Measures for Measure

TO JOY

Let our voyage together to Ithaca be slow, long and full of knowledge
With many a summer morning…

After Evangelis Sachperoglou's translation of C.P. Cafavy's *Ithaka*

Image courtesy of Joy Lawlor: *Untitled Still Life*. House paint on cardboard

Measures for Measure

Geology and the Industrial Revolution

How Carboniferous worlds changed ours forever

Mike Leeder

DUNEDIN

EDINBURGH ◆ LONDON

Published by
Dunedin Academic Press Ltd
Hudson House
8 Albany Street
Edinburgh EH1 3QB
Scotland

www.dunedinacademicpress.co.uk

ISBNs
9781780460819 (Hardback)
9781780466521 (ePub)
9781780466538 (Amazon Kindle)
9781780466545 (PDF)

British Library Cataloguing in Publication data
A catalogue record for this book is available from the British Library

Typeset by Kerrie Moncur Design & Typesetting

Printed in Poland by Hussar Books

Also by Mike Leeder and published by Dunedin:

GeoBritannica, Geological landscapes and the British peoples
With Joy Lawlor (2016)

For further details of these and other Dunedin
Earth and Environmental Sciences titles see
www.dunedinacademicpress.co.uk

Contents

PART 4 Landscapes of the Industrial Revolution

Acknowledgements

This is, of necessity, an interdisciplinary book, one that I believe breaks some new ground in considering essentially geological reasons behind the nature, location, timing and consequences of the first Industrial Revolution. My interest in the subject came initially from reading pioneering accounts of landscape and industrial archaeology by that polymath Dalesman, Arthur Raistrick, and from the work of W.G. Hoskins on the making of the broader British landscape.

My interests in geology and history were kindled first by my masters at the City of Norwich School, notably by an insistent perfectionist who insisted upon proper explanations for all phenomena: Mr Derek Lewis. That *leitmotif* carried on under tutelage and supervision at the Universities of Durham and Reading by Ronnie Girdler, Martin Bott, Tony Johnson and, most notably, Perce Allen.

During the book's three-year gestation I have received much assistance and advice, as follows:

From institutions and their chief contacts for granting and expediting my requests for permission to license images for reproduction: Ironbridge Gorge and Museum (Georgina Grant); Manchester Museum (David Gelthorpe, Kate Sherburn); National Galleries of Scotland (Beth Dunant).

From individuals for letting me reproduce their personal and published images and artwork: Julian Andrews (Parys Mountain); Chris Fielding (Fife Carboniferous strata); Joy Lawlor (*Black Fen Fantasy; Untitled Still Life*).

For general help and for logistic support during fieldwork: Brian Goggins (literature on Newry canal); Kate Johnston (help and accommodation in Lothian, Scotland); Joy Lawlor (organizer: 2019 'Pit-Stop Tour' of the North).

For ideas, general approval and critical reading of certain chapters (though remaining errors are my own): Julian Andrews (*Diagenesis; Carbon Cycling*); Hugh Falcon-Lang (he 'really enjoyed' my attempt at *Redress of Time*); Joy Lawlor (*Industrial Sublime*); Sam Lawlor (*Trees as Climate Modifiers*).

For providing me with the Cymru language for wetlands: Elgan Davies (Aberystwyth).

From Anthony Kinahan of Dunedin Academic Press for having faith in this project and his steady guidance and good advice. Kerrie Moncur and Anne Morton provided skilled expertise in design and editing. Sue Butterworth provided the excellent index and, in doing so, helpfully indicated some minor infelicities in the proofs that she worked on.

Preface

This book features once greatly disturbed landscapes – now largely remediated, physically at least, by post-industrial reconstruction. Yet the preserved machinery, industrial buildings and housing of the Industrial Revolution still dominate many parts of the British Isles and have done so for nearly two hundred and fifty years. They do so nowadays in the family-friendly and informative context of industrial museums, reconstructed industrial villages and in protected landscapes and townscapes still lived in by active communities. Coal, iron and other metallic ores and many other minerals and rocks had been taken out of Carboniferous strata and traded for over five hundred years before the Industrial Revolution, notably in the 'London Trade' of coal from Tyneside since the thirteenth century.

Yet by contrast, the neighbouring island of Ireland had no comparably great deposits of coal and never industrialized outside of coastal Belfast, though there was gold, copper, lead and zinc aplenty. What produced this abundance of fossil carbon preserved in the Carboniferous rocks of Britain? Why did the first Industrial Revolution originate there in the early- to mid-eighteenth century and not elsewhere in Europe where coal was also abundant?

As we shall see, a concatenation of geological, social and economic factors enabled the formation, preservation and exploitation of British coal- and iron-bearing strata. As economic historian Robert Allen puts it:

> The British were not more rational or prescient than the French in developing coal-based technologies. The British were simply luckier in their geology… there was only one route to the twentieth century – and it traversed northern Britain.[1]

Latterly other nations exploited their own Carboniferous treasure houses to establish, copy-cat style, the basis for their own industrial and urban growth – Belgium, France, Bismarck's united Germany, Poland, post-Civil War USA, Ukraine and the giants of modern day production, India and, supremely, China.

The simple fact is that, enlarging upon Allen's quote, problems and opportunities in the mining of coal and iron ore in Britain from Coal Measures strata determined the early Industrial Revolution's location, pace and extent. It gradually led to a thorough-going fossil fuel economy, the factory system and to the urbanization that would affect the entire nation in one way or another and, eventually, the whole world. I am not alone in thinking that at the present stage in human history it looks possible that the continued burning of fossil fuel at present levels will lead to global environmental catastrophe.

The British industrial legacy for wider humankind is thus profound and permanent. Great industrial cities became wonders of the world, their nouveau riche middle-class industrialists proud in themselves at what they had created. Industrial workers perhaps less so, for in their mines, foundries, forges, factories and mills they had to fight all the way for their economic, political and social rights. Our modern-day conserving culture has developed great care and concern to preserve the relics of a history that countless families shared down the ages. There were many personal and community traumas induced by the Industrial Revolution, and the memories of past suffering and opportunities taken and missed can echo down the generations, affecting individual psyches still.

The overriding aim in this book is therefore to trace the origins of the coal- and iron-bearing strata formed during the Carboniferous period within the context of social, economic and geological history and the fruits of the creative imaginations of people who lived through the Industrial Revolution. I have tried to do this with a lightness of touch, as cartoonist Gary Larson put it well when he has the felonious 'Brown Cow' in the dock, the judge appealing to the obviously guilty creature directly, with evident exasperation: 'Look. We know *how* you did it – *how* is no longer the issue. What we now want to know is …*why*. Why now brown cow?'

The book aims to explore all the adverbs, not only *how* and *why* but also the *where* and *when* of the early Industrial Revolution as it came about in eighteenth-century Britain. It does so mostly in everyday language, bar some necessary technical words and concepts whose definitions and usage are given in a glossary. The basic material should be understandable to anyone who enjoyed secondary education and who may/might have proceeded further, if not into a classroom or lecture theatre, then into the great outdoors of human imagination and experience.

Part 1 comprises a series of cameos. They are: pre-industrial landscapes explored by the intrepid Celia Fiennes and, a generation later, that social human Daniel Defoe; early resource exploitation by coal mining and the first rational attempt by George Sinclair in the seventeenth century to understand the course of coal-bearing strata in the landscape – the first account of structural geology; the invention by Thomas Newcomen of a steam engine that generated mechanical power from burning coal and efficiently drained mines; the use of coke by Abraham Darby as an advance over charcoal smelting in the world's first integrated iron-producing industrial complex in Shropshire; the new-built canal system pioneered by the Duke of Bridgwater, John Gilbert and James Brindley that enabled the bulk transport of industrial raw materials at prices far below those of pre-existing river or road transport monopolies.

Part 2 ventures into geology. It tries to explain why Carboniferous strata contained so much coal and ironstone compared to times before and after. It outlines the physical, chemical, biological and tectonic conditions that gave rise to the rocks and fossil floras of Carboniferous environments in a wider geological way. Also, since Carboniferous outcrops are not continuous, and since not all contain such once-valued economic deposits, the controls on their distribution in time and space are also examined.

Part 3 addresses social, economic, environmental and creative consequences of the early Industrial Revolution as the economics and technology of the factory system spread across the world. In the 1980s it finally reached the sleeping giants of populous East and South Asia, whose coal-burning economies now replicate certain of the miseries of the original one. More to the point, the continued burning of fossil carbon and hydrocarbon threatens global atmospheric stability through the inexorable workings of the carbon cycle. Finally, the creative legacy of early 'industrial sublime' and the reaction of artists and writers towards industrialization down to modern times is explored, again in a number of cameos and readings by the author.

Part 4 brings together geological, historical, social and cultural trends in the context of today's de-industrialized legacy of landscape, monuments and heritage in the former major regions of Britain that hosted the early Industrial Revolution: South Wales, England's West Country, North East Wales, English South Midlands, east of the South Pennines, west of the South Pennines, West Cumbria, North Pennines (Northumberland/ Durham) and midland Scotland.[2]

It seems appropriate here to recognize the universal downside for labouring humans over the millennia during the extraction of minerals and rocks useful or valuable to wider society, both in free-market and state-subsidized economies. It was memorably expressed fifty years ago with reference to gold mining in the Rand (though the symbolism is universal) by the late Mazisi Kunene, who became South Africa's and then Africa's poet laureate:

Towers rise to the skies,
Sounds echo their music,
Bells ring backwards and forwards
Awakening the crowds from the centre of fire.
Attendants at the feast glitter,
Wealth piles on the mountains.
But where are the people?
We stand by watching the parades
Walking the deserted halls
We who are locked in the pits of gold.[3]

PART 1

Economy in Motion

The English, Welsh and Scottish peoples and their national economies suffered greatly in the seventeenth century during the disastrous middle years of civil strife, extreme violence and widespread intolerance of others' religious and political beliefs. The first two countries subsequently expanded rapidly in the later seventeenth century, but in Scotland only slowly after the Union of 1707. Royal ownership of underground coal and mineral resources had by then been transferred to landed gentry who leased them on or charged royalties for extraction. The rights of freeholders were often questionable and frequently subject to litigation. This divestment of crown ownership stood in contrast to the situation in most of the rest of Europe and was an important but neglected factor contributing to that early increase in economic activity – it was obviously in landowners' financial self-interest to encourage increased output from minerals mined or quarried under their lands. Continued efficient and profitable expansion required an increasing scientific knowledge of underground geology and the principles of mine drainage. To provide this the profession of land or mineral surveyor evolved later in the seventeenth century, afterwards with increasing links to prospecting and general progress of the 'Enlightenment' science. Inventive practices and machines – coke-smelted iron and steam engines for mine drainage – generated unstoppable momentum. The first episode of British 'canal mania' from the 1740s to 1770s saw coal, metal ores, potter's clay and other bulk raw materials like roofing slates become markedly cheaper as producing and consuming districts were connected along their still waters – the first Industrial Revolution was well on its way. Steam-based mechanization utilizing rotative power courtesy of James Watt's later engines subsequently spread throughout transport and manufacturing, from locomotion to excavation and textiles.

1

Travellers' Tales 1: Celia Fiennes and Daniel Defoe in Pre-Industrial Britain

I was resolved to have a perfect knowledge of the most
remarkable things, and especially of the manufactures…

Daniel Defoe, *A Tour Through the Whole Island of Great Britain.*

Portraits of Britain at the Cusp of Change

There was a remarkable change in the economic and political system of Britain from the mid-seventeenth century with its upheavals caused by civil war. Over the next 270 or so years absolute monarchy, rural living and cottage economy gave way to universal suffrage, parliamentary democracy, urban living and industrial economy. It began as the northern European 'Age of Reason' produced discoveries in mathematics, astronomy and physics and underlined the right of individuals to have personal opinions independent of belief-dogma. There was a general increase in trading activities, urban renewal and home building, and a great expansion of mining in general and for coal and metals in particular.[1]

The onset of savage winters at the beginning of what has become known as the 'Little Ice Age'[2] around the early 1600s undoubtedly contributed to a greater demand for easily available and relatively cheap coal-based heating. This was seen in the expansion of the 'London Trade' for domestic coal, mostly from Tyneside (Fig. 1.1). The trade was in fact long established by then, for heating the great buildings of nobility, church and state and for pre-industrial commercial use (e.g. saltings, glass making, lime burning, dyeing) in the fuel-poor capital city. By the late-seventeenth century it had begun to grow rapidly on its trajectory to become the major and most populous mercantile port city of northern Europe, replacing first Antwerp and then Amsterdam after English victories in the Dutch naval wars. As cast iron ranges and fire grates became ever-more popular, brick chimney stacks rose to bristling predominance on urban skylines. The

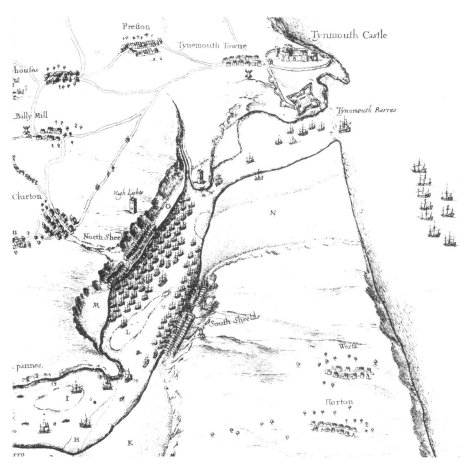

Figure 1.1 Part of Ralph Gardiner's extraordinary and unique map of 1655: *The River of Tyne leading from the Sea on the east to Newcastle on the west*, which forms the frontispiece to Nef (1932, 1972). This extract is a bird's-eye view about four miles wide of the great Tyneside collier fleet assembled at the 'Sheelds' quaysides in the sheltered outer estuary. Scores of collier vessels are lined up awaiting deliveries from the keelmen, whose lighters would bring them a couple of tens of tonnes per journey from coal depots further upstream along the narrow upper estuary that separated the walled towns of Newcastle and Gateshead. More vessels are waiting both outside and just inside the 'Tynmouth Barres' at the estuary entrance where the draught is just seven feet. The broad navigable outer estuarine channels had fifteen feet of water at high tide and just a couple of sandbanks to hinder navigation. The great salt 'pannes' of Holden are in view up-estuary on the left bank. Note the two lighthouses, 'High Lights' and 'Loo Lights' on the North Shields side: the latter features prominently in Joseph Turner's portrayal of the scene (his approximate viewpoint was on the left bank peninsula south of North Shields) by moonlight 180 years later (see cover image and text appropriate to Figure 15.2).

nation's trade expanded significantly after the eighteenth-century Seven Years War. Great mercantile concerns brought in tropical imports from India, Cathay and the Indies, with temperate items from Muscovy and Scandinavia. Scores of busy woollen factors set up large-scale cloth exports of the 'New Textiles' from the busy loom-clacking hinterlands of wider East Anglia.

For overviews of the immediately pre-industrial landscapes of Britain we have two lively primary sources to draw upon. The first is by Celia Fiennes in her 1690s account, *Through England On a Side Saddle in the Time of William and Mary*.[3] This was only brought to publication two hundred years later (in 1888) by her kinswoman Emily W. Griffiths, who writes in her brief introduction:

> The original MS, given to me by my father, has been copied verbatim, as I believe any correction or alteration would spoil its quaint originality. Celia Fiennes was daughter of Colonel Nathaniel Fiennes, a Parliamentarian Officer, by his marriage with Miss Whitehead, and was sister of the third Viscount Saye and Sele.

That Fiennes was a redoubtable, courageous and ever-observant daughter with a wide interest in the goings-on across English towns and the countryside becomes clear during the course of her 100,000 or so words. Her preface *To The Reader* makes clear that the book, warts and all, was of her own making, written for family reading. She reveals that the reasons behind her peregrinations were the need for bodily and mental exercise (in this and subsequent quotes, Fiennes' spelling and punctuation retain its 'quaint originality'):

> My Journeys as they were begun to regain my health by variety and change of aire and exercise, soe whatever promoted that was pursued; and those informations of things as could be obtein'd from jnns [inns] en passant, or from some acquaintance, inhabitants of such places could ffurnish me with for my diversion…

A knowledge of one's country, she writes, is essential to both men and women and the effort in gaining it by travel is a valuable antidote to ill-health, and she adds with (I imagine) enjoyable emphasis, 'Laziness'! The whole experience:

> would also fform such an Idea of England, add much to its Glory and Esteem in our minds and cure the evil Itch of overvalueing fforeign parts…

Cheering words for today's Brexiteers and home tourist industry! Fiennes' list of desirable features to observe while travelling included the obvious fashionables like 'pleasant prospects', 'good buildings' and 'the variety of sports and recreations' but also her emphasis on the items that most concern us in the present book – 'different produces and manufactures of each place' and, most emphatically:

ye nature of Land, ye genius of the Inhabitants, so as to promote and improve Manufacture and trade suitable to each…

She was the first economic landscape tourist!

The second primary source comes from the particularly detailed and researched observations made by a professional writer – in contrast to the vitality and freshness of Fiennes' highly personal and perhaps more human accounts. It was made in late age by Daniel Defoe in his 1720s *A Tour Through the Whole Island of Great Britain*.[4] Defoe provides us with a unique and broad-based documentary archive of the whole United Kingdom, rural and urban, unrivalled since. Defoe had a notably active life – stocking factor (he knew all the regional cloths of England), young rebel (fighting with Monmouth in the rising of 1686 against James II), novelist, journalist, government spy, debtor and indefatigable traveller. From his notes, recollections from as early as the 1680s, and with occasional scathing asides concerning rival travelling wordsmiths, he wrote a structured and consistently didactic narrative of several hundred thousand words. It combined accounts of visits, comments, reflections and generalized statements on a vast number of themes. Defoe also had an emotional and moral compass and socio-historical interests, as we might expect from the author of the *Robinson Crusoe* trilogy, *Moll Flanders* and *Roxanna*. This is most notable in his moving account of the working and domestic arrangements of a Derbyshire lead miner's family, featured later in this book.

Celia Fiennes: 'An Idea of England'

On her *Northern Journey* begun in May 1697 Fiennes first came across the reality of the Midlands mining districts after a meandering route through East Anglia, the East Midlands and parts of West Yorkshire. In north Derbyshire around Chesterfield we learn of the construction of mined levels or galleries excavated from a central shaft bottom outwards along a coal stratum – doubtless as far as the overlying stratum would resist collapse by judicious use of wooden props. A windlass arrangement let miners up and down and the coal out:

> Here we Entred Darbyshire and went to Chesterffield 6 mile, and Came by the Coale mines where they were digging. They make their mines at the Entrance Like a Well and so till they come to the Coale, then they dig all the Ground about where there is Coale and set pillars [wooden pit props] to support it, and so bring it to the well [shaft] where by basket Like a hand barrow by Cords they pull it up – so they let down and up the miners with a Cord.

The use of pit props to extend the lateral range of mining in such bell-pits is a unique observation, as far as I am aware (see Chapter 20 for contrasting information).

Fiennes had tastes and a sensibility that embraced fine ale, the occurrence of economic minerals, good building stone and the 'wisdom and benignitye of our greate Creator' as she generalizes her impressions of Derbyshire and of a specially happy Saturday market day spent by her in Chesterfield:

> In this town is the best ale in the Kingdom generally Esteem'd…You see neither hedge nor tree but only Low drye stone walls round some ground Else its only hills and Dales as thick as you Can Imagine, but tho' the Surface of the Earth Looks barren yet those hills are impregnated with Rich marble stone metals [solid freestone], Iron and Copper and Coal mines in their bowels, from whence we may see the wisdom and benignitye of our greate Creator to make up the deficiency of a place by an Equivolent, and also the diversity of the Creation which Encreaseth its Beauty.

She returns to this deistic theme a little later, first imagining that the freshly cleaved and silvery coloured galena (lead ore) that she sees brought up from a lead mine near Buxton was due to its high silver content. She goes on to give a down-to-earth description of the walled inner mine shaft down which the miners descend via a windlass:

> In that mine I saw there was 3 or 4 at work and all let down thro' the well [shaft]; they dig sometimes a great way before they Come to oar. There is also a sort of stuff they dig out mixt with the oar and all about the hills they call Sparr, it looks like Crystal or white sugar Candy, its pretty hard; the doctors use it in medicine for the Colloick; its smooth like glass but it looks all in Crack's all over [the mineral was calcite, complete with cleavage cracks]. They Wall round the Wells to the mines to Secure their Mold'ring [collapsing] in upon them, they generally Look very pale and yellow that work Underground, they are fforc'd to keep Lights with them and sometimes are forced to use Gunpowder to break the stones, and it is somtymes Hazardous to the people and destroys them at the work.

In her 1698 *My Great Journey to Newcastle and to Cornwall* Fiennes came across further pre-industrial districts that were later to feature greatly in the rise of industry and urbanization – the two Newcastles, Flintshire, South Lancashire, the Midlands and Bristol. We take a selective sample here. First was Newcastle-under-Lyme in north Staffordshire. Passing the great house of 'Mr Leveston Gore' [John Leveson-Gower, 1st Baron Gower, 1675–1709] at Trentham and riding thence up along the hillside road towards Newcastle, she writes lyrically of the meandering Trent in its floodplain below, which:

> ran and turn'd its silver streame forward and backward…Circleing about the fine meadows in their flourishing tyme…6 mile more to NewCastle under Line [Lyme] where there is the fine shining Channell [cannel] Coale; so the proverb to both the Newcastles of bringing Coales to them is a needless Labour, one being famous for this Coale that's Cloven [split] and makes

> white ashes as is this, and the Newcastle on the Tyne is for the sea Coale that
> Cakes and is what is Common and famillier to every smith in all villages.
> I went to this NewCastle in Staffordshire to see the making of the fine tea
> potts, Cups and saucers of the fine red earth [clays of the Late Carboniferous
> Warwickshire Group] in imitation and as Curious as that which Comes from
> China, but was defeated in my Design, they Comeing to an End of their Clay
> they made use for that sort of ware...

Seventy years later she would have looked down to see the beginnings of the Potteries industrial valley with Josiah Wedgwood's spanking new Etruria works and James Brindley's brave new Trent-Mersey Canal below her.

Coming into Flintshire via Chester she travels on from a relative's house at Harding where she is staying, to Holywell via Flint:

> a very Ragged place many villages in England are better, the houses all
> thatched and stone walls, but so decay'd that in many places Ready are to
> tumble down...

At Holywell the Welsh-speaking inhabitants fail to impress her; perhaps she was uncharacteristically shocked to see that they 'go barefoote and bare leg'd – a nasty sort of people.' She generalizes her impressions of coal mining around Flint:

> There are great Coale pitts of the Channell Coale that's Cloven huge great
> pieces: they have great wheeles that are turned with horses [gins] that draw
> up the water and so draine the Mines which would Else be over flowed so as
> they Could not dig the Coale...

Not only do the mines have the usual circular horse-powered 'gins' for pumping out water but the mined coal from the shafts is brought to the surface in barrow-like baskets powered by 'engines'. These would probably be courtesy of water-powered wheels turning a drum-wound rope. She goes on to document the common practice of moving goods (including coal) using horse-drawn wagons to and fro between Flintshire and Chester across the Dee at low tide with all its treacherous channels.

Moving into Lancashire and crossing the Mersey estuary by ferry 'an hour and halfe in the passage...[by] a sort of Hoy that I ferried over and my horses – the boat would have held 100 people.' Fiennes is impressed with 'Leverpoole', the smart, new-built town on the Mersey – full of Nonconformists:

> mostly new built houses of brick and stone after the London fashion; the
> first original was a few fishermens houses and now is grown to a large fine
> town and but a parish and one Church, tho' there be 24 streetes in it. There is
> Indeed a little Chappell and there are a great many dessenters in the town. It's
> a very Rich trading town...

She moves further across Lancashire past Ormskirk, avoiding treacherous marshes and mosses to re-acquaint herself with the cannel coal that seems to have become a favourite material:

> Thence to Wiggon [Wigan]…another pretty Market town built of stone and brick: here it is that the fine Channell Coales are in perfection – burns as light as a Candle – set the Coales together with some fire and it shall give a snap and burn up light. Of this Coale they make Saltcellars, Stand-dishes and many boxes and things…I bought some of them for Curiosity sake.

After Wigan there were Lancaster and the Lakes; then coming in towards Newcastle-on-Tyne from Carlisle and the Borders she gives us perhaps the earliest and canniest introduction to the London sea coal trade from Tyneside, the different types of coals available and the sulphurous nature of some of them:

> As I drew nearer and nearer to Newcastle I met with and saw abundance of Little Carriages with a yoke of oxen and a pair of horses together, which is to Convey the Coales from the pitts to the Barges [keels] on the river [Tyne]…I suppose they hold not above 2 or three Chaudron [the Newcastle chaldron was around two and a half tons]. This is the sea Coale which is pretty much small Coale tho' some is round Coales, yet none like the Cleft coales [cannel] and this is what the smiths use and it cakes in the ffire and makes a great heate, but it burns not up Light unless you put most round Coales which will burn Light, but then it is soon gone and that part of the Coales never Cakes, therefore the small sort is as good as any – if its black and shining [vitrinite-rich], that shows its goodness. This Country all about is full of this Coale, the sulphur of it taints the aire and it smells strongly to strangers, – upon a high hill 2 mile from Newcastle I could see all about the Country which was full of Coale pitts.

On her return south there was a close encounter with highwaymen (Fig. 1.2), which she graphically describes:

> here I may think I may say was the only tyme I had reason to suspect I was engaged with some Highway men; 2 fellows all of a suddain from the wood fell into the road, they look'd trussed up with great coates and as it were bundles about them which I believe was pistolls...

Heading for the south-west peninsula, Fiennes passed through the Kingswood coalfield that fed Bristol with its own needs, and also for what seems to have been a significant export by the many ships in the port:

> From Bath I went westward to Bristol over Landsdown 10 mile, and passed thro' Kingswood and was met with a great many horses passing

Figure 1.2 The splendid *Celia Fiennes Waypost* at No Mans Heath, Malpas, north of Whitchurch, Cheshire (53.0266, 2.7238) was a project initiated by Cheshire City Council and sculpted (out of what looks like Jurassic Portland limestone) by artist Jeff Aldridge of Chester. It was unveiled by a Fiennes descendant in 1998 to commemorate the 300th anniversary of her 'Great Journey'. Source: Wikimedia Commons File: No Fiennes Heath 484.jpg. Photographer: R.W. Haworth.

and returning Loaden with Coales Dug just thereabout; they give 12 pence a horse Load which carrys two Bushells, it makes very good ffires, this the Cakeing Coale. Bristol Lyes Low in a bottom the Greatest part of the town, tho' one End of it you have a pretty rise of ground. … This town is a very great tradeing Citty…I saw the harbour was full of shipps carrying Coales and all sorts of Commodityes to other parts.

Daniel Defoe: 'To Have a Perfect Knowledge of the Most Remarkable Things'

Here is a selection of Defoe's myriad observations, bearing in mind his general credo:

I was resolved to have a perfect knowledge of the most remarkable things, and especially of the manufactures…which I take to be as well worth a traveller's notice, as the most curious thing he can meet with, and which is so prodigious great…

His was therefore an analytical eye, though not parochial. As a rational individual, he refused to be cowed by antiquarian and other 'curiosities' or 'wonders' common in contemporary travelogues by less rounded and practically inclined authors. Though a Londoner by birth and habitation, and despite regarding the capital as the *sine qua non* of the nation's economic efforts, he also reflected deeply on the rising fortunes evident to him in the provinces, particularly those in the 'new' northern towns that might someday overtake the medieval south in trade:

we shall find them [the southern towns] matched, if not out-done, by the growing towns of Liverpool, Hull, Leeds, Newcastle, and Manchester, and the cities of Edinburgh and Glasgow…

This was good grandstanding for the times.

He was particularly attentive to the coal trade, both its mining and relevance to general commerce, and in its transport from pithead to market along the notoriously bad roads of the time (object of his most furious invective), by inland waterway and by coastal shipping. At this time coal's use was chiefly domestic and in trades requiring intense long-lasting heat, like smithy work, glass production, soap making, salt precipitation and the slow warming by coked coal of spreads of sprouting malt on brewery floors. He knew the broad extent and richness of coal reserves from Fife to South Wales (Fig. 1.3) and featured the notable ports that shipped it along navigable rivers to a vast internal market: Yarmouth (via the Yare to Norwich), King's Lynn (via the extensive Ouse network), Boston (via the Witham), Hull (via the Humber and Trent), Bristol (via the Severn to the English Midlands), Swansea and Neath (to the south-west and Ireland), and Whitehaven (to northern Ireland).

This regional trade was dwarfed by the aforementioned 'London Trade' from Tyneside, centred at the daily Billingsgate coal market that supplied the City and all of the east and south of England:

> This trade is so considerable, that it is esteemed the great nursery of our best seamen. The quantity of coals, which it is supposed are, *communibus annis*, burnt and consumed in and about this city, is supposed to be about five hundred thousand chalder…generally weighing about thirty hundred weight [the London chaldron measure].

It represented an annual trade in what had become known as 'sea-coal' of at least three-quarters of a million tonnes. The term 'sea coal' simply refers to the means by which the material was brought to places like London, emphatically not on its being casually picked or dug from northern beaches and outcrops, as often assumed. As Defoe notes, sea-coal voyaging was also a major factor in the development of maritime expertise in the British navy – passage along the stormy and dangerous shores of the eastern seaboard (usually a leeward coast, but full-on windward during many winter gales from the north-east quadrant) were full of challenges to navigation and seamanship. He analyses the particular vulnerability of collier vessels to such gales in the stretch from Flamborough Head to north-east Norfolk. Hence we can read with new insight why Samuel Pepys was so delighted when he scribbled in his diary in the 1660s: 'This afternoon came my great store of Coles in, being 10 Chaldron.' That would be about fifteen tonnes to fill his reconstructed cellars, of which he records on 8 February 1661:

> All the morning in the cellar with the colliers, removing the coles out of the old cole hole into the new one.

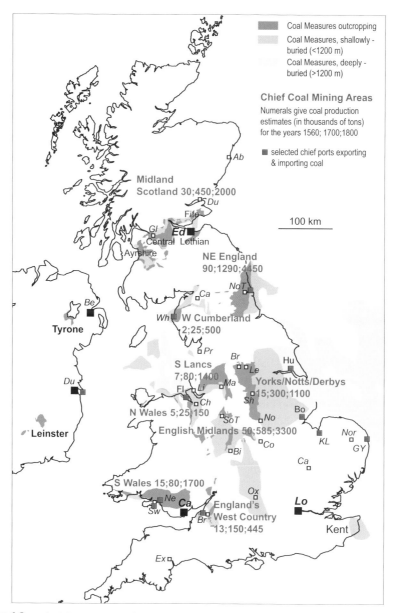

Figure 1.3 Both Celia Fiennes and Daniel Defoe were aware at first hand of the wide extent of coal-bearing rocks throughout England, Wales and Scotland. Here are shown the chief coalfields, including those very few in Ireland. They occur both as exposed Coal Measures at outcrop and as 'concealed' Measures overlain by younger strata. The coalfields mapped here are all Late Carboniferous in age, but production also came from Early and Middle Carboniferous coal-bearing strata in northern Britain, particularly in NE Northumberland and, more importantly, in midland Scotland. Sources: redrawn, modified and crafted from British and Irish Geological Survey map sources and the primary scientific literature. Coal production estimates for *c.*1560, 1700 and 1800 are garnered from sources cited by Allen (2009, his Table 4.1).

Amused at the thought of Mr Pepys in shirtsleeves shovelling coal in his own cellars, one understands how this huge amount for just one well-off family house in London underscores estimates of the scope and volume of the whole trade. Defoe also sheds light on its wider economics and organization. The east coast ports of Whitby, Great Yarmouth and Ipswich, as well as Newcastle itself, are named as the main shipbuilding centres and the home ports for the collier fleet. Ipswich is particularly highlighted from his own memories as a young man in the 1680s. It was:

> the greatest town in England for large colliers or coal-ships, employed between New Castle and London…built so prodigious strong…to reign for forty or fifty years, and more…above a hundred sail of them…the least of which carried…300 chaldrons of coals…this about the year 1688 (when I first knew the place). This made the town at that time to be so populous, for those masters, as they had good ships at sea, so they had large families, who lived plentifully, and in very good houses in the town, and several streets were chiefly inhabited by such.

He goes on to remark that the trade there had been much diminished since the Anglo-Dutch wars when many captured Dutch 'flyboats' entered the trade, used by both London and Great Yarmouth merchants.

The increased price of sea-coal in London, compared to that at the pit-heads of the Tyne, included not only the costs of transport but also a stiff tax on import taken at the incoming port. This gave great opportunity for non-Tyneside coal producers that could use untaxed riverine transport. Such was the case in West Yorkshire and adjacent counties along the Humber-Trent navigable river network. This served chiefly to export woollen goods out of the West Riding down the newly navigable rivers Aire and Calder to Hull and beyond into northern European markets, but also:

> they carry coals down from Wakefield (especially) and also from Leeds… down into the Humber and then up the Ouse to York, and up the Trent, and other rivers, where there are abundance of large towns who they supply with coals; with this advantage too, that whereas the Newcastle coals pay four shillings per chaldron duty to the public; these being only called river borne coal, are exempted and pay nothing…

Defoe also gives relevant information concerning metal mining and smelting in many districts: iron in the Weald, West Midlands, Ross-on-Wye, Bristol, Sheffield, 'Black' Barnsley (as he calls it), Rotherham and, recently set up, at Newcastle-on-Tyne; lead from Cornwall, west (Mendip) and north Somerset, Cardiganshire and Derbyshire; tin from many localities in Cornwall; copper also from Cornwall, in particular west of Truro where:

> being lately found in large quantities…and which is much improved since the several mills are erected at Bristol, and other parts, for the manufactures

of battery ware [pots and pans made by beating malleable sheet metal] or, as 'tis called, brass, which is made out of English copper, [and zinc, from Mendip in Somerset] most of it dug in these parts; the ore itself also being found very rich and good.

Brass featured again on the Thames east of Reading (also known from near Bath, at Saltford) where water-powered trip hammers pounded the brass into the various shapes required:

by the help of *lapis caliminaris* [calamine] they convert copper into brass in large broad plates, they beat them out by force of great hammers, wrought by the water mills, into what shape they think fit for sale…

He is particularly informative about the coal-fired salt works ('pans') in Cheshire, Lothians and Fife, especially the long-established works at North Shields on the outer Tyne estuary (*see* Fig. 1.1) where, on a journey up to Durham, he notes:

It is a prodigious quantity [of] coals which these salt works consume, and the fires make such a smoke, that we saw it ascend in clouds over the hills, four miles before we came to Durham, which is at least sixteen miles from the place.

Coda

Celia Fiennes and Daniel Defoe skilfully mix scenic, manufacturing, architectural, historic and social themes in their own unique and creative ways. Defoe's book became a best seller in its time and through repeated later editions, right down to the abridged (but still thick) Penguin paperback of today. Since its publication in 1888, Fiennes' lively squib of a book has captured the hearts and minds of many. Yet both works were but samplers of Britain from the 1680s to the 1720s, containing echoes of painful memories, both human-induced and natural: of civil war and regicide; of civil unrest; the narrow escape in 1688 from renewed tyranny; the calamitous financial consequences of the South Sea Bubble; the 'Great Storm' of 1703, to name but a few.

From our own point in history we recognize the uniqueness of Celia Fiennes' perambulations in the safer times of William and Mary, with Defoe writing between wars in very early Georgian Britain – the two defeated Jacobite rebellions of 1715 and 1745–46. Both books are real antecedents to J.B. Priestley's *English Journey* of 1934, also conducted and written between wars.[5] The eighteenth-century civil conflicts solidified the Hanoverian dynasty's somewhat shaky start and gave a degree of political stability through successive Whig regimes for almost the entire century. The two books were completed just a generation or two before the common adoption of technological advances that would change the face of Britain and, eventually, the world.

One wonders how the ever-curious and game Fiennes and the industry-enthusiast Defoe would have reacted to smoking coke heaps and ovens, flame-belching Bessemer converters, steam-powered textile mills, general mechanization and cramped urbanization. Perhaps a combination of the thrills of the 'industrial sublime' as pictured by artists Joseph Turner and Joseph Wright with the emotions of William Blake – those of dreadful horror at the 'dark, satanic mills' populating England's 'green and pleasant land'?

2

Beginning of Coallery:
George Sinclair, 'Scoto-Lothiani'

to give you my judgement in a matter wherein I have been so little conversant myself, and have had the steps of no other to follow, never one having hitherto touched that subject in writing; I mean of Coals, and other Minerals of that nature, their Course, and other things relating thereunto; the observation whereof (I grant) wants not its own pleasure and usefulness.

George Sinclair beginning his tract, the *Short History of Coal* (1672)

Perspectives

As a consequence of the great increase in coal usage for domestic heating, by the late seventeenth century large and deep mines had developed, particularly along Tyneside, as noted by both Fiennes and Defoe. In the aforementioned monumental history of the coal mining industry, J.U. Nef (1932) writes of the knock-on consequences to this expansion:

There were other specialists in mining operations, who did not attach themselves permanently to the staff of any particular enterprise, but put themselves at the disposal of all adventurers [prospectors, investors] who wanted advice…The increase in the number of mine officials and mining experts during the period between 1550 and 1700 justifies us in speaking of the formation of a new social class within the coal industry; a class which we may call professional because its members gained their living principally by brain work rather than by manual labour.

Which kind of 'brain work' Nef does not specify, but it would have included surveying, prospecting and mine drainage. The former would reveal the spatial extent of underground workings under a landowner's property to be shown on a surface map plan.

This established the extent of workings in relation to its surface boundaries with those of adjacent estates. It is not difficult to imagine why frequent and often interminable litigation arose that plagued mining enterprises during the rise of early mining activities. Neighbours accused neighbours of underground trespass, theft, physical harassment, drainage dislocation and other acts of sabotage. Both Nef and J. Hatcher (1993) give many examples of such shenanigans – some readers may be more familiar with them from episodes of *Poldark*.

The central figure in the present chapter, George Sinclair (*c.*1630–1696), was the first and most distinguished and original of the professional mineral surveyors of the Stuart business world. This self-styled 'Scoto-Lothiani' published a contribution in 1672 that was part of a larger work, *Hydrostaticks*.[1] It was subtitled a *Short History of Coal*, a wholly original work concerning mineral surveying in the natural landscape and the mining of coal, together with the ancillary problems of ventilation and drainage. He describes his work as dealing with 'coallery' i.e. all manner of things concerning coal. His geological insights have been largely ignored by economic historians and neglected by geoscience historians. In what follows, Sinclair's approach and discoveries are placed in the context of his *milieu*, emphasizing the practical side to Enlightenment philosophy and its legacy.

Life, Times, Background

Sinclair was a contemporary of such eminent English natural philosophers and anti-quarians as John Evelyn, Thomas Browne and John Aubrey – also of practical men, early civil engineers like Andrew Yarranton, William Sandys and Sir Richard Weston.[2] Sinclair had intellectual clout, holding a chair in philosophy (later known as 'natural philosophy', physics as we now call it) at the University of Glasgow from 1654 to 1666. He was a lifelong polymath, writing (often in Latin) on geometry, applied physics (Aristotelian hydrostatics, pneumatics) and religion. He also conducted some daring geophysical field experiments – the 'outdoor physics' of Lord Rayleigh – determining the approximate altitude of Scottish mountains using a barometer, and descending off Mull in an original diving-bell to examine the contents of wreckage from the Spanish Armada.[3]

The turmoil that ensued after the Restoration of the 1660s determined the historical background to Sinclair's professional life as a mineral surveyor. This was during the dangerous times of Covenanting and a Scottish near-obsession with witchcraft and superstition inherited from the sometimes deranged James VI, and to which Sinclair himself later contributed. Scotland had come under the sovereignty of England's Charles II through the events of its own Restoration – the dissolution of the Scottish parliament and the governance of the country by Charles's (secretly) Catholic brother James. As an unbending Presbyterian, Sinclair was dismissed from his Glasgow post in 1666 by the Episcopalian authorities for refusing to swear an oath of allegiance to either prelate or King.

An East Lothian by birth, Sinclair lived latterly in Leith and owned property in Haddington, though we know nothing of his personal or family life. He must have taken up mineral surveying to keep body and soul together during the period of his dismissal from Glasgow. He also taught and practised hydraulic engineering at this time, being entrusted by the Edinburgh magistrates (despite his Glasgow sacking) both to lecture and to successfully bring a piped supply of copious fresh spring water into the city from adjacent uplands (certain outlets for his scheme are still preserved in the city today). He became deeply knowledgeable about Carboniferous stratal patterns in the coal-bearing Lothians and elsewhere in central Scotland and northern England. After the 'Glorious Revolution' of 1688 he was able to resume his chair of philosophy at the University of Glasgow in 1689. He was subsequently made the first professor of mathematics there, a post he held till his death in 1696. His professorial chair is still extant: the George Sinclair Chair of Mathematics.

Sinclair's Special Contribution

The *Short History of Coal* (Fig. 2.1) is more than 12,000 words long with wood-block diagrams that include the first recognizable diagrammatic geological section. He begins with a modest prologue to an unnamed patron (*see* chapter motto). Sinclair was a rational observer not prepared to speculate beyond his immediate experience. When trying to decide on the true origins of coal, minerals and rock there were, according to him, two possibilities; either they were created *ab initio* (once, from starters) or that they are continually being created naturally within the Earth. In his eloquent words:

> if Coal, and Free-stone…have been created in the beginning, in their perfection, as we now find them, and since that time only preserved, as they were created for the use of men, to whom all sublunary [earthly] things were made subservient. Or, if they have been but produced gradually, as they speak of Gold and other Minerals, by the influence of the Sun, in the bowels of the Earth? And if their production be of that nature, out of what matter they are formed? These things being above my reach, I shall leave their inquiry, to those that are knowing in the secrets of Nature, and shall therefore give you a narration, of what either I have observed of these things, which occur in the Winning of Coal in my own experience, or by conversing with others of more experience than myself in doing whereof, I shall follow this Method.

The last sentence has echoes of European Renaissance philosophy from Georgius Agricola in his preface to *De Re Metallica* (Sinclair would have read him in the original Latin) paraphrasing Julius Scaliger as he bluntly states his devotion to empirical enquiry:

> I have omitted all those things which I have not myself seen, or have not read or heard of from persons upon whom I can rely. That which I have neither seen, nor carefully considered after reading or hearing of, I have not written about.[4]

Perusing Sinclair's writings on the subject we realize that he had reduced the chief determining features of rock at outcrop and also underground in the coal-rich Lothian

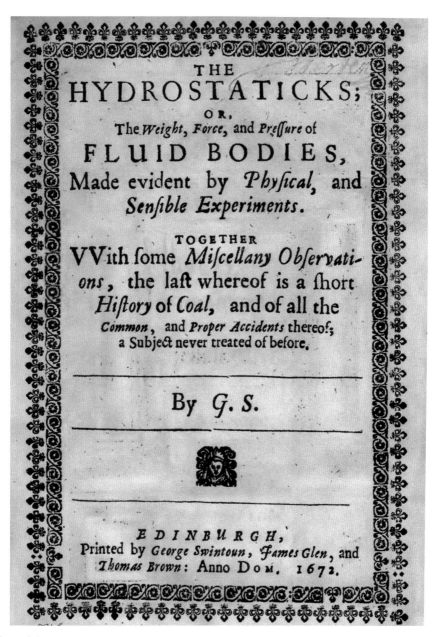

Figure 2.1 The title page to George Sinclair's *The Hydrostaticks*, in which his revelations concerning the nature of the emerging subject of 'coallery' (and, as we can now recognize, structural geology) are advertized in the long subordinate subtitle. Not the best blurb for a book that drew little attention from later geologically attuned readers, with the notable exceptions of Sir Edward Bailey's brief and emphatic 1952 resumé and the attention of John Hatcher (1993) to Sinclair's writings on prospecting by borehole and mine drainage. As far as I am aware his geological insights have been previously missed by historians of that science.

landscape to mental constructs. In his mathematical mind they were geometric entities –
Euclidian objects like our Meccano or Lego, their dimensions and orientations occupying
space in the way that Descartes had developed for graphical coordinates. Though Sinclair
was apparently a bitter enemy of philosophical rationalists like Descartes, Hobbes and
Spinoza, his view of strata was itself, as we have said, *ab initio*. He defined coal-bearing
rock as more-or-less continuous solid layers: that strata could be so-divided was an
entirely new development. He placed an imaginary grid of Cartesian coordinates from
the surface of the Earth downwards and described the course of strata on them as they
might appear underground. It was an astonishing and original achievement that laid the
basis for all subsequent geological mapping activity, from plane tables and entangled
survey chains right down to modern GPS coordinates and real-time mapping on laptop
computer.

By comparison, Nicholas Steno's famous 'section' illustrating the geological evolution
of Tuscany in his *Prodromus* (aka *De Solido)* of 1669 was a composite three-dimensional
pictograph.[5] Sinclair's work was published just four years after Steno's – I feel it
unlikely that he was aware of it at the time of his own writing. Given Sinclair's religious
proclivities it would be doubtful if the Presbyterian Scot would have entertained Steno's
wider philosophy, especially the justification of its geological aspects (chiefly history and
fossilization) as being all at one with the teaching of the Catholic Church.

But let us be clear about what Sinclair did not achieve in comparison with Steno.
He has nothing to say to us on the true succession in time represented by rock layers,
never mentioning that underlying examples in a succession might be older in their
formation than overlying ones. As he freely admits from the previous long quotation, he
has no intellectual theory available for the actual creation of rock. That they might have
sedimented as successive layers one on top of the other had not occurred to him – it was
simply beyond his ken at the time. But it was not beyond the ken of Nicholas Steno, and
that was the difference between the two men as natural philosophers. Steno was clearly
one of those 'knowing in the secrets of Nature…', as Sinclair mellifluously (or perhaps
slightly sarcastically) put it, just as James Hutton of Edinburgh was to become supreme
in the art seventy or so years in the future.

Context: Dip and Course

Beneath the surface soil, Sinclair recognized that various types of 'freestone' may
underlie and overlie coal. These have the same spatial orientation and regularity as the
coal itself, except where they are found to be absent for some reason or other. Given the
Lothian context of his field observations, he might have been referring to sandstones or,
less likely, limestones as 'freestone'[6]. The spatial characteristic of both coal and freestone
(strata) he defined as their 'Course' – their combined geometrical attributes of dip – any
sloping of the coal or freestone down or up – and 'streek' (dialect for our strike), their
planar orientation. He was in fact describing strata and stratification (Fig. 2.2), the first
person to give geometric precision to the description of rock. In his own words:

A Schematic section of successive coal and rock strata dipping similarly

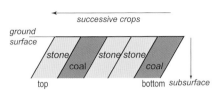

B Variation in angle of stratal dip: more coal is mined per unit depth for shallower dips

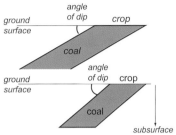

C Contrary stratal dips as seen across the Midlothian coalfield (the first described synclinal fold)

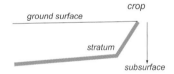

D Contrary dips with a central reversal (a syncline & subsidiary anticline

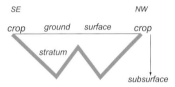

E Lateral variation in stratal dip (part of a monocline)

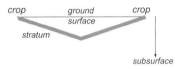

F Fife strata '...made to ly like a bowe...' (an anticlinal fold intruded by a dyke

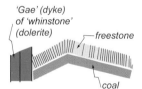

G Sinclair's original 1672 woodcut of **F**

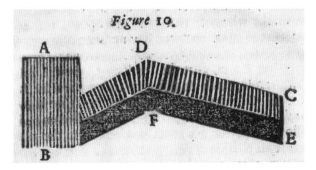

Figure 2.2 **A–E** Modern versions of Sinclair's geometrical stratal diagrams, somewhat simplified and with his alphabetical notations removed. **F, G** The Fife anticline in a recoloured version and the original woodcut based on Sinclair's field observations taken along the Fife coast. These are the first of their kind ever published. Annotations are by the present author.

> The *Dipp*, and *Rise*, are nothing but a declining of the whole body of the *Metalls* [bedrock strata]…This declining or *Dipping*, of the *Coal*, is sometimes greater, and sensible, sometimes lesser, and almost insensible. There being some, that if you consider the declination, it will not be found one foot in ten; some one foot in twenty, or one in thirty. Whereas in others it will be one foot in three, or one in five…this general Rule, that, having found your *Dipp* and *Rise*, to what ever Points that *Course* is directed, the *Streek* is to the quite contrary. For supposing a *Coal Dipp* SE…Then it must needs follow, that the *Streek* must run SW, and NE, which two *courses* [i.e. dip and strike] divides the Compass, at *right Angles*.

He found that whole groups of interstratified coals and strata share the same spatial orientations at particular locations. He made the caveat that their 'regular course' – tabular bodies of given extent and thickness – might be interrupted by 'accidental' features such as 'troubles' (dykes and faults; *see* further below).

He observed that coal and strata together dip underground from the near-surface, or, putting it conversely, rise from the depths to the surface, where the material was said to 'crop out' or, more quaintly, to 'rise to grass'. As the cropping is approached he noticed that strata were altered to softer material; as sand from sandstone or as 'inflammable dross' in the case of coal. He illustrated (Fig. 2.2A) a succession of coals and strata lying in course one above another. Any stratum immediately above a coal defines the coal's 'roof', i.e. the top of the working space that a coal-hewer would work to. His figures also geometrically illustrated the notion that the spacing of coals at outcrop depended upon both the original separation (thickness) of coal and roof stone as well as the amount of their dip. Since dip can vary from near-horizontal to near-vertical, the lower the dip the greater the amount of coal that can be obtained to a given depth (Fig. 2.2B). Simple trigonometry using three illustrative gradients was applied to solve the depth needed to dig a shaft in order to mine coal found at outcrop. Field examples of each type of dipping coal are given for the district around Tranent. Changing surface elevation was not generally accompanied by any similar proportional rise and fall of dipping coal. This implies that whatever the reason for changing surface elevation, it was not the same reason as that which controls the dipping.

Contrary Dips: Structural Mapping of a Whole Coalfield Syncline

Sinclair made the major discovery that certain dipping coal-bearing strata may, after some distance, reverse their dip direction to rise once more from a low-point towards the surface. The opposing outcrops were of the same material continuously joined at depth (Fig. 2.2C). He so defines the geometry of a synclinal fold – a downfold. His underground observations and measurements proved that the trend towards contrary dips occurs in a regular manner when unaffected by 'troubles' (faulting). Such observations obviously needed to be made where physical obstructions to measurement were absent. His phrase that 'troubles…hath cut them off…' implies that features such as faults and dykes physically cut through strata, placing them elsewhere.

In other examples of opposing dip it was indirect evidence that held sway, the nature of strata and the coal itself being exactly the same across an extent of coalfield where there were no intermediate exposures, the rock lying above and below the coals having the same particulars and course. Today we call this 'mapping by extrapolation', a basic tenet of geological mapping in areas of incomplete exposure. In doing so he defined what we now call the Midlothian syncline – the first account of a regional geological structure ever made. An extract from his description of its course (a litany of Lothian place names and aristocratic landowners), together with a modern-age map and section, may be found in Chapter 24.

Other Contrary Courses

Sinclair used the term 'coallery' for both the technical vocabulary in coal prospecting and for the rational approach necessarily used in that profession. It is the first printed use of geological language. He referred to the persons who dig out the coal not as miners but 'Coal-hewers' – some of his coallery terminology most probably gained from the vernacular words used by them in their labours. He uses recognizable terminologies for the first time, as in 'the body of the Coal…' with the implication that such bodies had three dimensions – two of lateral extent and one of thickness. The extent of coal beneath the surface defines what he terms as a coal 'field' – also his use of 'bed' for body as in 'flat-bed Coal' for a horizontal course. He emphasized that when contrary dips are found (i.e folding has taken place) the strike must necessarily rotate in a curvilinear fashion as it gradually changes to a contrary dip. Further, 'troubles' may also cause local abrupt changes to the strike, as he observed around Prestonpans. He also describes a small upfold within an overall downfold (Fig. 2.2D) and a structure that is just the half limb of an anticline, what we would nowadays term a monocline (Fig. 2.2E).

He reasons that since coal is a solid physical body contained between concordant strata, those strata above the coal define its 'roof' and those below its 'pavement' (or 'thill' in dialect). Tantalizingly he does not then say that, as in a house, the pavement is built before the roof! He notes that the depth to a coal is constant along the strike on level ground or when the strike is parallel to the gradient and trend of sloping ground surface. Outcrop of coal or freestone depends upon the angular difference between surface slope and strike, occurring when elevation of ground surface coincides with elevation of the coal surface – a geometric concept that underlies our modern understanding of structure contours. Although Sinclair saw no general orientation of the dip and strike of coal-bearing rocks over midland Scotland as a whole, in all examples he had seen (in Midlothian especially; but he admits he has not seen all) the strike is commonly to the southern and northern quadrants with NE–SW strikes giving much longer outcrops than SE–NW ones; seven and four miles respectively.

On 'Trouble' and 'Gaes'

'Trouble' is the general term Sinclair introduced from the coal-hewer's vernacular for 'gaes' and 'dykes' (respectively as faults and dykes of 'whinstone'). These are nuisance features for miners since they alter the course of coals and their associated strata and juxtapose worthless hard rock against coal or ore, which somehow need to be relocated by the forward-excavating hewers. He often uses the two terms synonymously. To us, faults are fractures that displace rock masses upwards, downwards or sideways, while dykes are intrusions of hard igneous dolerite ('whin' in dialect, from the gorse that commonly infests them at outcrop). Dykes might occur along faults, but not necessarily so. Sinclair ascertained that both may have any orientation, parallel to dip or strike or oblique to either. He thought there was no indication of their own course of dip or strike from within the dyke rocks themselves, though at one point he was able to follow a dyke at surface for a very great distance along its own strike, from Prestonpans to Leith. On the modern geological map such dykes are seen as part of an east–west trending late-Carboniferous swarm across midland Scotland. They are technically known as quartz-dolerites, his continuous dyke nowadays mapped more hesitantly as disconnected segments.

He noted that dykes and faults affect the nature of coals and sandstone in their proximity. Faults exhibit pulverized rock along their course (our fault-rock) and he seems to include such altered rock in his use of the term 'gae'. More significantly, he observes that dykes may render adjacent coals crumbly, 'as if they were already burnt…' – a tantalizing link forward to James Hutton here, and the Plutonist strand of arguments for the molten origins of crystalline rocks like dolerite. More generally, he writes that faults and dykes alter the local dip and also the strike of coal and freestone, causing an increase or decrease of dip as they are approached (we would say causing a 'roll-over' or 'reverse drag'), providing a warning of an impending 'trouble' ahead. In one case he is able to trace a fault dying out so that the displacement of a coal seam becomes zero at its tip:

> the gae wearing out towards the West, the two parts of the Coal that was separated by it, joynes themselves again, and continues in one body, as they were before separation.

Sinclair's technique for solving the practical day-to-day problems arising from discovering a fault or dyke cutting across a coal-face came from a knowledge of stratal dip, and the succession of coal-bearing strata. For thin dykes with small displacements the main coal might be found again by simply hewing through and finding identifiable stata again some short distance away. But as he points out, faults and dykes vary greatly in their scale, and the trick and practice of 'Coallery' is to recognize that large dykes usually show no signs of a particular mined coal seam and its attendant strata anywhere near their other side. To counter this Sinclair describes what we now know as the fault or dyke surface, which he describes as being like a 'vise' [dialect for 'fissure'], the line of fracture along a margin, which he says should be examined before mining into its 'other side'. He makes the margin itself a spatial object with its own strike orientation and dip. Referring to this dipping surface he describes its dip as 'downward' when it slopes away from the mined level and descends below the mine floor. In modern parlance the working miners as they work through such a normal fault must dig downwards to find the displaced coal seam.

Finally, in a remarkable paragraph, Sinclair describes and illustrates a coastal outcrop at Miln-Stone, between Burntisland and Kinghorn in Fife. This shows a domal, convex-upwards form, the coal and freestone strata having been '…made to ly like a Bowe…'. Such an arrangement defines an anticlinal fold, the structure in question bounded by a whinstone dyke whose contact on the fold's northern side is said to '…quite cut off…' the coal and freestone strata. His sketch illustrating all this is the first ever geological section made from a field outcrop (Fig. 2.2F, G).

Dud Dudley: A Near-Contemporary

We have one other person contemporary to Sinclair whose written testimony on matters of coallery may serve for brief comparison. In 1665 Dud Dudley, then around sixty-five years old, the natural and favourite son of Edward, Lord Dudley, published a short booklet, *Metallum Martis*, on his life's thwarted ambitions to patent and put into mass production a secret method of smelting iron ore using coal instead of charcoal.[7] We shall return briefly to his central claim in the next chapter; but he also presents us with his knowledge of the succession of coals in the wider Dudley area of the South Staffordshire coalfield (*see* Fig. 19.2) and the manner of their mining:

> the uppermost or top measures of Coles are ten, eleven, and some twelve yards thick; the Coles Ascending, Basseting, or as the Colliers term it, Cropping up even unto the superfices of the Earth, and there the Colliers formerly got the Coles…But now the Colliers working more in the deep of these Works, they are constrained to sink Pits some of which Pits are from eight unto twenty yards deep, and some are near twenty fathome deep…I might here give you the names, and partly the nature of every measure, or parting of each cole lying upon each other; the three uppermost measures are called the white measures for his Arcenical, Salsuginos and Sulphurious which is in that Cole; the next measure, is the shoulder-cole, the toe-cole, the foote-cole, the yard-cole, the slipper-cole, the sawyer-cole, and the frisly-cole…I might give you other names of coles but desire not prolixity…

We see that Dudley knew his local coal succession, but in contrast to Sinclair has nothing to say concerning accompanying strata or the reasons for the coal's 'ascending', i.e. its spatial orientation and organization, of sometimes being shallow and sometimes deep.

Coda: 'these unhappy Wretches…'

It is easy to find sympathy for George Sinclair as a consulting geologist in pre-industrial times. He would have descended along a gently sloping adit or down a shallow shaft, making observations underground and writing down his compass measurements and notes. All in the flickering light of a tallow candle or rushlight held by one of the 'hewers' or perhaps by one of his Edinburgh students acting as a field assistant. All mine work calls for a cool head and the ability to work in dangerous spaces quickly and with due care for safety. The advent of gunpowder led to the explosive removal of rock material when the mine personnel would evacuate the workings before the shots were fired, the dangers well appreciated by Celia Fiennes.

But what of the anonymous 'hewers', the men who dug the coal underground, and whose families often undertook the essential ancillary duties of shovelling, creel-loading, creel-dragging and ladder-climbing to the pit head with their father's daily production? Such largely invisible existences were lived out at the very margins of safety, endurance, health, nutrition, suffering and human decency – a state of affairs that must be taken as a common thread throughout the early development of the British mining industry in general.[8] It met with eloquent protest from the likes of Edmund Burke in1756:

> I suppose that there are in Great Britain upwards of an hundred thousand people employed in Lead, Tin, Iron, Copper and Coal Mines; these unhappy Wretches scarce ever see the Light of the Sun; they are buried in the Bowels of the Earth; there they work at a severe and dismal task, without the least Prospect of being delivered from it; they subsist upon the coarsest and worst sort of fare; they have their health miserably impaired, and their Lives cut short, by being perpetually confined in the close Vapour of these malignant Minerals.[9]

Worse, under Scotland's merciless Poor Laws, many miners were *de facto* legal serfs (*see* Chapter 24). That state of affairs lasted until the advent of late-Georgian legislation and mid-Victorian reforms (Chapter 13) that led to some amelioration of conditions for children and females – later unionization helping to cement a stratum of human decency.

New Reductions: Iron and 'Charking Coles'

Every Farm has one Forge or more; so that the Farmers carry on two very different Businesses, working at their Forges as Smiths, when they are not employed in the Fields as farmers. And all their work they bring to market, where the great Tradesmen buy it up and send to London…We cannot travel far in any direction out of sound of the hammer.

Daniel Defoe in *A Tour Through the Whole Island of Great Britain*. Him travelling through the Dudley area of the West Midlands.

Introduction

There are a few occasions in Defoe's 'Tour' where a ready opinion of his flies in the face of both contemporary and later judgement. One such involves the rising expense and scarcity of charcoal, used since early Iron Age times for the chemical reduction of ores in iron smelting. Here is the offending passage, written in Book 2 of the *Tour* around 1722–1723 as he leaves Battle in Sussex and rides northwards for the High Weald and the fleshpots of Tunbridge Wells. This part of the Weald was the long-established centre of iron smelting in southern England, making use of abundant Early Cretaceous clay ironstone ores, identical in their make-up and mode of occurrence to the Carboniferous ores to be discussed below. They had formed the basis for iron working in the Weald for more than two and a half thousand years. Defoe, of course, knew nothing of the geology, but in keeping with his general philosophy of dutiful enquiry, he:

had the curiosity to see the great foundries, or iron-works, which are in this country, and where they are carried on at such a prodigious expense of wood, that even in a country almost over-run with timber, they begin to complain of the consuming of it for those furnaces, and leaving the next age to want timber for building their navies. I must own however, that I had found that complaint perfectly groundless, the three counties of Kent, Sussex and Hampshire, (all which lie contiguous to one another) being one inexhaustible store-house of timber never to be destroyed, but by a general conflagration, and able at this time to supply timber to rebuild all the royal navies in Europe, if they were all to be destroyed, and set about the building them together.[1]

The modern reader juggles this opinion with the bare historical facts gained by many economic historians over the years – that there was indeed a crisis in the availability and price of charcoal at this time. Indeed, almost exactly one hundred years before, the shortage of timber for iron-smelting in England was featured in a 1620 prospectus by the nascent Virginia Company as one of the reasons why North American colonization was a splendid thing:

> The iron [smelting industry], which hath so wasted our English woods, that itself in short time must decay together with them, is to be had in Virginia… for all good conditions answerable to the best in the world.

Such signs of crisis became particularly evident in the early eighteenth-century flight of ironmasters from the Sussex Weald to other sites in the country where clay ironstone deposits were available to be quarried and mined from still heavily forested private demesnes with plentiful swift-running water for working hydraulic machinery. These places were all located on Coal Measures bedrock and included the Forest of Dean, north-east Glamorgan, north-west Monmouthshire, the South Riding of Yorkshire, north Derbyshire and eastern midland Scotland around Falkirk; also to the lowlands of Furness and Egremont in southern and western Cumberland, where high-grade hematite ore was available from Early Carboniferous limestone strata.

Many of the small-scale iron furnaces that grew up in these 'new' iron-founding districts were in fact former monastic smelting sites of the Cistercian order – pre-Dissolution 'bloomeries'. These smelted at a lower temperature than blast furnaces and were so-named from the cauliflower-like masses of smelted iron that dribbled out of them to the ground ready to be annealed into wrought (bar) iron by water-powered trip hammers. Blast furnaces (a Chinese invention of over 2000 years before) were introduced to the Sussex Weald from Namur in Belgium in the late sixteenth century, from there spreading to all iron smelting districts. Blast-furnace smelting soon melded with forge- and casting-work and the conversion of raw pig-iron by charcoal-refining and hammer-working to rod and bar iron. These now supplied scores of independent ironmongers across the land who specialized in iron products such as plough shares, gun barrels, equestrian tackle, railings, fire grates, locks, nails, chains, wire and much else. That many of these determinedly independent and canny men and women were Nonconformists and Quakers is well recorded, particularly in the West Midlands. Yet everywhere charcoal was becoming dearer as demand for iron grew and charcoaleries spread further and further from the furnaces – the enhanced transport costs further increasing the price of the raw material and also, inevitably, that of the smelted iron. But this was soon to change.

Abraham Darby I: Early Achievements in Bristol

In the first decade of the eighteenth century, at a time when Daniel Defoe had been charged in London with sedition and made to stand in pillory, one young, ambitious and inventive man of Quaker calling, Abraham Darby, a native of Dudley near Birmingham, was making a living in Bristol as a brassmaster.[2] The road to this golden occupation came via a local and family history of forge iron work mixed with farming in the general Dudley area[3], as Defoe observed. Darby had developed experience of forge work with his father while growing up, yet he seemed to have viewed the growing crisis of charcoal supply to iron furnaces to be a superable problem. This confidence doubtless came from his experiences of working in Bristol, first apprenticed to a manufacturer of iron-ware for malt brewing equipment, and subsequently, from around 1699, as a manufacturer of malt mill engineering parts in his own right. He most certainly would have known of the production of coke from certain coals since the substance allowed the slow, virtually smokeless, odourless and regulated heat necessary to lightly roast sprouting barley on maltings floors. This Derbyshire-invented process of brewing was considered as an '…alteration which all England admired' – the resulting pale ale so lovingly tasted and toasted by Celia Fiennes in Chesterfield in 1697.

Also, probably, he would have heard of, or read for himself Dud Dudley's aforementioned *Metallum Martis* with its claims for iron smelting using coal and of the patents taken out for it. The earliest of these dated from 1620 when Lord Dudley obtained a royal patent from James I for his twenty-year-old son, who was then running his ironworks at Pensnet in Worcestershire. No previous claim or patent seems to have been successfully seen through, and though Dudley claimed he had produced limited supplies of pig-iron by his own method he never gave proof or details of the processes involved, so it is not known whether he had 'charked' his favoured 'smal coal' into coke or not. A favourable view of his claims recognizes that his knowledge of South Staffordshire coal and ironstone successions includes specific mention that the slipper-, sawyer- and frisly-coals 'are the best for the making of Iron, yet other coles may be made use of.' A generous view might indicate that Dudley knew from experience that perhaps the majority of coal seams were not suitable for smelting because they could not make coke. A tantalizing account in *Metallum Martis* of the long, slow, underground burning of 'smal coles' might confirm this view, but the hard evidence is not there. Should we then doubt or accept Dud Dudley's veracity, given his lifetime of bad luck in the iron industry (major floods, sabotage, obstruction, civil war)? It is possible that he may have smelted clay ironstone with coke, but even so there is no documentation that the product was ever successfully marketed, probably because it was too brittle.

Darby's Bristol brass career had none of Dudley's disasters. In 1702 when he left off the manufacture of brewing ironware, he co-founded his own brass company with several Quaker friends. Before this time the Dutch had pioneered and monopolized direct casting of brassware into dry sand moulds. After a visit to the Netherlands and the hiring of Dutch casters in 1704, he began his new venture. As we have seen from Defoe's observations, a growing market existed amongst the better-off for brass merchandise for both domestic and ornamental use. The business became successful, yet Darby returned to his first love of ironworking, probably in his spare time and making use of furnaces provided by friendly co-religionists in the Bristol iron trade. He showed his talent for lateral thinking and improvization by getting his Netherlanders to try casting iron directly from a blast furnace into sand moulds. His aim was to produce domestic ironware that was cheaper and better than brass. For, as any modern *le Creuset* owner knows, cast iron, though always brittle and failing under tension, is a superior material for long, slow and uniform cooking. Though the process was initially unsuccessful, he persisted with help from a young fellow-Quaker, one John Thomas, who during an apprenticeship and after many trials succeeded in direct casting. Thomas's daughter wrote late in life that, after the failure of the Dutchmen's efforts, her father had:

> asked his master [Darby] to let him try, so with his leave he did it, and afterwards his Master and him were bound in Articles in the year 1707 that John Thomas should be bound to work at that business and keep it a secret and not teach anybody else, for three years. They were so private [in their experiments] as to stop the keyhole of the door.[4]

What sparse but gripping lines that tell the story of an extended and secretive effort by the two men! The outcome was that within a short time a patent was issued for the casting methodology and Darby, perhaps not encouraged by his fellow brass-company directors to branch out into ironware, made the 'momentous' decision (as Arthur Raistrick put it[2]) to sell his brass company shares and move from Bristol to the heart of Shropshire. Here he was to use his capital to take over and refit for his own purposes an existing blast furnace and foundry in Coalbrookdale.

Coalbrookdale and the Art of Coking

With the whole of the mineral-rich English West Midlands to choose from, why did Darby choose Coalbrookdale? It was a left-bank tributary to the navigable Severn, at the time one of Europe's busiest waterways, leading downstream to Gloucester and Bristol with important markets (and communities of Friends) on the way and also upstream via Welshpool, Shrewsbury, Stourbridge, Wolverhampton, Birmingham, Dudley and many other towns. Their markets were all within easy or reasonable travelling distance, and where household cast-ironware of high quality and relatively low cost could and would soon be sold, often by Darby himself. Apart from the obvious advantages of this major navigable river location, we can infer that the intensely-focused and rational Darby was drawn to Coalbrookdale for several linked reasons in addition to a pre-existing furnace site. The valley had variably steep topography with perennial streamflow for generating ample water power and adequate woodlands should the use of charcoal prove unavoidable. Its bedrock of Carboniferous and older strata provided a complete range of cheap, reliable raw materials for hearthstone and furnace linings and for the furnace itself – clay ironstone, coking coal, and limestone fragments. The coal and iron ore were cheap to mine on account of their simple geological structure further discussed in Chapter 19.

It seems clear that from his arrival at Coalbrookdale in 1708 until at least the second blast of 1709, possibly earlier, during and after the months of refitting the plant, Darby experimented with small-scale, open-air coking.[4] We have no idea exactly how he proceeded, but it was presumably 'charked' *in situ* in smouldering piles covered with earth in the works grounds (*see* Fig. 3.1). A letter written sixty years later by Darby's daughter-in-law, Abiah, drew on both the memories of her husband (Darby's son Abraham II) and also that of a workman who had been at the furnace works from the outset. After arriving in Coalbrookdale, Darby and his partners:

> cast iron Goods in sand out of the Blast Furnace that blow'd with wood charcoal; for it was not yet thought of to blow with Pit Coal. Sometime after he suggested the thought, that it might be practicable to smelt the Iron from the ore in the Blast Furnace with Pit Coal: Upon this he first try'd with raw coal as it came out of the Mines, but it did not answer. He not discouraged, had the coal coak'd into Cynder, as is done for drying Malt, and it then succeeded to his satisfaction. But he found that only one sort of pit Coal [ie from one particular seam] would suit best for the purpose of making good iron…He then erected another Blast Furnace, and enlarged the Works. The discovery soon got abroad and became of great utility.[5]

Darby chose the most suitable coal for coking from the nearest uphill mines around Madelely, probably for the first blast but certainly for the second. He had mastered the choice of both coal seam – that now known as the Clod Coal, though others would also coke, notably the Fungous, Top and Double Coals[6] – and the coking process itself, so that enough first-rate coke (strong, porous and clean) was available.

In January 1709 the stage was set for the historic first blast in the reconditioned furnace (Fig. 3.2) – the world's first commercial iron smelting with coke. All was in place – furnace hearths and water-wheel driven bellows had been repaired, the casting-floors set up by young John Thomas with moulding sand and storage facilities provisioned with ample fuel, ore and limestone. The preserved Coalbrookdale accounts examined by Arthur Raistrick[4] show that no charcoal was brought into the furnace works at this time. They clearly state that in late-1708 and early-1709, prior to the first blast and a few months afterwards, 'Charking coles' (ie cakeable coal suitable for coke-making) were supplied in small quantities. For the second blast, the accounts list 'coles' supplied from the adjacent Madeley Wood collieries by the owner, one Richard Hartshorne, in July 1709. There was a great deal of such coal supplied, worth over sixty-five pounds: if the pithead price was around three shillings per tonne that makes several hundred tonnes.

Figure 3.1 *The Late Works at Coalbrookdale* by Francois Vivares, 1758. This hand-coloured engraving after artwork by Thomas Smith and George Perry is a historic view of mid-century Coalbrookdale. The foreground has a file of six horses towing a cart bearing a huge cast iron steam engine cylinder. Not perhaps an everyday sight on the country lanes of Shropshire but a regular one here, for by this time the Coalbrookdale Company had set up adjacent ironworks that were producing parts for enhanced and powerful Newcomen-type steam engines. The one shown here, its bored cylinder looking to be over a metre in diameter with a stroke of two or so metres, is in transit uphill to the works at Horsehay or Ketley. The view is south-west over the Coalbrookdale Upper Works with the Upper Furnace Reservoir to the right (at this time water-wheels drove the furnace bellows) with four adjacent smoking coke heaps. A roofed pleasure barge on the reservoir, rowed by an oarsman in the prow with two couples in it, drifts innocuously through the industrial landscape. Down the steep incline above the boat a carter is braking his two-horse wagon full of baskets of either coal or iron ore, perhaps from Coalmoor to the north-west. In the works itself the two central tall iron-hooped chimneys issue from reverberatory (hot-air) furnaces used to remelt coke-smelted pig-iron to convert it to bar-iron in a process invented by Abraham Darby II a couple of decades earlier. The Upper Furnace is the low, square building just to the right of the air furnace chimneys with a wider effusion of smoke from its ongoing blast. The possible position of the artist is shown in Figure 19.3. Source: Reproduced by permission from the copyright holder: The Sir Arthur Elton Collection, The Ironbridge Gorge Museum Trust.

The coking process involved slow burning to drive off volatile constituents from the bituminous coal – gases like methane, sulphur dioxide (several of the local coal seams were sulphurous), carbon dioxide and water vapour.[7] Also the coke had to cake and compact into brittle but strong biscuit-like vesicular lumps or nuggets so that while burning and emitting carbon monoxide (the reducing agent) in the furnace it could also withstand the weight of the overlying charge of ironstone and flux. Not every coal had

Figure 3.2 Coalbrookdale Museum (52.639, 2.492). The original Coalbrookdale furnace of 1658 was restored for operation by Abrahan Darby I in the early 1700s and rebuilt by his grandson in 1777. It is nowadays preserved as a hallowed industrial monument sheltered by the beautiful austerity of its protective shield of glass and steel, part of the Coalbrookdale Museum complex, around the outer shell of the Upper Works. The enshrined furnace is seen through the glass from the hearth side, one of whose original openings served as a tapping point for the molten iron into an adjoining casting house; the other opening was for the tuyère – the entry point for air from bellows worked by water wheel and (later) compression cylinders. Source: Wikimedia Commons via File: Darby Furnace UK.jpg. Photographer: Helen Simonsson.

the required composition and make-up to achieve this compacting ability. Later studies[8] determined that the best coking coal was from soft, 'bright' seams rich in vitrinite (*see* Chapter 10) with a low volatile content of less than thirty per cent.

Coda

Abraham Darby deserves history's recognition as the person responsible for the all-important practical measures necessary to smelt iron with coke, ensuring at the same time that coal gained added value when so transformed. Yet at the time of his premature death in 1717 the perfection of coke smelting to produce other than pig-iron for casting would take more than the lifetimes of both his son and grandson to achieve. The reasons for this delay were due to the chemical and metallurgical complexity of the coke-based smelting process when applied to clay ironstone ore from the Coal Measures. The pig-iron so produced had a relatively high phosphorous content that rendered it difficult to work by forge tilt-hammers to obtain malleable, fibrous-textured bar (or wrought) iron. The same strictures did not apply to the low phosphorous hematite ores of West Cumberland.

Despite these drawbacks, Darby's Coalbrookdale pig-iron reigned supreme when cast into pots, pans, fire grates, kitchen ranges and steam engine parts (*see* Fig. 3.1), to name but a few products. For other essential uses in the iron industry there was an active trade in superior bar iron that permeated Europe and Russia. This was the fabled low-phosphorous *öregrund* iron from Sweden, whose use in steel making had no equal. It came especially from the magnetite ores mined and smelted in the Uppsala area of southern Sweden – the famous Dannemora Mine, the Vallonbruk ironworks, and in particular the Äkerby and Lövsta forges.[9] Most was imported into Britain through the ports of Bristol, Hull and Liverpool to steel producers in the South Midlands, Lancashire and Yorkshire. Therein lies the tale of the rise of Sheffield, which will emerge when regional industrial history is considered later in this book.

Many iron manufacturers in Britain persisted with the use of charcoal using Cumberland hematite ore, the Rawlinsons of Backbarrow doing so for over a hundred years. From the 1750s onwards the pace of technological progress quickened in Britain: furnaces used hot air blasts; indirect heating was used to burn off impurities in reverberatory furnaces; 'puddling' techniques were re-invented (the original efforts date from Song dynasty times in China) by Henry Cort. There was also large-scale importation of Cumbrian ore, particularly to the blossoming South Wales iron industry, where it was known as 'red ore'. Finally in the mid-nineteenth century Henry Bessemer's eponymous 'converter' was initially able to produce copious mild steel from pig-iron smelted from hematite ore – the steel age arrived speedily, and on steel rails. Then, as economic historian Robert Allen comments: 'the French, Germans and Americans [hitherto all charcoal users] moved directly to the most advanced technology…'.[10] For over a hundred years British pioneers had worked everything out from scratch.

4

Steam Engine Works: Newcomen to Watt and Boulton

I have a letter from Watt – he has brought his curious wheel
to go, it works by steam – he says it appears to be right in all
essentials and goes with an equitable motion and great power.

Letter from James Hutton to George Clerk-Maxwell, August 1774

An Economy Set Free by Coal

Following Abraham Darby's pioneering efforts, the metal masters were at last able to
break free from the restrictions placed by charcoal availability and price. Further, once
the canal system was in place by the 1770s, cokeable coal was cheaply available at near
pithead prices from mines in most of the many British coalfields. From the most calorific
energy source on Earth at the time sprang an abundance of cheap pig-iron that trickled
and glowed from the tappings of scores, then hundreds and thousands of blast furnaces.
As a consequence, production of coal rose ever upward with the opening of larger and
deeper pits. Late-medieval villages were transformed almost overnight with rows of
new-build brick terraces cheek-by-jowl with medieval timber-frames, and even Anglo-
Saxon churches, as at Escomb in County Durham.

The parallel boom in shaft mining of non-ferrous metalliferous ores (tin, copper, lead
and zinc), most of which could eventually be coal-smelted in reverberatory furnaces,
brought renewed mining ventures into the Early Carboniferous and older bedrock of upland fells and moors in south-west England, the south and north Pennines, North Wales and
the Southern Uplands of Scotland. These fed a new generation of large smelters that grew up
in both coastal (Swansea, Avonmouth, Chester) and many inland near-coalfield locations.
Enlarged settlements led to the growth of the highest market towns and working villages in
the realm, as at Alston in Cumbria and Leadhills/Wanlockhead in southern Lanarkshire.

The consumption of coal was further increased as new inventions utilizing steam-
driven power offered keen competition to water power in many industries, most notably

in cotton, linen and woollen textiles. Coal also satisfied an entirely new market – the stoking of the vast boilers of steam-driven engines, whose evolution we now trace.

The Mine Drainage Problem

As mine shafts were sunk ever deeper and working levels branched out to unheard-of distances from shaft bottoms, a natural check to the efficiency and economic margins of the mining industry came about. This was the unceasing egress of groundwater into workings, together with the need for more thorough ventilation: the two chief engineering problems of deep shaft-mining. Although throughflow of water is ever present below ground surface, it is usually containable until the excavation of rock reaches a certain depth and the permanent water table is penetrated, leading to the continuous (though usually slow) incoming of water. Traditional hand- and animal-propelled chain pumps could not cope with such water volumes, or at least they became uneconomic to use. The one remedy to hand was an ancient one based on the recognition that shallow groundwater could drain into gently sloping adit tunnels (aka soughs) and flow off downhill by gravity, like an underground stream. The downside was that labour had to be hired to blast and dig out the rock: ventures that might lead to no immediate financial reward for mine-owners and shareholders, unless of course new mineral reserves were discovered *en route*, as they sometimes were.

Adits were driven upwards on a gentle incline from valley sides or coastal cliffs through bedrock towards the often distant mine workings, as unnamed Roman military engineers, Agricola and George Sinclair were well aware. Such drainage conduits often doubled up as access tunnels for miners, pit ponies and ore if water flow was sufficiently low. But a mine can be drained by gravity alone only to a certain limit, defined by the deepest possible open 'drainage adit'. Above this depth ('above adit') the mine workings remain unflooded; below, flooding would occur unless pumps were operating. Adits were still useful in deeper mines because waters pumped from below needed only to be raised to the level of the drainage adit, not necessarily all the way to the surface. Great cooperative schemes were set up to share the benefits and savings of adit drainage by driving them through groups of adjacent mine properties and their coal seams and ore veins. Such were the sixteen-kilometre Milwr tunnel system of the North Wales (Halkyn) lead/zinc mining field (Chapter 18) and the sixty-five-kilometre Great County Adit in Cornwall that drained the entire Gwennap district tin and copper mines.

Yet by the early eighteenth century the inexorable exhaustion of coal and metal ores at shallow depths led to crisis as mine shafts reached below the limit set by adit drainage. Alternatives to draining by gravity were now urgently needed.

Denis Papin and Newcomen's Great Engine

Denis Papin[1] was a brilliant French (Huguenot) emigré and inventor, a friend of Boyle, a correspondent of Newton, who had studied under Huygens and Leibniz. He

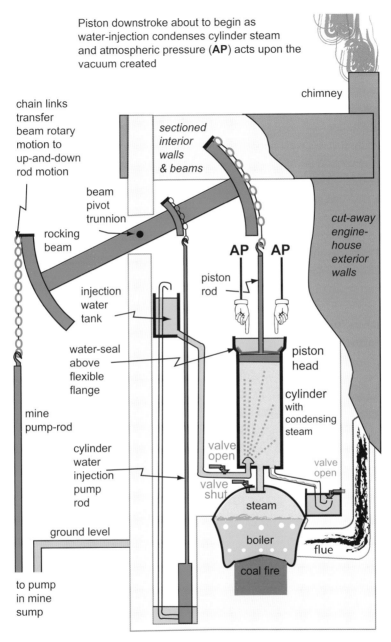

Piston downstroke about to begin as water-injection condenses cylinder steam and atmospheric pressure (**AP**) acts upon the vacuum created

chimney

chain links transfer beam rotary motion to up-and-down rod motion

sectioned interior walls & beams

beam pivot trunnion

rocking beam

cut-away engine-house exterior walls

AP   AP

piston rod

injection water tank

water-seal above flexible flange

piston head

cylinder with condensing steam

mine pump-rod

cylinder water injection pump rod

valve open

valve open

valve shut

steam

ground level

boiler

flue

coal fire

to pump in mine sump

Figure 4.1 Diagrammatic version of a Newcomen steam engine. Redrawn, modified and colour-crafted from an original cutaway drawing by Cossons (1993). For typical scale compare with the Figure 4.2.

envisaged the motive power of steam being channelled against the base of a rod tight-fitted inside a cylindrical tube, a bit like a muzzle-loading gun or cannon barrel. Such an arrangement allowed a sliding motion to produce transferable mechanical energy that could do useful work by moving superimposed loads – the piston-cylinder mechanism. First demonstrated in 1690, twenty years later Papin deposited accounts of his piston mechanism and related inventions (including the first pressure cooker) with the Royal Society of London.

Papin's discoveries were open to inspection by fellows of the Society, and word must somehow eventually have reached the ears of that inventive and practical Devonshire man, one-time ironmonger, Thomas Newcomen in Dartmouth. After many years of effort his 1712 invention utilized the Papin cylinder idea (or perhaps Newcomen made an independent and parallel discovery of his own – we do not know) in a completely novel context – that of a coal-powered, reciprocating beam-balance steam engine (Fig. 4.1).[2] This obtained its mechanical energy from steam by moving a pivoting beam via a smooth piston within a brass cylinder so that the piston's natural upstroke due to its own weight was reversed as the steam in its cylinder was evacuated. The vacuum that drew down the piston was induced by water-injection, quenching steam already let in from the coal-fired boiler at atmospheric pressure. Re-introduction of more steam into the cylinder followed the upstroke of the beam member. The whole cycle repeated endlessly, and led to reciprocal up-and-down motions translated to the pump rods by the ever-rocking overhead beam. This was emphatically not a pressurized steam engine (hence its description as 'atmospheric') like that of Papin's experiment or the later invention by Cornishman, Richard Trevithick. At the time it was much the safer for that, since metal welding technology was insufficiently developed at that time to safely maintain highly pressurized vessels.

However, in using power from a coal-fired boiler to produce steam and condensing it to pump water, Newcomen had duplicated the bare principles of a previously-patented (but practically useless) invention – a sort of siphonic thermal pump that did not generate mechanical power. It was designed more as a 'concept model' by military engineer (ex-Captain) Thomas Savery. In the words of the 1698 patent, it was:

> A new invention for raising of water and occasioning motion to all sorts of mill work by the impellent force of fire, which will be of great use and advantage for draining mines, serving towns with water, and for the working of all sorts of mills where they have not the benefit of water nor constant winds.

Such puff was wildly optimistic, and with hindsight it is clear that the Savery patent should never have been extended to the Newcomen mechanical engine. The Savery 'pump' was also a totally impractical object to use underground, yet the words of the patent describing the 'raising of water…by… fire…' were broad enough at the time to stop Newcomen patenting his new, thoroughly practical, safe, novel and ingenious solution to the world's mine drainage problems. It forced him to go into partnership with

Savery who, doubtless with connections in high places, was able in 1699 to get the term of his own patent massively extended until 1733, by which time Newcomen was four years dead. This was done by an Act of Parliament, no less – the 'Fire Engine Act'. James Watt and Matthew Boulton were to go to similar lengths to protect their own patented engine seventy-six years later.

The Cost of Coal

By the time of the patent's expiry there were scores of Newcomen engines pumping water out of British mines; many were also exported abroad. The immense superstructures of the rocking engine heads and cylinders supported on thick stone-built gable-ends of tower-like engine housings with their adjacent chimneyed boiler houses were (and are) iconic emblems of available steam power (Fig. 4.2). We are lucky in Britain that some have been preserved in their working landscapes. Yet their undoubted efficacy over and above manual- and horse-driven pumps and the engineering robustness of the low-pressure steam principle came at a steep price for south-west England mining companies because of the high cost of coal shipped into a region that had no indigenous coal deposits of its own. Another pricing problem was circumvented when, as we saw previously (Fig. 3.1) the Darbys in Coalbrookdale began casting steam engine parts of iron rather than expensive brass.

The expensively dewatered copper mines of the south-west proved vulnerable to competition from lower-cost enterprises. This was brought home by the large fall in world copper prices (almost 50% by 1780) that followed the 1768 discovery of an enormous new extension to the Roman-worked Parys Mountain copper deposit in Anglesey. The three-dimensional deposit could be quarried from the surface, obviating the immediate need for shaft mining and deep drainage (Fig. 4.3). Although large, it was overall a low-grade deposit, the raw ore averaging 4% or so of copper. Yet after clever and extensive surface processing, the mined ore concentrate reached 25% copper – very rich indeed.

James Hutton, the eminent Edinburgh geologist, visited Anglesey whilst on fieldwork in the late summer of 1774, and in a letter to his friend James Watt gives an early account of the Parys workings, the ore mineralogy and enrichment techniques for the treated ore. He wrote to Watt (punctuation is as in the original – Hutton was a free and spontaneous writer):

> Mr Rouse the manager a Derbyshire man is a very sensible clever fellow and the process is admirable – the mine is an immense lump of Pyrites iron mixed with copper they break and pick out the bits of yellow copper pyrites the whiter and poorer sort is kindled in great heaps like lime kilns and burns of itself with the smell & quality of hell fire for 3 months sometimes, when thus calcined the iron pyrites is reduced to a calx the bits of copper pyrites interspersed is consolidated like a kind of reguluss it is put into long ponds made like tarn holes and covered with water for a day or so then taken up

Figure 4.2 A–B The business ends of two Newcomen-type beam engines used for mine dewatering. **A** The Elsecar pumping engine of 1787 a few miles south-east of Barnsley, West Yorkshire is preserved within the wider Elsecar Heritage Site, a former industrial village with forges and a coal mine that worked the famous Barnsley/Top Hard seam (53.293923, 1.251166). It is the only such engine surviving in the world (many hundreds were exported) that is still installed at its original location, residing in an elegant sandstone-built, ashlar-quoined, engine house (*c*.9 m high) with a datestone. The beam-end linked to the vertical pump shaft fed 100 metres underground to the terminal wide-bore perforated iron sump-drainer laid out in the foreground. Refurbishment of the engine in 1983 was by the National Coal Board engineer, Peter Clayton. **B** The pumping engine at Prestongrange mine, Prestonpans, East Lothian (55.570593, 3.003200) is also contained in an ashlar sandstone engine house (*c*.14 m high), this time a substantially more massive structure with a two-metre thick retaining front wall to support the pivot-bearing of its huge iron beam. It is the only example in Scotland of an *in situ* engine in its workplace. It is a Cornish engine with a detached furnace (long demolished), a close cousin of the Newcomen and manufactured by J.E. Mare & Co of Plymouth to the design of engineers Hocking & Loam. It was used in three different metal mines in Cornwall until finally bought by Harvey and Co. of Hayle, Cornwall, who sold it on complete with a new beam of their own manufacture (their moniker clearly seen on the protruding right-hand side of the beam) to Prestongrange Coal and Iron Company in 1874 and shipped north. It operated in the notoriously wet colliery until 1954, when electric pumps took over till the colliery closed in 1962. Sources: Author.

Figure 4.3 Parys Mountain opencast copper mine is perhaps the most iconic and sublime visual remnant of the metal mining heritage of early Industrial Revolution Britain. This view (53.225949, 4.195969) is to the north-east from the lip of the crater from which the copper ore was quarried from the late-1770s and later mined below surface in the nineteenth century. The distance to the disused windmill stack on the horizon is around a kilometre and the crater is *c*. fifty metres deep. Remnants of Bronze Age and Roman mining are also known. The kaleidoscope of ochreous colouring is the result of chemical weathering of the copper-iron sulphide minerals that made up the deposit. Modern runoff into the holding lakes that surround the opencast are classified as 'acid-mine' with a measured pH of as low as 2. Source and data courtesy of Professor Julian Andrews, UEA Norwich.

> buckered or beaten and washed like ore whereby the heavy and hard reguline parts are separated…[so they] concentrate the copper into one which makes the mine a most valuable thing…they send off 3 or 4000 tuns a year to Liverpool.[3]

Watt's Idea

The notion that the Newcomen engine might be 're-invented' to a new patent in a more economic form first occurred to James Watt in Glasgow around 1761 where he was employed as the University's instrument-maker. Examination of a scale model of Newcomen's engine that he had been asked to repair led him to think deeply about how he might be able to increase the machine's efficiency (Fig. 4.4). He pondered over this critical question (it was a sideline to his day-job), until in the spring of 1765, on a walk by the College Green, he had an inspirational idea:

> I was thinking upon the engine…when the idea came into my mind, that as steam was an elastic body it would rush into a vacuum, and if a communication

Figure 4.4 *James Watt and the Steam Engine: the Dawn of the Nineteenth Century* by James Eckford Lauder (1855). Oil on canvas. 147x239 cm. Unique for the great engineers who spurred on the later development of the Industrial Revolution, James Watt had his portrait painted posthumously. It shows him in thoughtful mood in somewhat surreal and eclectic surroundings. Side-lit from the left by an intensely bright but unseen source, he ponders a collage of sometimes shadowy objects. From his definite hand-held dividers and resting hammer, one pans across to the silhouette of the beams to a Newcomen engine, a boiling kettle, a pumping cylinder and, enclosed in a large box, a separate condenser receiving a definite ejaculate of water. The artist is clearly trying to capture the moment of discovery of the 'separate condenser' mechanism that improved the efficiency of the Newcomen engine. The lighting and symbolism should be considered as one with the earlier 'sublime' industrial and volcanic imagery of Joseph Wright of Derby. Source: reproduced by licence courtesy of the National Gallery of Scotland.

> was made between the cylinder and an exhausted vessel, it would rush into it, and might there be condensed without cooling the cylinder…[4]

But how could this be done practically? Realization proved a nightmare for Watt. Money problems over the next five years saw him finally earning his family's main living as a skilled surveyor of canal constructions all over Scotland – the time of the 'canal mania'. Meanwhile his supremely competent, positive and supportive wife, Margaret, looked after the instrument business back in Glasgow. Yet he continued with his engine work in his spare time regardless, and by 1769 had reached the point of a positive outcome sufficient to patent his invention of 'separate condensation'. Yet, unlike the Newcomen engine, now long out of patent, his engine could not yet work economically because of steam leakage from the piston margins. This despite the specially cast iron cylinder provided courtesy of his financial sponsor and friend John Roebuck, proprietor of the pioneering Carron Ironworks at Falkirk in midland Scotland. His patented invention prototype was the 'Kinneil' steam engine, named after Kinneil House, Roebuck's home near Bo'ness, its place of manufacture. But in 1773 disaster struck – Roebuck was declared bankrupt.

The Low Road to England

So it was that in May of 1774 Watt travelled south from Edinburgh to the West Midlands accompanied by his old friend James Hutton. In the next fifteen years Hutton was to revolutionize geology as Watt was to revolutionize available mechanical power from burning coal. Watt had previously written to a Birmingham friend, William Small:

> I have persuaded my friend Dr Hutton, the famous fossil philosopher, to make the jaunt with me, and there are hopes that Dr Black's coming also.

Why was Watt going south with Hutton at this time (Black did not in the end accompany them)? The Kinneil engine itself had already been sent to the Midlands the previous year since Roebuck's bankruptcy. It was a doubly sad and troubled year for Watt; his beloved Peggy, mother of their two young children, had died suddenly while in late pregnancy. His unhappy exodus from Scotland ('I am heartsick of this cursed country', Watt wrote to William Small) to the English Midlands was full of familial grief and the stress of his impossible financial situation.

Watt's migration was the result of years of patient encouragement by one of the canniest and most forceful of men in the early Industrial Revolution. This was Matthew Boulton, a West Midlands polymath and fanatic for coal-generated steam who, as one of Roebuck's chief creditors, took over his share of the Kinneil patent, and in doing so became Watts' saviour.

The spring journey south with Hutton went well, and after a spot of geologizing in the Cheshire basin's salt mines (featured later in Hutton's great opus, *Theory of the Earth*, 1788) both men arrived safely in Birmingham after a journey that lasted two weeks. Watt wrote to his father that it had been: 'a pleasant journey in which nothing remarkable happened.' Nothing for Watt, perhaps, but observations at the salt mines convinced Hutton that their deposits of rock salt were not in fact precipitates from sea water, as widely thought: rather they were the product of crystallization from the molten state – one of the great man's few misreadings of the origin of geological materials. Hutton stayed for six weeks or so in the Midlands, geologizing locally and meeting Boulton, Erasmus Darwin (Charles's grandfather) and other members of the renowned Birmingham intelligensia that formed the natural philosophy club, the Lunar Society. This was so-named from their meetings held monthly at full moon (for personal safety in those pre-street lamp days), the members known affectionately amongst themselves as 'Lunaticks'.

Watt at his new home was now more happily reconsidering his patented engine's design and performance. Hutton left in mid-July to do fieldwork across Wales and south-west

England, a long and tiring stint, on horseback all the way. In a letter written at Bath in August to his friend George Clerk-Maxwell back in Edinburgh, he bemoans four times having to repair his riding 'breeks' and pities himself as a saddlesore seeker after rocks: 'Lord pity the arse that's clagged [stuck] to a head that will hunt stones…'. In addition to further tales of fieldwork and the joys of lodging in civilized Bath, he also writes of momentous news from Birmingham:

> I have a letter from Watt – he has brought his curious wheel to go, it works by steam – he says it appears to be right in all essentials and goes with an equitable motion and great power. In short I believe it will answer (in his own words) – this will raise his fame yonder it being so new a thing for that is what catches the multitude; tho I think the improvement of the reciprocating fire engine [Newcomen steam engine] is the thing of greater utility but everyone will not be sensible of the merits of that great improvement.[5]

He added that Clerk should: 'tell Dr Black of Watts success.' The letter is the earliest confirmation that Watt's separate condenser mechanism was now fully working – what a moment for Hutton to have been able to describe for history! Meanwhile, under the expert eye of Boulton's chief mechanic, Yorkshireman John Harrison, the original Kinneil steam engine was re-assembled in short order and spent the summer happily pumping water back up to the reservoir that fed a water-wheel helping to power Boulton's Soho Works near Birmingham.

A tremendous demand had built up in mining districts in expectation of the new engine's performance. The now-shortened term of its original patent – only eight years were left – meant that the ever-canny Boulton foresaw potential problems for longer term profitability of the new engine, which was still not completely engineered to either his or Watt's satisfaction. His solution was to 'do a Savery': lobbying support for a Private Bill through Parliament to extend the patent for a further twenty-five years. Despite furious opposition to such blatant monopolism from other inventors (even from Edmund Burke), the Bill was so extended, also to cover Scotland, and received assent in May 1775.

Before full-scale production of engines from the Soho Works could commence, a final and all-important technical problem had to be overcome – a perfectly straight, uniform diameter, smooth-bored cylinder was needed, along which a precisely cast piston head could move with no leakage or impediment whatsoever. The only way this could be achieved, as Boulton now knew, was to use a cannon-boring machine of the type recently invented by John 'Iron-Mad' Wilkinson, a Cumbrian ironmaster of great technical prowess now settled in the Coalbrookdale area. After negotiations, two giant cylinders thus bored were breathtakingly realized in Wilkinson's forge at New Willey, Shropshire. One had a thirty-eight inch bore cylinder designed for providing blast at Wilkinson's own iron-smelting furnace; the other giant was all of fifty inches in diameter, and was to pump water from the deep Bloomfield colliery at Tipton, north of Dudley, in what was then Staffordshire. Here in the vivid words of Jenny Uglow, distinguished biographer of Lunar Society members, is the climax to the finished engine's trials at the Bloomfield pit in March 1776:

> From the first thrust the great beam-engine, standing high and proud in the March winds, managed about fifteen strokes a minute and drained the pit of nearly sixty feet of water in less than an hour.[6]

Recollections of Boulton in the same year are by James Boswell:

> Mr Hector was so good as to accompany me to see the great works of Mr Bolton (*sic*), at a place which he has called Soho, about two miles from Birmingham, which the very ingenious proprietor showed me himself to the best advantage. I wish Johnson had been with us: for it was a scene which I should have been glad to contemplate by his light. The vastness and the contrivance of some of the machinery would have 'matched his mighty mind'. I shall never forget Mr Bolton's expression to me: 'I sell here, Sir, what all the world desires to have, – POWER.' He had about seven hundred people at work. I contemplated him as an iron chieftain, and he seemed to be a father to his tribe. One of them came to him, complaining grievously of his landlord for having distrained his goods. 'Your landlord is in the right, Smith, (said Bolton), But I'll tell you what: find you a friend who will lay down one half of your rent, and I'll lay down the other half; and you shall have your goods again.'[7]

Later years were to see the spread of Boulton and Watt's ever more efficient, powerful, compact and adaptable (providing rotary motion) steam engines and their many successors. They powered all manner of tasks in mines, factories, textile mills, agricultural and construction machinery and transport (steam locomotives on steel rails, steamships) over the face of Britain and in exported guise across the entire globe.

Coda

In the universal *Système Internationale* (SI system) of scientific units the unit of power is defined as 'the time rate of the liberation of energy'. It is named the Watt in tribute to the Scot's dogged persistence and ultimate success in harnessing steam power to work diverse machinery throughout his long career as an inventor of genius. But, unlike Newton (similarly honoured in the naming of the unit of force), he really had stood on the shoulders of a giant to get there. One cannot help but think that the quantity for the rate of doing work ought really to have been named the Newcomen, for it was on his broad shoulders that James Watt stood, reaching upwards.

5

Still Waters Run Shallow: Canal Mania

I found my love by the gaslight crawls
Dreamed a dream by the old canal
Kissed my girl by the factory wall
Dirty old town, dirty old town

I heard a siren from the docks
Saw a train set the night on fire
Smelled the smoke on the Salford wind
Dirty old town, dirty old town…

Dirty Old Town. Beginning of the lyrics written in 1949 by Ewan MacColl. He was Salford born and raised – the Bridgewater coal canal ran through Salford into Manchester.

Navigable Rivers

Prior to canalization, navigable rivers were the quickest and most economical way to transport bulky goods to and from coastal ports. In mainland northern Europe they had been employed for millennia centred on the well-used Rhine and Meuse. In the early 1600s England had around 700 miles of navigable rivers upstream from coastal estuaries (Fig. 5.1), the majority deemed 'open' in that no major engineering works had taken place to restrict access or enhance the navigability of their natural channels.[1] Such were the Severn to Shrewsbury, the Trent to Nottingham, the Ouse to York and the Yare to Norwich from Great Yarmouth. Some idea of the scope of riverine trade by the 1720s is given by Defoe in part of his North Midlands and Yorkshire Tour, written from Nottingham:

> The Trent is navigable here for vessels or barges of great burthern, by which all their heavy and bulky goods are brought from the Humber, and even from Hull; such as iron, block-tin, salt, hops, grocery, dyers' wares, wine, oil,

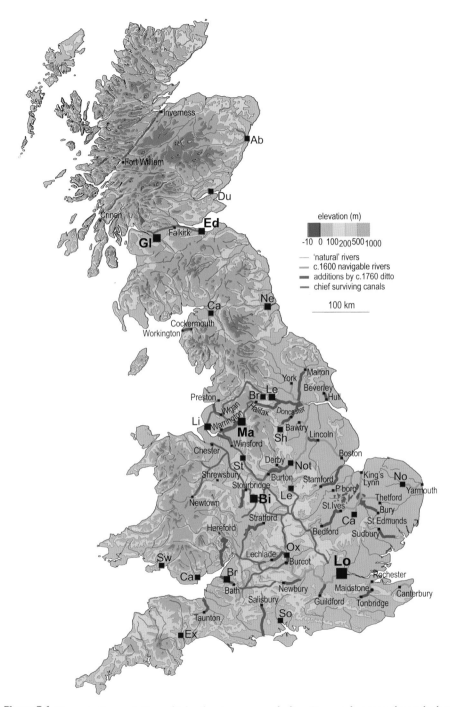

Figure 5.1 Stages in the evolution of inland waterways, excluding Roman drainage channels that may have been used as navigable cuts. Redrawn, modified and crafted by the author from data in the Times Atlas (1987), Cossons (1993) and Skempton (1996).

tar, hemp, flax, etc. and the same vessels carry down lead, coal, wood, corn; as also cheese in great quantities, from Warwickshire and Staffordshire.[2]

The Thames was open to sea-going vessels only as far as London Bridge; cargoes were then taken on as far as Windsor by large barges. Upstream as far as Oxford there were numerous mill weirs with a system of specially designed 'flash' locks (single-gated, channel-wide structures) to facilitate an upstream head of water to work the mill wheels. These had been in existence since at least the thirteenth century; once opened when milling work had finished, they enabled waiting boats to quickly pass the weirs by.

Nevertheless, natural rivers were and are unsteady things – seasonal flooding, drought, sedimentation, log jams and vicarious bed and bank erosion could make river transport a risky and unpredictable business. River training to facilitate aqueous portage took place subsequently along several major waterways. More importantly, several growing commercial and industrial centres, hitherto far from existing navigable rivers, petitioned Parliament to engineer navigability of local tributary rivers to major water courses. Such were the proposals (Fig. 5.1) to link the nascent Sheffield metal trades to the Trent via its own River Don, the coal mines and clothiers of West Yorkshire to the Humber via the Rivers Aire and Calder, the Wey navigation from the Thames to Guildford, and the Warwick Avon scheme from the Severn up to Stratford. These and many other schemes increased the mileage of navigable rivers by some six hundred miles by the mid-eighteenth century, at the dawning of the indigenous British canal boom.

The increase in navigability into river reaches of steeper gradient could only proceed by the construction of pound locks. These structures, with masonry-walled chambers of water between two successive pairs of gates, had been developed in several European countries. They were first introduced to Britain as early as 1567 by the Welshman John Trew on a five-mile lateral cut on the River Exe upstream from Exeter. His scheme also featured a large holding basin for vessels during low tide in the river estuary.

The Impetus for Canals

The early purveyors of industrial capital had to carefully consider the location of plant in their investments so as to minimize transport of raw materials and finished products. Sources of many industrial raw materials existed far away from existing natural waterways so that industrialists became frustrated and filled with vexation as efforts to expand their businesses stalled or were made impossible by typically high land transport costs. The raw materials that made the early Industrial Revolution possible were variable in their chemical and physical states – wool, raw cotton, timber, grain, metallic and non-metallic ores, mineral fluxes, coal, coke, stone, sand, gravel, potters clay, to name but a few. Yet they had one thing in common – economics determined they needed to be transported in large volumes. Low frictional resistance determined that a horse-drawn barge on water could transport around 250 times the weight that the animal could physically carry, or four times the weight it could pull in a wagon on iron rails.

Unlike rivers, true canals are entirely engineered, cut perfectly horizontal with their contained water topped up from upslope reservoirs. They are largely unaffected by weather (bar freezing), the natural fall in gradient of their course taken up entirely by the arrangement of pound locks. Prior to their establishment in Britain, canals were commonplace in northern Europe. Most spectacular was that stunning show-off of engineering skills in the 'French century' under Louie XIVs finance minister Colbert, the 240-kilometre-long Canal du Midi (aka Canal de Deux Mers). It was began in 1660 and engineered by the Baron Pierre-Paul Riquet and a team of skilled engineers with a workforce of 12 000. Taking only fifteen years to build, it passed below the foothills of the Montagne Noir in south-west France, joining the Atlantic and Mediterranean coasts. The entire system is still working after over 350 years and remains in the French public domain. Yet in England and Scotland during the entire eighteenth century no capital outlays were ever proposed by Parliament for canal building (but see below for Ireland); they were all subsequently financed from private capital put into limited company stocks and shares. That much of the Georgian canal network survives today after the stiff competition from railways and, later, tarmacked roads and the internal combustion engine is nothing short of miraculous.

We saw earlier two contrasting examples of rapidly expanding Stuart and early Georgian trade and industry based upon distinctive solutions to the problem of transporting raw materials. One was the 'London Trade' in coal from the Tyne – the material loaded from keels (lighters) into sea-going colliers. The other was Darby's careful selection of a vertically integrated site for his iron business in Coalbrookdale. There were enough other energetic and far-sighted individuals from the dawn of the 'New Manufacturers' – Matthew Boulton of Birmingham, Josiah Wedgwood of Burslem; John Roebuck, Samuel Garbett and William Cadell of Falkirk – who could see the great benefits that canals would bring in transporting industrial materials and manufactured products throughout the Anglo-Scottish lowlands, where many coal basins were situated. Eventually they were also cut through the steep valleys of upland Glamorgan and across the Pennines. Josiah Wedgwood, for example, lobbied for a canal to run through the Potteries, and once James Brindley's planned routes were approved by parliamentary act, he promptly built his Etruria factory-cum-studio within a stone's throw of its planned course. Such was also the outlook and situation of aristocratic (and technocratic) coal owners like the 3rd Duke of Bridgewater, wishing to sell their subterranean products at a decent profit to warm the houses of rapidly growing industrial cities like Manchester and Salford. In fact, most canals were specifically laid out to connect distant coalfields and their associated industries with ports or centres of population and industry.

The first in this regard was not, as commonly supposed, in England but Ireland – the Newry Canal, which opened in 1741.[3] It was also primarily motivated by the coal trade, for it connected the newly discovered (but geologically complex) Tyrone coalfield (for location *see* Fig. 1.2) above Coalisland along the Tyrone navigation (aka Coalisland Canal) and the River Blackwater via the Maghery Cut into Loch Neagh. Boats crossed the loch to the upper reaches of the River Bann and hence to the Newry Canal onwards to Newry, situated by Carlingford Lough bordering the western Irish Sea. Apart from it being the earliest true canal in Britain and Ireland, it is noteworthy for having been publicly funded and instigated by the Irish Parliament who, as early as 1717, had offered a prize of 1000 pounds to the first person to ship 500 tonnes of Irish coal to the port of Dublin.

Not only was the Newry Canal the first such; it was also a 'summit-level' cutting, as was the Canal de Midi, in that it passed up, over and down the drainage divide between adjacent river catchments. In this case the total fall towards Carlingford Lough was only twenty-four metres with a total of fourteen individual stone-built pound locks (originally brick-lined, but these failed in places) along its length. In 1769 the three-kilometre-long Newry Ship Canal completed the circuit for sea-going craft to offload barge cargoes. In fact, though originally constructed entirely for coal traffic, the supply source from the Tyrone pits proved limited – steep stratal dips up to fifty degrees and numerous faults cut up the seams and made the mines difficult to work at a profit. Within seventy-five years, passenger traffic, general merchandise and produce from the rich agricultural hinterland comprised the majority of cargo.

Canal Ways and Lancashire Coal Mining

The 3rd Duke of Bridgewater in the mid-1750s was an eager and independent young man with a passion for applied science and engineering.[4] He was also stupendously rich from an unlikely inheritance – unlikely because he was a seventh son – his elder male siblings all predeceasing him through the scourge of consumption. On his 'Grand Tour' he spent time doing what we might call 'engineering fieldwork'. In particular, he viewed and admired the building, working and subsequent maintenance of the Canal de Midi, with its use of tunnels and barge-carrying aqueducts as well as the usual pound-locks and loading basins.[5] On return to England the Duke left his London home after a failed betrothal and set up as a life-long bachelor in his inherited South Lancashire seat, the Egerton family's Worsley Old Hall.

The Duke's intentions were immediate and clear: he became a capitalist-cum-adventurer in mining and canal works, doubtless eagerly travelling the fifteen kilometres or so to witness the Sankey canal scheme. This was the first in England, being carried out to completion in 1757 between the South Lancashire coalfield's Haydock collieries east of the town of St Helens to the right bank of the River Mersey, with its later extension to the then village of Widnes. The Act proposing this initiative was the first such issued from Parliament and in following years was followed by a flood of others – the era of 'Canal Mania'. No objections to the Sankey canal were filed for parliamentary scrutiny, a phenomenon never repeated. Canal building subsequently became a long-drawn-out and often litigious process, as Wedgwood, Boulton, Brindley and others would discover. The absence of objections in this case was perhaps due to the disingenuous wording of the Act, which stated simply that the river was going to be improved by '… making [it] navigable…', not that in doing so, a separate canal with locks would be cut out of the adjacent natural riparian landscape. At any rate there was no legal objection from any landowners concerning the works, so we must assume they all profited from their coal leases after construction.

The canal was no 'summit-level' effort, yet the initial southern course down the Sankey Valley to the floodplain of the Mersey had a drop of some twenty-five metres over twelve or so kilometres, which required construction of several pound locks. A subsequent extension eight kilometres or so westwards to the village of Widnes was cut a few years later after more serious parliamentary opposition was overcome by a simple 'contour-course' cut with a smaller drop. The Mersey–Irwell Navigation (chief and vociferous objectors to the extension) then took the barge-coal on to Liverpool and Chester for use in saltworks and in the nascent Merseyside and Deeside chemical and smelting industries, and for domestic use in those districts. Interestingly in view of the history of British and Irish canal building, one of the two Sankey surveyors, Henry Berry, had also worked on the Newry canal under Thomas Steers – a man described by A.W. Skempton as England's first professional civil engineer. Berry himself carried on this work, designing and constructing much of mid-Georgian Liverpool's docklands – he became the second chief port engineer after Steers' death in 1750.

The success of the Sankey initiative caused Bridgewater to reflect on the problems of promoting the coal trade from his own mines on the Worsley estate. He had hitherto faced the heightening expense of transporting thousands of tonnes of mined coal annually to the rapidly growing Manchester market by cartage and then by boat via the meandering and unreliable drought- and flood-prone River Irwell that drained the wet and hilly Rossendale catchment north of the city. His thoughts turned to the advantages of canals. Around 1758 the Duke and his canny junior land agent, John Gilbert, fixed the siting of one along a course from Worsley to Ordsall in Salford for the documentation required for a parliamentary act that received acceptance in 1759.

Some construction had already begun when around 1760 Bridgewater contacted and hired the adept, forceful and rising James Brindley to act as land surveyor for the canal. Brindley was also a proven expert on adit and shaft mine drainage, and subsequently spent two months living at Worsley Old Hall as Bridgewater's house guest on a detailed survey. It has been popularly assumed that this included work on both the new canal extension and the Worsley mine levels (Fig. 5.2). But the research of Hugh Malet indicates that it was the Duke himself in conjunction with John Gilbert who was responsible for suggesting both the mine plan and the revised contour-course ideas. This comes directly from the pen of Sir Joseph Banks, who visited the construction works in 1767–68 and who categorically states that it was the Duke in person who was the originator, with Gilbert acting the responsible agent '…overlooking every part, and trusting scarce the smallest thing to be done except under his own eye'.[4]

In the event, the canal required no locks, the water surface running at a constant twenty-five metres above OD from Worsley Delph into a canal basin and warehouse

Figure 5.2 The entrance (right) and exit (left, obscured by slippages) to the Worsley canal levels at Worsley Delph (53.300334, 2.225242) separated by a mooring portage abutment for transfer of coal to the main Bridgewater Canal. Figure 21.3 shows a historic engraving of the scene for comparison. The weathered and partly overgrown Worsley Delph Sandstone along the quarry faces exhibits channel-like structures cut by the Late Carboniferous delta distributaries that deposited it. Source: Wikimedia Commons, photographer: *Parrot of Doom* (Tom Jeffs).

complex in central Manchester. However, it did require an aqueduct for the canal (the first in Britain) over the River Irwell at Barton. It was Brindley who successfully presented the Duke's revised proposals to the relevant parliamentary committee: a second parliamentary act soon followed. We can see nowadays from the geological map that from Worsley the canal was first cut through glacial-age boulder clay until Triassic bedrock (useful for foundations) was reached north of the Irwell crossing. From then on it passed through Irwell fluvio-glacial terrace sands and gravels into central Manchester. The Barton aqueduct, with its three Romanesque arches that took the canal twelve metres above the Irwell, was constructed from Worsley's very own strong sandstone (Fig. 5.3), which we now know as 'Worsley Delf Rock' – 'Delf' or 'Delph' is Old English for a quarry or pit.

Also on the agenda during preparation of the revised canal plan was a pressing issue concerning the interminable and vexatious problem of Worsley mine drainage and the longer-term issue of mine development. The gripping story of their solution to the Worsley mines drainage problem, the construction of the Worsley Navigable Levels, is taken further in Chapter 21.

Brindley and Wedgwood: Canal Action

A Stoke-on-Trent journalist wrote the following words in the 1840s about the longest canal project in Georgian England, writing at the dawn of the later industrial era of steam railway engines, iron-hulled steamships and steam shovels:

> The first clod of the Grand Trunk Canal was dug the 26th July, 1766, on the declivity of Brownhills, in a piece of land, now belonging to Mr. Wood, within a few yards of the bridge which crosses the canal, by Mr. Josiah Wedgwood, then of Burslem, (the gentleman who afterwards rose to such eminence as a Potter), in the presence of Brindley, the Engineer, and many respectable persons of the neighbourhood, who each cut a sod to felicitate the work.

What was the pre-eminent potter doing at Brownhills in North Staffordshire initiating the work on a canal to be constructed by the same James Brindley who had surveyed in the Bridgewater Canal a few years previously? Perhaps it is explicit in the description

A. The navigation.
B. The mouth of the tunnel, with large doors to open and fhut.
C. The Quarry.

D. A crane of a very curious conftruction, for heaving the ftones out of the quarry into the barges

Figure 5.3 Historical engraving (from c.1759–60?) of the Worsley Delph sandstone quarry with an interesting crane that lifted and rotated quarried sandstone blocks onto barges for transport down the canal, notably during construction of the famous aqueduct over the Irwell Navigation (later the Manchester Ship Canal) at Barton. The view shows no exit tunnel coming out of the levels to the left (cf. Fig. 5.2) and would appear to be an early stage in the cutting of the levels network northwards. Source: Malet (1990).

and definite article '…the Engineer…', for Brindley must have been a happy man as he took the sod turfs away in his wheelbarrow. He was now engaged on the largest project of an intensely busy life – out and about in the countryside on horseback and mapping with assistants and surveying kit in all weathers. The occasion was when members of the Lunar Society of Birmingham, with their multifarious inventions and initiatives, went universal – the initiation of the Trent–Mersey Canal, planned to span the entire breadth of the English Midlands from sea to sea.[6]

It had taken years to get to this stage – the Parliamentary Act allowing a complete navigable canal connection between the northern Europe-facing port of Hull in the east and Atlantic-facing Liverpool had just recently passed through its final reading. Judging by all accounts, it is fair to say that it was both Brindley's and Wedgwood's vision, energy and dogged determination, plus inspiration and encouragement from several other 'Lunaticks', chiefly Darwin, Edgeworth and Boulton, to get to this position. There had been years of wrangling between different groups of rival financiers concerning the specific route that such a canal might take. There had also been strong and prolonged opposition from existing navigable river concerns, chiefly on the lower and middle reaches of the Trent up to Burton and along the Weaver navigation up in the Cheshire Plain from Winsford. The river navigation companies had monopolistic rights that encouraged steep prices for carriage of goods that the new canals were now savagely undercutting. The price of carriage by the Trent–Mersey Canal was stipulated in advance amongst the clauses laid out in the parliamentary act responsible, so that the canal's advantages were plain for all to see.

It fell to noblemen politicians like the Earl Gower, soon to become Lord President of the Privy Council, to help push the canal-building movement forwards. He had far-reaching tentacles of influence in the all-controlling Whig political circles of the time. His home and estate lands were around Trentham, located across the upper Trent valley from the canal's route past the six Potteries towns (we previously noted Celia Fiennes in the area at the time of his grandfather). The canal bug had bitten him early, as it infected all British landowners with property on Coal Measures outcrops – his lands around Newcastle-Under-Lyme were in what came to be known as the North Staffordshire or 'Potteries' Coalfield. He had previously witnessed and supported the relevant parliamentary act that led the success of the Sankey experiment. More importantly, he also supported the Duke of Bridgewater's subsequent proposal. There were close family links involved here, for the Duke's sister, Lady Louisa Egerton, was his second wife. Gower had in fact asked Brindley as early as 1758 to be the surveyor of a route to link the upper Trent valley with the River Tern, the English headwater tributary to the Severn at Market Drayton. Wedgwood vigorously lobbied Gower with Brindley's planned routes through the Potteries, with the results of a survey for a section from Burslem downriver almost to the last navigable stretch of the Trent at Wilden. This included a branch to Lichfield close to Erasmus Darwin's house and, more to the point, close by a locality where the Doctor had planned to invest in a new wrought ironworks. So it went on –

canal mania was affecting the titled and affluent classes, middle-class entrepreneurs, and the intellectuals of the Lunar Society willy-nilly.

Following parliamentary approval for the Trent–Mersey scheme, the burgeoning industrial-scale pottery makers of north Staffordshire could look forward to the day when their delicate hand-painted forms, cushioned and packed carefully within bulky crates, would have some certitude of being delivered to customers and to markets both cheaply and in one piece. For the moment, though, the old expensive long haul by packhorse or rattling wagon on rutted roads to the nearest navigable reaches of the River Trent at Burton, the River Weaver at Winsford, or the Mersey at Runcorn would have to do. It was to take eleven years for that day to pass – years that would see grindingly slow progress across the Coal Measures core of the North Staffordshire syncline above Tunstall, the most northerly of the Potteries towns.

The U-shaped course of the Trent–Mersey joins the Mesozoic basins of East Yorkshire, Nottinghamshire, Derbyshire, Staffordshire and Cheshire that bound the Coal Measures of the South Pennine massif. The basins are low-lying because they have a substrate of soft Triassic marls (Mercia Mudstone), with slightly higher land over more resistant Sherwood Sandstone. Moorland catchments today briskly feed their winter runoff into the Trent, but its course had been cut into older alluvium formed by powerful discharges induced by late-Pleistocene ice melt.

Within these broad constraints the first requirement of Brindley's surveyors and overseers was to choose the elevation of the canal above any adjacent river channel. This was a tricky one: on one hand it had to be higher than extreme flood level, but also commensurate with the nature of the geological substrate and to fit with the landowner's preferences. In the case of the Trent, this usually involved a winding course along the slightly raised boundary between the modern floodplain and the first elevated late-Pleistocene river terrace above highest flood-prone level. The substrate below soil level was mostly sand and gravel, more easily excavated by pick and hand shovel than the more plastic and cohesive clays deposited as a direct result of glacial action.

It was on this permeable substrate, using local brick clays quarried from borrow pits in outcropping clay or from the harder Mercia Mudstone that Brindley perfected his 'puddling' technique. This involved soft clay and/or pulverized mudrock mixed with just enough water to make a thick sticky mess and scraped like marzipan over the permeable terrace substrate. Thus sealed, it stopped water loss by percolation. However, some water loss by evaporation and minor seepage could not be prevented, and so reservoirs along the bordering uplands were needed for periodic replenishment; vice versa drainages for excess precipitation or other inflows.

All this meant that Brindley was forced to follow the general trend of the long Trent valley west of the already navigable reaches at Burton-on-Trent all the way to Tunstall in the northern Potteries, where geological problems along his choosen route, a long tunnel under Goldenhill, proved near-impossible to surmount, and which delayed completion of the finished canal by several years (this intriguing story is continued in Chapter 21). Thereafter the canal course tracked irregularly across several low points in the Cheshire basin to its Mersey termination.

Feat of Clay: From Cornwall to Etruria by Water

Looking out in 1777 from the front of his two-year-old Etruria works, Josiah Wedgwood could satisfyingly contemplate the busy scene – the canal basin full; a pleasing line of waiting barges queuing to offload raw materials – minerals for glazes, china clay, coal – or loading up with finished pottery for export across the nation and to the world at large. The output ranged from simple, durable everyday household wares to complete finely decorated individual dinner services for state occasions of the wealthy and royalty of Europe and beyond. Large orders had come or would come in from the likes of Queen Charlotte and the wider British nobility, with a great trade eastwards out of Hull into Europe and beyond – to include Catherine the Great in St Petersburg.

With the opening of the canal, the slump in transport charges for the transport of bulk materials affected not just the price of coal but also, of even greater importance to the Potteries, that of potter's clay. Josiah Wedgwood and Josiah Spode were able at last to realize the once-impossible – the industrial-scale importation of Cornish china-clay, soon to be cheaper than ever before. It was sailed up to the Merseyside entrance of the great canal from the harbours of Cornwall and Devon, eventually, by the nineteenth century, to include Par, Fowey, Charlestown, Pentewan, Newquay and Wadebridge.[7] Offloaded into barges, the whiter-than-white cargoes bagged-up in hessian sacks ghosted silently across the Cheshire plain, up the set of locks at Kidsgrove, into the Harecastle tunnel and out into the smoky air of the Potteries.

The background to all this was that Wedgwood, always on the lookout for mineral perfection or novelty in the commercial usage of glazes and fluxes, had been alerted to the discovery of significant quantities of fine china clay in the area north of St Austell. In fact it was the indefatigable James Watt, down in Cornwall in the late-1770s and early 1780s looking after his and Boulton's steam engine sales, who tipped him the wink as to the location of existing St Austell excavations. This came as a godsend to Wedgwood, for production of such clay in the countryside of Germany between Bohemia and Saxony had led to the establishment of the rival Meissen white porcelain works. He had wished for his own home-grown source of the clay, though not so much for porcelain-making as for his own totally original wares – jasper and basalt (stoneware) pottery. Although he did not know it at the time, the geology and mineralization of Cornwall was similar in some aspects to that of Saxony – granitic rock shot through with metal- and quartz-rich veins and with kaolinitic clay alteration products.

The original discoverer of the Cornish clays was one William Cookworthy, proprietor of its early quarries and an aspiring ceramicist. He was a Quaker minister and pharmacist who came across the material in pre-existing tin opencasts on Tregonning Hill, Germoe. Obtaining monopoly rights to the St Austell deposits, he made porcelain at Plymouth until 1772 before giving up the business and selling his interests and monopoly to a friend, one Richard Champion of Bristol. Both men made classic high-fired, hard-paste porcelain, but neither could make a sustained profit from it. Champion tried to retain his monopoly over use of the St Austell clay, but after Wedgwood's 1775 visit to Cornwall, with the help of Watt and another Staffordshire master-potter, John Turner, a successful legal challenge was issued against the monopoly. Wedgwood wrote after his visit that:

> Mr Turner and I have concluded to set about washing clay for ourselves and for others as soon as we can, if they chuse to have it, but at the same time to have the raw clay open to all who chuse rather to prepare it themselves, for I am firmly perswaded that an exclusive Company, or rather an exclusive right in the Clay in any Company, or under regulations whatever, would soon degenerate into a pernicious monopoly.[8]

Wedgwood thereafter bought the lease of Carloggas pit and shares in other concerns. These sold the washed and matured clay on to the Staffordshire potteries via transport by sea, river and canal barge. Watt took a share in the profits from these enterprises, reflecting his role as mineral prospector.

Wedgwood and Spode as Experimental Geochemists

[9]Wedgwood was a particularly keen amateur mineralogist – modern x-ray fluorescence analysis reveals that the secret recipe for his famous jasperware included a large proportion of barium. The chief source for this element, then and now, is the mineral barytes (aka barite), a barium sulphate readily and cheaply available at the time in large quantities from nearby Derbyshire lead mines. Another source is the much rarer barium carbonate. A specimen of the latter obtained from a Cumberland lead mine by Matthew Boulton and formerly attributed to barytes was carefully analysed by the medical doctor, botanist, mineralogist and fellow Lunatick, William Withering. He found it to be the true carbonate of barium and, five years after publication of his results, the great Gottlieb Werner formally named the mineral in his honour in 1789. Witherite, referred to as 'fusible spar', or *terra ponderosa* (heavy earth), is specifically mentioned in letters between Lunaticks. It was subsequently mined from a single Northumberland location, at Settlingstones, a unique vein deposit of pure witherite that the author was lucky to visit underground as a student just before it closed down.

Wedgwood's jasperware took on characteristic colours by the judicious addition to his glazes of chemicals derived from various mineral ores: chromium oxide to give sage green; cobalt oxide to give a deliciously popular blue; manganese oxide to give lilac; a salt of antimony to give yellow, and iron oxide to give black. It goes without saying that Wedgwood undertook thousands of high-temperature experiments to obtain his results. He can certainly be viewed as the original intuitive experimental petrologist, a worthy predecessor to James Hall of Edinburgh who made experimental basalt.

China clay is pure kaolinite, a silicate of aluminium, particularly vital to Josiah Spode, fellow pupil with Wedgwood of the master potter Thomas Wheildon of Fenton. The mineral was a major ingredient (25% by weight) in his original successful and lucrative bone china ware of the late 1780s. This was a translucent and strong form of soft-paste china that utilized finely ground and calcined animal bones (45–50% by weight) together with the mineral feldspar (Cornish stone) (25–30%) to provide a source of calcium and phosphate fired at moderate temperatures of around 1250 °C. Onto bone china and other wares Spode transferred gorgeous scenes and motifs from his own original technique of underglaze blue transfer printing, which make his wares so emphatically beautiful and creative.

Coda

By the time that the Trent and Mersey Canal was fully opened, there was a substantial regional network of canals across southern England and the Midlands – an English national system linking the major navigable Trent, Mersey, Severn and Thames rivers. In the autumn of 1772, Matthew Boulton could truly write of the existence of an inland merchant marine: 'Our navigation goes prosperously…we already sail from Birmingham to Bristol and to Hull', though perhaps his use of 'sail' is over-egging the cake. The canal-building mania reached midland Scotland at the same time where, as we saw previously, James Watt was part-time surveyor for one (unsuccessful) route for the Forth–Clyde Canal. All this and later canal building in Wales and Ireland was accomplished by the concerted efforts of an interlinked band of thousands of navvies and numerous surveyors, engineers, financiers and aristocratic/manorial (mostly) landowners who had coal under their lands. The greatest beneficiaries outside of the canal shareholders (though rewarded, profits were never excessive) were such mine owners and leasees and the bosses and shareholders of new industries that depended upon a radically reduced price of their chief fuel and energy source. A vastly expanded market for coal had also opened up for domestic heating amongst the burgeoning middle classes. All this was unique on the world stage: a single small nation at fifty-two degrees of latitude north with what seemed like an inexhaustible source of relatively cheap home fuel during a time of dramatic climatic deterioration. But within seventy years or so the canal ways would become outdated for many purposes as steam locomotives on mild steel rails raced their passengers and loads of manufactures and raw materials from one end of the island to another ten times and more faster than any canal barge could achieve – the Age of Steam went into overdrive by the 1840s for a long interval of over one hundred years.

To end: a contemporary newspaper epitaph[10] for James Brindley, worthy in its execution and phrasing of William Topaz McGonagall, but anticipating that great poet's style by a century:

JAMES BRINDLEY lies amongst these Rocks,
He made Canals, Bridges, and Locks,
To convey Water; he made Tunnels
For Barges, Boats, and Air-Vessels;
He erected several Banks,
Mills, Pumps, Machines, with Wheels and Cranks;
He was famous t'invent Engines,
Calculated for working Mines;
He knew Water, its Weight and Strength,
Turn'd Brooks, made Soughs to a great Length;
While he used the Miners' Blast,
There ne'er was paid such Attention
As he did to Navigation.
But while busy with Pit or Well,
His Spirits sunk below Level;
And, when too late, his Doctor found,
Water sent him to the Ground.

The last lines refer to Brindley's sad end after his doctor, Erasmus Darwin, diagnosed the advanced development of the diabetic condition that eventually killed him, aged fifty-six, with his greatest project, the Trent–Mersey Canal, unfinished.

6

Travellers' Tales 2:
Louis Simond Under Tyneside, 1811

As I went up Sandgate, up Sandgate, up Sandgate,
As I went up Sandgate I heard a lassie sing –

Weel may the keel row, the keel row, the keel row,
Weel may the keel row that my laddie's in!

He wears a blue bonnet, blue bonnet, blue bonnet,
He wears a blue bonnet, a dimple in his chin:

And weel may the keel row, the keel row, the keel row,
And weel may the keel row that my laddie's in!

The Keel Row (*c*.1770). Traditional lyrics that owe more to Border Scots speech than Tyneside (for background *see* Chapter 23). Bagpipe versions and Kathleen Ferrier's resonant take may be found online.

Introduction

The grand scale of the Tyneside coal mines was established by the mid-seventeenth century. For example, the Grand Lease colliery probably employed around 500 colliers to produce its 75 to 100,000 tonnes of coal per year. It had rapidly expanded (as did much other trade) after the disasters of plague (the devastating 1636 Newcastle plague, *see* Chapter 23) and the Civil Wars when Sir George Vane became chief shareholder – an early example for historian John Nef of a capitalist coalmining entrepreneur.[1] Even before the Wars it featured the first recorded wagonway to transport coal to the wharves of the Tyne – the Whickham Grand Lease Way of 1620.

In common with other early modern mining areas, written accounts of underground scenes are few and far between, especially by those visitors with an interest in mining methods or in the working miners. One such later visitor was the Abbé Le Blanc, whose appreciation of Tyneside skills in mechanization led him to write in 1745 that they:

> really multiply men by lessening their work…Thus in the coal pits of NEWCASTLE, a single person can, by means of an engine [probably a Newcomen] equally surprising and simple, raise five hundred tons of water to the height of a hundred and eighty feet.[2]

Sixty years later an adventurous, observant and fluent American traveller of French extraction, Louis Simond, travelled the length and breadth of England, Wales and Scotland with his wife for nearly two years between 1809 and 1811, recording their often acute and precise observations. These became part of a journal first published anonymously in 1815; a modern (but not well-known) edition is *An American in Regency England* (1968).[3] The journal deals with life, society, landscape and many aspects of industry, commerce and governance in what was a pivotal time for Britain during the later Napoleonic Wars: Wellington's victorious but bloody battles of Bussaco and Albuera were fought in these years. Simond's account of a visit to an unnamed Newcastle coal mine in March 1811 is unique for its time, and a valuable historical document in its own right concerning the manner of working and underground arrangements of what must have been one of Tyneside's larger mines.

Louis Simond's Narrative

Simond starts by referring to the great Tyneside mines as underground 'farms' leased by the same landlords who own and lease conventional surface farms, a slant that places the peculiar British law of underground mineral rights vested with the landowner rather than the crown or state under immediate inspection:

> The name of Newcastle is identified with that of coals, the country about containing immense strata of this mineral, which is the object of a great trade. There are farms under ground as well as on the surface, and leased separately. I know of a subterranean farm of this kind of 5000 acres, for which 3000 sterling a-year is paid, and a percentage depending on the quantity of coals extracted, which may double that rent.

The beginning of his underground excursion reveals the stark reality of the initial descent, the writer generating a good deal of dramatic tension:

> I accepted with pleasure an invitation to descend in a coal-mine. The mode is rather alarming. The extremity of the rope which works up and down the shaft being formed into a loop, you pass one leg through it, so as to sit, or be almost astride on the rope; then, hugging it with both arms, you are turned off from the platform over a dark abyss, where you would hardly venture if

the depth was seen. This was 63 fathoms deep [378 feet]. One of the workmen bestrode the loop by the side of me, and down we went with considerable rapidity. The wall of rock seemed to rush upwards – the darkness increased – the mouth above appeared a mere speck of light. I shut my eyes for fear of growing giddy; the motion soon diminished, and we touched the ground.

These were the days before lift cages. We are not told of the mode of transmission that drove the descent, but given the time and place, it would be in order to assume that it was by winding gear driven by a Watt steam engine:

Here we stopped for two other persons. Each of us had a flannel dress and a candle, and thus proceeded through a long passage – rock above, rock below – and a shining black wall of coal on each side; a railway in the middle for horses (for there are fifty or sixty horses living in this subterraneous world), to draw two four-wheel carriages, with each eight large baskets of coal; these baskets are brought one at a time by diminutive wagons, on four little wheels, drawn or pushed by boys along other rail-ways, coming down side streets to this main horse-road, the ceiling of which is cut in the main rock, high enough for a man to stand upright, while the side streets are no higher than the stratum of coals (four and a half feet), therefore you must walk stooping.

The whole extent of the mine is worked in streets, intersecting each other at right angles, 24 feet wide and 36 feet asunder, leaving therefore solid blocks 36 feet every way.

He is describing the 'pillar-and-stall' method of mining ('stoop-and-room' in Scotland) a thick seam sandwiched between solid rock (sandstone) above and below (see Figure 20.2). In its original guise, the method left coal pillars whole and abandoned, acting as permanent roof supports as the mine headings advanced, but as we shall see this was not the case here. Simond continues on the necessity of thorough ventilation:

The miners have two enemies to contend with, air and water; that air is hydrogen gas [actually, methane, as eventually discovered by Humphry Davy], continually emitted by the coals, with an audible hissing noise. The contact of the lights necessary to be used would infallibly set fire to the hydrogen gas, if allowed to accumulate, and either blow up or singe the miners severely; it is therefore necessary that there should be a continual current of air going in and out by two different issues. At the beginning of the works, and while there is only one shaft, this is effected by means of a wooden partition, carried down along the middle of the shaft, then along the first street opened, and so disposed afterwards, that the air which comes down the shaft on one side of the partition, may circulate successively through each and every street before it returns up the other division of the shaft, a small fire establishing and keeping up the draught. As to water, the dip, or inclination of the stratum of coals being known, all the art consists in making the first shaft in the lowest part of the tract; a steam engine at the top drains up the water, and draws up the coals. Wherever the shaft comes in contact with any stratum yielding water, it must be kept out by means of a drum, or lining of timber, made tight round the inside of the shaft. I saw a small spring of clear water issueing from the bed of coal below, near the stables where the horses are kept, and serving water to them. These horses are in very good order; their coats soft and glossy, like the skin of a mole: they are conveyed down, or taken out, with great care and expedition, by means of a great net or bag.

Such ventilation techniques were pretty much standard, but in practice were made more difficult in single-shaft mines than in the double shaft system forced on colliery owners by later mine-safety legislation. The comments on the fine state of the pit ponies echoes the long tradition of care given to such underground animals, though the modern mind troubles at the circumstances of their labours only little less than the unfortunate children working the stalls in such conditions. Simond then gives a rare (and hair-raising), possibly eye-witness account of back-retreat that enabled the near-complete removal of all coal along any mined level:

Some of the mines are more extensive than the city of Philadelphia, and their streets are as regular. When the whole area is thus excavated in streets, it must not be supposed that the solid blocks [pillars] are abandoned; but, beginning at the furthest extremity, the miners proceed to pull down all the blocks one after the other. When a space two or three hundred foot square has been left thus unsupported, the ceiling of solid rock begins to sag and crack, with a hideous noise; the workmen go on notwithstanding, trusting that the ceiling will not break down close to the blocks, but some way behind; and such is the case – the cracks grow wider and wider – the rocks bend down, coming at last in contact with the floor – and the whole extent is thus filled up. On the surface of the ground, however, nothing is perceived; the rocks are left to manage the business among themselves below. Houses – and stone houses too – remain standing, and their inhabitants sleep in peace all the while.

Together with the aforementioned information concerning mine depth and coal thickness, the following snippet ought to enable the modern geologist to have a fair guess as to the likely stratal horizon the coal occupied in this unknown mine in the Newcastle area north of the river:

The miners know, by the nature of the rocks they meet while sinking the shaft, when they approach the coal, which is generally found between two beds of white sandstone. They sink the shaft at the rate of about two fathoms a-week.

Unfortunately in this case it is difficult to pinpoint a precise seam, as none of those documented in the authoritative British Geological Survey memoir of the area[4] have such a combination of depth from surface, thickness and stratigraphy.

The following nautical details confirm Defoe's comments a century earlier on the wider significance of the collier trade to British naval affairs. Such a pedigree had been much reinforced by Cook's famous voyage with Joseph Banks of 1768–71 in the barque *Endeavour*, a converted Whitby collier whose tough hull built for breasting East Coast gales survived severe and almost calamitous grounding on the Great Barrier Reef off Queensland.[5] The system of downslope transport of coal down railed carriageways from pit to waterway that Simond describes was by then almost 150 years old, whereas the closing comment about the fine state of miners' health seems risible:

> The consumption of London [from Tyneside] has increased one-fourth in the course of the last few years; it amounts now to about 1,000,000 chaldrons, or 1,200,000 tons annually, forming 6000 cargoes of vessels 200 tons each; and as they perform twelve voyages a-year, the trade employs 500 ships; the crew consists of two old sailors, for captain and mate, and seven or eight apprentices, all protected from impressment; the two old men have £9 each voyage. The mere coal trade between Newcastle and London is, therefore, a nursery for 4000 young sailors, and a preferment for 1000 old ones. The celebrated navigator, Captain Cook, had served his time on board a collier.

> The coal drawn to the surface of the soil is conveyed to the lighters by means of low carriages, on four small wheels fixed to their axis, that there motion may be perfectly equal. They travel on rail-ways, which are composed of two bars of iron, upon which the wheels, which have grooves at their circumference, run without impediment. Ninety-two bushels, weighing about two tons, besides the wagon, are drawn by a single horse, with so much ease, that the driver is obliged, on the least descent of the road, to press on the wheel with a sort of lever, to retard its motion by the friction, that the carriage may not run too much on the horse. The lighters, called keels, of about fifteen tons, carry coals on board vessels waiting in deep water. It is remarked, that the men employed under ground enjoy better health than those on the surface; the regularity of temperature securing them against many disorders, and the air constantly renewed being sufficiently pure.

PART 2

Redress of Time:
Carboniferous Worlds Reconstructed

We turn to geology to explore the reasons for the origins of the materials that enabled the Industrial Revolution to take place. By the beginning of the Carboniferous Period the planet's course had run for over ninety per cent of geological time. By then a vigorous tree-like flora dominated by the spore-bearing Lepidodendrales had evolved, perfectly adapted to wetland conditions. By the Middle Carboniferous a unique set of factors came together to form and preserve abundant, thick and extensive swamp forest peats composed of the remains of these trees, our future coals, together with copious iron-rich mudrock strata, our future iron ores. Over this time the Rheic Ocean closed up and the northern continent of Laurussia slid and scraped eastwards along the western edges of its southern neighbour, Gondwana. As a result the northern part of Laurussia was stretched and sheared. Vast subsiding equatorial coastal plains developed, fed by innumerable river networks whose mountainous catchments bestrode the tropics. Massive discharges of monsoonal rainwater brought in copious sediment to feed riverine and deltaic plains. These were colonized as dense forest swamps, chiefly by the Lepidodendrales but also other forms, and an accompanying diverse underflora.

Over fifteen or so million years widespread tectonic subsidence across the sedimentary basins of Euro-America – from the south-west USA to Donetsk and in the separate terranes of north-east India, north China and east Australia – allowed preservation of carbon abstracted by photosynthesis and of microbial iron minerals precipitated beneath the forest swamps. During all this time Gondwana's great southern polar ice cap continued its periodic freezing and thawing in response to orbitally controlled climatic fluctuations. Around the fringes of bordering Panthalassa and Palaeotethys, icebergs streamed out into oceans subject to yo-yoing global sea levels.

End-Carboniferous times saw the assemblage of supercontinental Pangea, riven across by the high dry fastnesses of the rapidly eroding Appalachian–Variscan mountains. Significant amounts of coal and iron ore deposits survived this mountain building and the ravages of subsequent geological events of Atlantic rifting, with volcanic outpourings and thorough-going Pleistocene glaciations. The sources of cheap energy and iron ore for the world's first Industrial Revolution were abundantly preserved in the downfolded sedimentary basins of the British Isles.

7

Devonian Prequel: Scintillas, then Splashes of Green

It's everywhere, and if we are
Its archaeologists and dig, stale
Excrement lies under
Excrement, and further down

Most ancient and compact
Dried excrement; and under that
Lies coprolite, so many thousand
Excremental years B.P. B.C.

The beginning of Geoffrey Grigson's *I Did Not Say Content* from *Selected Poems* (2017). Grigson's geological sensibility reveals itself in many of his poems.

Introduction

The biological colonization of land by higher plants during Devonian and Carboniferous times around 420 to 300 million years ago (Ma) heralded major changes in the way that the continental surface, created by eons of plate tectonics, volcanism, erosion and sedimentation, reacted with the atmosphere. Bacteria and photosynthesizing cellular forms (algae) had long colonized shallow oceans and, probably, lakes. Their activity had created an oxygenated atmosphere – plant cover would add to its abundance. It is not known when rock weathering (the physical and chemical breakdown of bedrock by frost, rain and dew) first involved organisms, but it seems likely that sometime in the later Precambrian (1000–600 Ma), filamentous forms of green algae (like modern *Spirogyra* on tidal flats) colonized liminal environments on coastal plains, where their cell walls protected protoplasm from drying out and where photosynthesis could take place.

Important clues as to the timing of colonization comes from acid-digestion of sedimentary mudrocks – crushed, washed, sieved and centrifuged residues yield plant microfossils.

These comprise thick-walled spores (cryptospores: 'hidden' spores) shed from primitive plants found in rocks as old as the early Middle Ordovician (in Argentina), around 470 million years ago (Ma), with numerous other examples from later Ordovician and Silurian sedimentary rocks worldwide.[1]

Tectonic Geography

By Early Devonian times (420–393 Ma), the majority of continental shorelines were lapped by the Panthalassa Ocean (*thalassa* is Greek for sea, so, 'all-enveloping sea': Fig. 7.1). They formed two major groupings in the southern hemisphere. One was the cluster of long-lived Gondwana terranes, named after the Gondi homelands of the northern Deccan, India. The other was the younger Laurussia, its name taken from Pre-cambrian rocks along the St Lawrence Seaway, Canada and from coeval rocks in Russia. They were separated by the Rheic Ocean, which covered nearly ninety degrees of latitude at its widest, rimmed by an outer archipelago that included the Siberia and North China terranes at the northern limits to the tropics.

Things were to hot up on the tectonic front between Gondwana and the other continental masses during Devonian times. In the face of circum-global plate consumption (subduction) the Rheic Ocean began to close around its entire perimeter, much as the Pacific does today. As a consequence western Gondwana and Laurussia converged, with the former drifting and rotating generally southwards to eventually occupy much of the high southern latitudes. Meanwhile a vigorous new mid-ocean ridge of the Palaeotethys Ocean propagated eastwards along northern Gondwana. This rifted a cluster of terranes away from the main continent and set them on a rapid and momentous western journey, each bearing clear geological evidence for its Pan-African geological heritage – a late Precambrian mountain-building event. We shall call these the 'Euroterranes'. They included the nuclei of modern south-west England and formerly disconnected bits of what is now continuous mainland European crust – Iberia, Armorica, the Massif Central, central Germany and Bohemia. These played key roles in the eventual tectonic joining and consolidation of supercontinental Pangea along what was to become the Variscan (aka Hercynian) mountain belt by the end-Carboniferous, around 300 Ma.

During the Early and Middle Devonian, north-west-subducting Rheic ocean plate interacted in a complex way with south-east Laurussia. It created a great arc of compression that resulted in what is known as the Acadian mountain-building event, named after its importance in Maritime Canada. In northern Britain it records the impact of the Midlands–Ardennes terrane whose rocks include the Gondwanan volcanic and sedimentary assemblages visible today in the uplands of Charnwood Forest, Leicestershire. Further spectacular consequences of the Acadian were the formation of numerous 'Newer Granite' igneous bodies in northern Britain and the Great Glen Fault that slashed across Highland Scotland with the Orcadian (Orkney–Caithness) lake basin and its vegetated margins developing over its northern splays.

A. EARLY DEVONIAN TECTONIC GEOGRAPHY

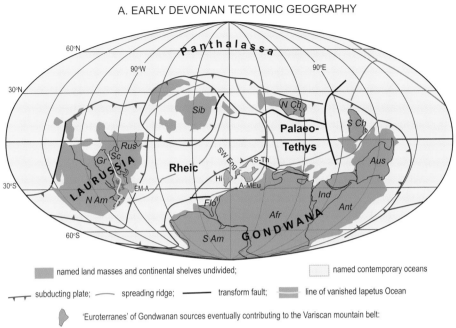

named land masses and continental shelves undivided;   named contemporary oceans

⊢⊢⊢ subducting plate;   —— spreading ridge;   —— transform fault;   ▓▓ line of vanished Iapetus Ocean

'Euroterranes' of Gondwanan sources eventually contributing to the Variscan mountain belt:

S-Th - Saxo-Thuringia; A-MEu - Armorica-MittelEuropa; SW Eng - South West England; Hi - Hispania; EM-A - English Midlands/Ardennes

B. LATE DEVONIAN TECTONIC GEOGRAPHY

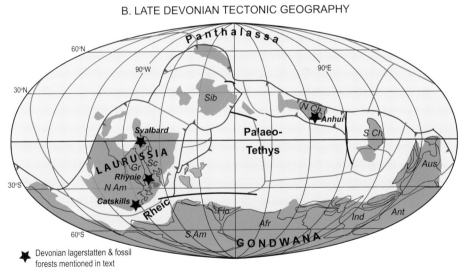

★ Devonian lagerstatten & fossil forests mentioned in text

Figure 7.1 **A–B** Early and Late Devonian tectonic geography. Redrawn and crafted chiefly from material in Torsvik and Cocks (2017) and from relevant ideas on the Acadian event and the impact of a Midlands–Brabant terrane in northern Britain presented by Woodcock and Strachan (2012).

Devonian Climate

The kind of global tectonic geography just described had markedly different land–ocean distributions compared to the modern world – especially the concentration of landmasses in the southern hemisphere. A possible general climatic scenario for Late Devonian times is sketched out in Figure 7.2. The majority of Gondwana would have been under the thrall of mid-latitude moist winter westerlies blowing off widening Palaeotethys and the closing Rheic oceans. Given the high latitude of eastern Gondwana, we can infer the likely presence of ice sheets there, nourished by snowfall from the interaction of the westerlies with cold air masses along the polar front.

Large areas of central Laurussia, eastern Gondwana and the South China terrane would have been favourably placed with respect to the north-east trade winds to have experienced a strong north west monsoon in the southern hemisphere summer. This would have intensified as the trades ascended the flanks of the Acadian mountains of eastern Laurussia. In the subtropical northern hemisphere, northern Laurussia and the South China terrane would have been prone to summer south-west monsoonal conditions as the south-east trades crossed that sector of the Equator.

Western Laurussia seems likely to have been arid to semi-arid due to the dry south-east trades associated with subtropical high pressure. Around the southern tropics, summer subtropical trade winds rose towards the Acadian mountains, with winter aridity under the same high-pressure cells. Such summer-wet conditions were perfect for development of red-brown, iron-rich sediments and monsoonal forest soils analogous to those of the modern Indo-Gangetic plains that are recorded in the late Devonian deposits of Munster, Ireland[2] and the Catskill riverine environments of upstate New York and Pennsylvania.[3] The calcic soils found widely in the Upper Old Red Sandstone of

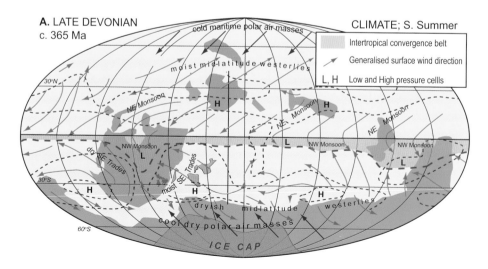

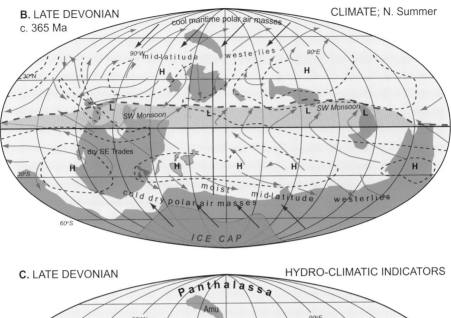

B. LATE DEVONIAN
c. 365 Ma

CLIMATE; N. Summer

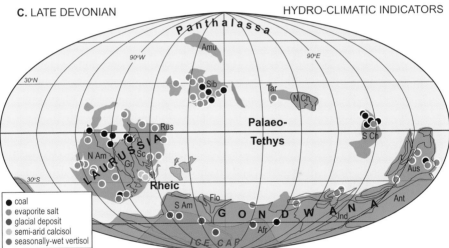

C. LATE DEVONIAN

HYDRO-CLIMATIC INDICATORS

- ● coal
- ● evaporite salt
- ● glacial deposit
- ○ semi-arid calcisol
- ● seasonally-wet vertisol

Figure 7.2 A–B Reconstructed Late Devonian summer and winter climatic regimes using the geographic map of Figure 7.1B. Reconstructions inspired by present-day general atmospheric circulation and world climatic belts. **C** Hydro-climatic indicators; largely after Torsvik and Cocks (2017) using the prior work of Scotese (2000) and Siberian data from Volkova (1994). Coals indicate wetland peat accumulations in any climatic zone; evaporite salts indicate seasonal water body evaporation or coastal sabkhas typical of hot, arid tropical latitudes; glacial deposits from ice-erosion and deposition indicate snowfall accumulations at high latitudes and/or high altitudes; calcrete soils (calcisols) are indicative of low levels of seasonal rainfall in semi-arid climates (as in the south-west USA today); iron-rich vertisols with clay horizons are characteristic of summer-wet/winter-dry climates, often associated with a monsoonal regime (as in the Indo-Gangetic Plain of modern northern India).

Britain[4] indicate a somewhat more arid climate in central-eastern Laurussia, analogous to that of the modern-day semi-arid South West USA.

From Pre-Devonian Holdfasts to First Forests

The sources for the cryptospores noted above are thought to have been pre-Devonian plant forms without roots, leaves or vascular (conducting tubes) stem tissue. They were probably related to the plant grouping of modern bryophytes. Such 'living fossils' include the familiar mosses and perhaps not-so-familiar worts: liverworts and hornworts, abundant in damp climates, whose spores are dispersed by rain droplet impact (Fig. 7.3). For around forty million years, the ancestors of such plants provided a decent scintilla of greenery and shed their spores into pre-Devonian surface waters and atmospheric winds. Yet the low preservation potential of the bryophyte plant itself meant that no actual 'body' fossils of them are recorded until Middle Devonian times (around 385 Ma), discovered fairly recently in strata of the Catskill Mountains.[5]

The oldest remains of land plants with water-conducting stem tissue (vascular plants) are curved and double-branched forms named *Cooksonia*. They bear millimetric-size kidney-shaped spore-bearing parts – sporangia. The discovery of these in 1980 as carbonaceous fragments in micaceous siltstone strata from County Tipperary in Ireland by Diane Edwards and John Feehan[6] created a botanical sensation, for they were dated from associated fossils as Middle Silurian in age, around 425–430 Ma. Further discoveries, published in 1992, of coalified remains from the Early Devonian, again by Edwards and co-workers[7], revealed exceptionally well-preserved vascular structures in *Cooksonia*. Although not (yet) represented by complete remains, it seems likely that the plant was small in stature (decimetric) and leafless, featuring bryophyte-like holdfasts. Thanks to the voluminous production of their characteristic multiple (four-fold) spore-aggregates, they would have spread rapidly across soft, moist fine-grained muddy/silty sediment substrates – river-bank alluvium and lake shoreline deposits. Exceptional records of Late-Silurian (*c.*423 Ma) fossil charcoal[8] imply the action of lightning strikes upon clumped and desiccated (?autumnal) accumulations along seasonal watercourses.

A factor governing the rarity of fossilized early vascular plants was that of preservation potential. Post-mortem survival of terrestrial remains requires sudden, but not destructive, deposition by sediment influxes, the fallout of fine volcanic ash or mineral precipitation around them. Luckily the latter was the case in the Early Devonian lake and riverine basins of south-east Laurussia in what is now north-east Scotland. The volcanically active Rhynie basin in Aberdeenshire provided remarkable conditions of fossilization for the most outstanding early terrestrial biological community yet discovered[9], dated to around 410 Ma. Palaeontologists use the German term *Lagerstätte* ('dwelling-place') for such richly fossilized *in situ* assemblages, an apposite term due to the brilliant and witty pioneering palaeoecologist, the late Adolf Seilacher. The floral remains include the spore-bearing vascular genus, *Rhynia,* whose branched stems with terminal sporangia issued from a basal rhizome (Fig. 7.4). This and related plant forms

Figure 7.3 **A–B** Living fossils mimicking a bryophyte-encrusted Palaeozoic land surface. **A** Thriving bryophyte mats (*Lunularia sp.*) on sheltered, bare, damp and compacted garden soil in maritime, moist-temperate south-west Ireland. Field of view is *c.*150 mm wide. **B** Close-up (scale in millimetres) of *Lunularia* with its interlocking and protective 'leaves' (thalli), the cup-like examples containing unattached spores awaiting raindrop impacts for their dispersal in splashes. The plants are attached to the soil substrate by net-like holdfasts, not true vascular roots. Source: Author.

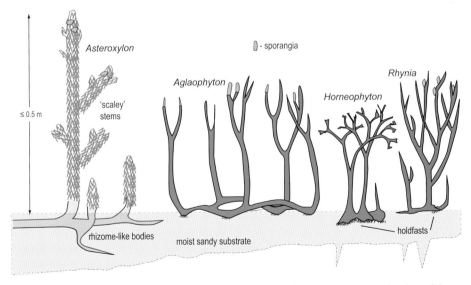

Figure 7.4 Early Devonian plants from the Rhynie terrestrial *Lagerstätten*. Four sketches of the chief Rhynie plants to show their holdfast anchorages, sporangia and exposed or buried rhizomes colonizing a moist sand substrate. Recrafted from material in Trewin, (2001).

are exquisitely preserved in chert, a flinty, hard, siliceous sedimentary rock. An entire ecosystem was fossilized here[10] by mineral precipitation from successive percolating and enveloping influxes of warm silica-rich, alkaline waters gently debouching from overflowing pools of a nearby sinter cone (as seen around modern Icelandic and Yellowstone geysers). In addition to several other distinct forms of vascular plants, the rich community also featured algal charophytes (like the modern limey stonecrops), fungae and lichens that hosted pioneering plant-eating insects including the first known spiders, mites and springtail-like creatures.

The first true *in situ* branching lateral roots from a primitive member of the Lepidodendrales (clubmoss) were found recently in the Early Devonian (*c*.411 Ma) of Wyoming.[11] These are of very similar age to the Rhynie flora – indicative of widespread botanical radiation. That it continued in Middle Devonian times (*c*.392 Ma) is spectacularly witnessed by long-known[12] fossil *Lagerstätten* and by more recent discoveries[13,14] from the Catskills. Here the south-west coastal plains of Laurussia were lapped by the shores of the oceanic strait joining Panthalassa with the Rheic Ocean. Sedimentary deposits reveal evidence for seasonal tropical storms periodically driving waves over coastal barrier and lagoon complexes to deposit 'washover' sands and silts onto riverine vegetated wetlands.[15]

That the Devonian Catskill coastal plains were colonized by dense forest stands was established nearly one hundred years ago by Winifred Goldring of the New York State Museum – known from its place of discovery as the Gilboa *Lagerstätte*.[12] Her 'oldest fossil forest', though now usurped by more recent discoveries, has since been proven to be a true mixed forest.[13] It had underground growths featuring rhizomes and surface sprouts of woody ancestral conifer-like forms and primitive lepidodendrales. Goldring's discovery concentrated on the impressively large tree-like form *Eospermaptopteris*, spore-bearing ancestors to the extant clubmosses (Fig. 7.5A). They were tall (*c*.30 m) tree fern-like constructs with stout unbranched boles up to 1 m in diameter preserved as sandstone casts with rounded bulbous bottoms, features that led Goldring to suggest that they 'might be expected of certain trees growing under swampy conditions'. The boles have a spaghetti-like bristle of numerous thin, sometimes long rootlets that may have repeatedly died back and regrown rather than being permanent features.[14]

The tree-like forms were leafless, and their boles branchless. It is thought they photosynthesized from short-lived, constantly renewing crown branches via hand-like arrays with terminal appendages containing spore-bearing apparatus. Such dense forests erupted massive clouds of spores into the winds of the Devonian atmosphere. There are now many records of thick Middle Devonian coals (some in the Barzas area of the Siberian Kuznetsk basin are over three metres thick) whose constituent parts are dominated by plant cuticle remains and spores.[16] The former predominate, their waxy residues yielding coals rich in volatile (oil- or gas-prone) constituents termed liptinite.

The 2019 discovery at Cairo Quarry in upstate New York[14] has spectacularly moved back the record of the traces of the 'woody ancestral conifer-like forms' found at Gilboa

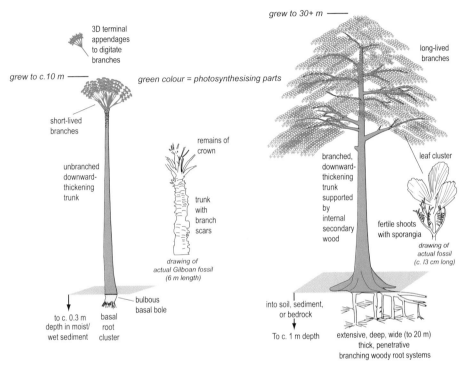

A MIDDLE DEVONIAN

GILBOAN 'TREE-FERN' MODE *(Eospermatopteris)*

FROND-LIKE; TRUNK UNBRANCHED; NARROW-CROWNED; SHALLOW-ROOTED
PROVIDED LITTLE SHELTER, SHADE OR ROOT-INDUCED WEATHERING

B MIDDLE DEVONIAN

ANCESTRAL 'CONIFER' MODE *(Archaeopteris)*

LEAFED; TRUNK BRANCHED; WIDE-CROWNED; DEEP- & WIDE-ROOTED
PROVIDED SHELTER, SHADE & ROOT-INDUCED WEATHERING

grew to 30+ m

3D terminal appendages to digitate branches

long-lived branches

grew to c.10 m

green colour = photosynthesising parts

short-lived branches

remains of crown

leaf cluster

unbranched downward-thickening trunk

trunk with branch scars

branched, downward-thickening trunk supported by internal secondary wood

fertile shoots with sporangia

drawing of actual fossil (c. 13 cm long)

drawing of actual Gilboan fossil (6 m length)

bulbous basal bole

into soil, sediment, or bedrock

to c. 0.3 m depth in moist/ wet sediment

basal root cluster

To c. 1 m depth

extensive, deep, wide (to 20 m) thick, penetrative branching woody root systems

Figure 7.5 **A–B** Reconstructions showing contrasting characteristics of two of the larger trees that evolved to populate the first forests in the Middle to Late Devonian. These illustrate the diverse evolutionary strategies available at these times. **A** Reconstruction of the 'Gilboa Forest' architectural lineage that developed in Middle Devonian 'short-lived' moistland habitats: a form that later evolved separately into modern tree-ferns and palms. More permanent equatorial riverine and deltaic wetland swamp environments, soon to dominate the tropics, awaited their own explosive vegetational development and colonization by descendants of the less-common arborescent Lepidodendrales, also found in the Gilboa *Lagerstätten*. Data assembled from Stein *et al.* (2007, 2012) and the accompanying commentaries of Meyer-Berthaud and Decombeix (2007, 2012). **B** The *Archaeopteris* lineage grew high, their superstructures supported by a thick growth of secondary wood. They colonized better-drained Middle Devonian substrates and infiltrated them with dense, wide and deep branching woody rooting systems. It later founded the gymnosperm line (modern conifers and cycads are descendants) and that of most seed-plants. Data from Stewart and Rothwell (1993), Algeo and Schechler (1998) and Stein *et al.* (2019).

in 2012. This is the beefy genus *Archaeopteris* (Fig. 7.5B) that gave rise to all modern seed plants, up until then thought to have evolved only in Late Devonian times. The Cairo discovery was found in Middle Devonian strata, two to three million years older than the Gilboa forest. It comprised a soil horizon in which were anchored abundant large, *in situ* radial woody roots (≤11 m long) of many *Archaeopteris* individuals. These trees had diversified tremendously by this time (385 Ma), with specialized, deep-penetrating, branching, lignin-rich roots, crown outgrowths and substantial long-lived, leaved and branched trunks fortified by thick bark lignin. Cleverly assembled sedimentary evidence indicates the Cairo coastal plain forest floor and its widespread and distinctive 'vertisol' soil (formed in a climate with strong wet/dry seasonal alternations) were flooded and preserved by a coastal storm-surge event. This deposited detrital silts, an assemblage of broken-up fish remains, and enough seawater sulphate to enable iron pyrites (an iron sulphide) mineralization around some of the buried and dead root structures.

By late-Devonian times the ancestral forms of many long-lived plant lines had diversified spectacularly. We might single out the recent discoveries of seed-related plants with multi-strand, water-conducting vascular tissues[17] and early Lepidodendrales frond-forest stands in both Svalbard, Spitsbergen[18] and Anhui Province, China.[19] Lepidodendrales ancestors were in the undergrowth of the Gilboan forest[13] – their successors destined to dominate the plant communities of the swampy Carboniferous equatorial world.

Coda: of Trees, Soils and Atmospheres

Vegetation had long-lasting and pervasive effects on terrestrial environments during Devonian times. Leaf crown growths provided canopy sheltering and rain droplet break-up (the umbrella effect), the interception lessening soil erosion. Wild fires could erupt after lightning strikes, particularly after drought, the potassium-rich ash renewing soil fertility. Deeper, stronger, thicker and more densely branched roots of the Archaeopterid line meant greater soil weathering and bedrock breakdown. The evolutionary impetus in gaining more light for photosynthesis gave a parallel gigantism in stem bark and crown – the entity we know as a forest came into being. These gave niches for increasingly diverse fungal and insect populations that fed on living and dead vegetation, with knock-on feeding opportunities for early amphibians, and, in their turn, predatory fish. By late-Devonian times the greening of the continents had proceeded apace, coastal and inland floodplains of the world and their interior margins appearing as green-lined riverside glades, branching and widening lengthways from inland catchment edges to their salty coastal termini. Magnified photosynthetic activity would have produced a steady and progressive enrichment of oxygen in the planet's atmosphere, with concomitant carbon dioxide depletion, so reducing the planet's greenhouse effect and causing global cooling.[20] It would seem no coincidence that the first evidence for growth of ice sheets and upland valley glaciers (Fig. 7.2) over high-latitude Gondwana was in the Late Devonian and Early Carboniferous periods.[21]

8

Carboniferous Tectonic Geography and Climate

Dionysius of Halicarnassus once likened
Aeschylus' poetry to this Cyclopean
wall beneath Apollo's temple before us,
this wall I always gaze on whenever in Delphi,
blocks shaped like continents pre-early Jurassic
where capers cascade down landlocked Pangea,
polygonal Gondwanaland, in tasselly swathes…

Opening lines of Tony Harrison's *Polygons* (London Review of Books, February 2015),
a long almost-lament for the classical Greece of his theatrical imagination and for his
friends, Ted Hughes and Seamus Heaney.

Defining the Carboniferous and its Boundaries

The Carboniferous Period (Fig. 8.1) spanned the interval of geological time from around
359 Ma to *c*.299 Ma. A momentous sequence of events – geographic, tectonic, climatic and
evolutionary – occurred over this time. The nature of these has been pieced together by
close examination of Carboniferous strata in different parts of the world recognized by their
common fossil content. The chief fossil groups used for matching strata over such large areas either rode the currents of the three great contemporary oceans, Panthalassa, Rhea and
Palaeotethys, or were blown hither and thither by planetary winds. The groups comprise:

- …goniatites – free-swimming and coiled marine molluscs, ancestors
  to the well-known Mesozoic ammonites and the modern *Nautilus* (see
  Chapter 9);

- … conodonts – phosphatic tooth-like remains of the soft-bodied, worm-
  like marine creature *Clydagnathus*; and

- … spores – airborne plant reproductive structures shed, for example, by
  liverworts, ferns and the tree-like Lepidodendrales.

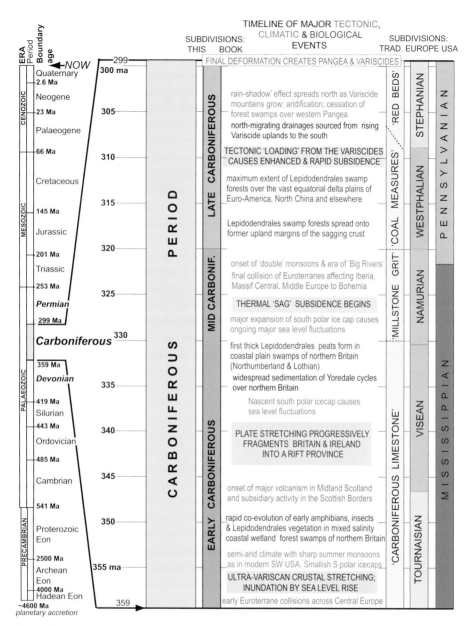

Figure 8.1 Timescales, timelines and various geological events enacted during the Carboniferous period in UK/Ireland, northern Europe and globally. Details follow in the text. The timescale follows the International Commission on Stratigraphy (2012) as reinforced by subsequent radiometric dates made on zircon crystals found in volcanic ash strata of Carboniferous age.

All these life forms evolved sufficiently rapidly during the 60 million years of the Carboniferous for their fossils, found at different levels of stratal successions, to appear distinctive, allowing global matching and linkage between them. By way of contrast, other regional or locally occurring fossil creatures are of lesser use in such correlation, since they evolved under place-specific conditions in local or regional habitats. Such were certain organisms that lived in and around sub-tropical coral reefs, shell-sand beaches or the freshwater clams of shallow lakes and ponded mires in riverine and deltaic lakes and lagoons. Over the centuries there have been heated debates and arguments amongst palaeontologists concerning resolution of these issues in Carboniferous stratigraphy.

Tectonic Geography

Figure 8.2 shows two snapshots of Carboniferous time. Together with those for Devonian times in Figure 7.1 they show the supercontinent Pangea assembling by the slow conjoining of Gondwana and Laurussia along the remnants of the Rheic Ocean. Sandwiched between them were the Euroterranes noted previously. In addition to a general northern shift of all these, there was also a significant longitudinal shear adjacent to their margins. This involved both compression and tension along the undulating plate boundary between Laurussia and Gondwana and a significant sliding motion (wrenching) with a rightward (dextral) sense. In other words, Laurussia moved relatively eastwards, Gondwana westwards. It was this combination of sliding and compression, known as transpression, that sent north-west Gondwana (including modern Amazonia, Florida and north-west Africa) scraping and buffeting along former southern Laurussia for a thousand or so kilometres over the sixty million years of Carboniferous time. The mean rate of this motion was around 15 mm per year, close to that of modern Pacific sea-floor spreading. By the latest Carboniferous, the process had docked north-west Gondwana firmly into Laurussia, forming a compressive mountain belt dominated by thrust faults and related folding. Its surviving roots may be mapped out in North America from the Ouachitas of west Arkansas and south-east Oklahoma along the Allegheny ranges from south-west Virginia to north central Pennsylvania. These latter comprise the eastern USA's chief coal regions. In Europe it is defined by the Variscan mountain belt spanning thirty degrees of present-day longitude between western Iberia and Bohemia.

Early Carboniferous times (Fig. 8.2A) saw the rapidly opening central equatorial Palaeotethys Ocean in familiar guise. Along the Laurussian margin, in what is now Maritime Canada, a powerful offshoot of the plate-sliding regime led to the formation of a plethora of narrow rift basins and broader rift sags. These were criss-crossed by dextral wrench faults, and by thrust and normal faults according to the orientation of major bends along the main fault line and its many splays and offshoots. This Maritime transpressional belt faced directly across to the western (Ulster) part of the Ultra-Variscide rift province (ultra: Latin for 'beyond') of Ireland, Britain and the southern North Sea. All these places were as yet untouched by any hint of the mountain-building storm that was hitting Europe to the south from collision of the Euroterranes with

A EARLY CARBONIFEROUS
TECTONIC GEOGRAPHY

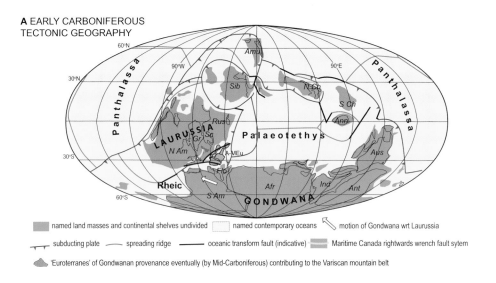

B LATE CARBONIFEROUS
TECTONIC GEOGRAPHY

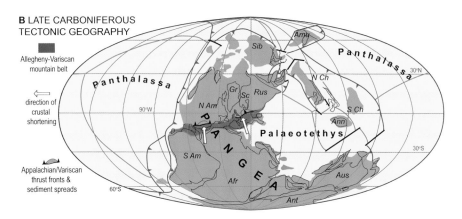

Figure 8.2 **A–B** Carboniferous tectonic geographies. **A** Early Carboniferous **B** Late Carboniferous. Redrawn and crafted chiefly from material in Torsvik and Cocks (2017).

south-east Laurussia. By Middle Carboniferous times active rifting had ceased in the Ultra-Variscides and there was continued and general northwards drift of Laurussia and Gondwana, shifting them fifteen or so degrees northwards towards the Equator. During the Late Carboniferous (Fig. 8.2B) Pangea gradually assembled from the tectonic collision of Gondwana and Laurussia as they both continued their relative drift along different vectors.

Eventually, crustal shortening produced by the collision propagated northwards to engulf all of the Ultra-Variscan basins, inverting them into mountains and uplands riven across by thrust faults and related folding. Associated crustal thickening caused melting in south-west England, from the Isles of Scilly to Dartmoor, where during the Early Permian a ridge-like mass (batholith) of granitic magma intruded into the upper crust. Basic magma was generated in the earliest Permian in northern Britain, perhaps by a plume-like upwelling of magma. It rose to form large sills (strata-parallel intrusions) with associated dykes in the upper crust – the Whin Sill of north-east England and the Forth Sill of Lothian and Fife.

Equatorial Double Monsoons

The assembly of Pangea had major climatic consequences (Figs 8.3, 8.4). It placed the Acadian and older mountains, the newly forming Variscide mountains and the Ultra-Variscan rift province in the equatorial climatic engine-house. The positioning of equatorial atmosphere and ocean with wide mountainous land masses would have encouraged monsoonal conditions to dominate tropical climates. In southern high latitudes western Gondwana rotated clockwise with respect to the pole, shifting climatic zones and the locus of polar ice correspondingly.

During the Early Carboniferous the gradual northward drift of Laurussia brought more and more of the continent under the thrall of both south-west and north-west monsoons. Equatorial forest swamps with thick peaty substrates spread to higher latitudes in what is now Arctic Canada, and expanded further southwards over much of Laurussia.

Middle to Late Carboniferous times saw large tracts of Laurussia positioned to experience both summer and winter monsoons. The vigorous Lepidodendrales floras, waiting in the wings since the Early Carboniferous, responded with ecological and evolutionary vigour. It was into this subsiding 'Bay of Middle Europe' (a Carboniferous Bay of Bengal) and other examples worldwide, chiefly north China and north-east Australia, that great rivers developed, spreading prodigious discharges of water and sediment. The mightily fecund Lepidodendrales colonized several million square kilometres of wetland channel sand banks, levees, bayou and floodplains with swamp forest, their reproductive spores being spread widely by water and wind.

Carboniferous Tectonics and Sedimentation in Britain

Tectonic stretching of the Ultra-Variscide crust in the Early Carboniferous (Fig. 8.5A) fractured it into tilted blocks and platforms that subsided rapidly, with widespread basaltic volcanism in northern areas.[1] Earliest Carboniferous sedimentary deposits record a trend to rising global sea level that submerged former coastal plains under rapidly deepening marine waters. This marine flooding was much accentuated by the rifting itself as the stretched crust subsided.

Shallow riverine lagoons and lakes on coastal plains with wildly fluctuating salinities developed widely in the early Carboniferous of northern Britain.[2] In these water bodies, small early amphibians waddled and thrived, feeding off equally flourishing insect

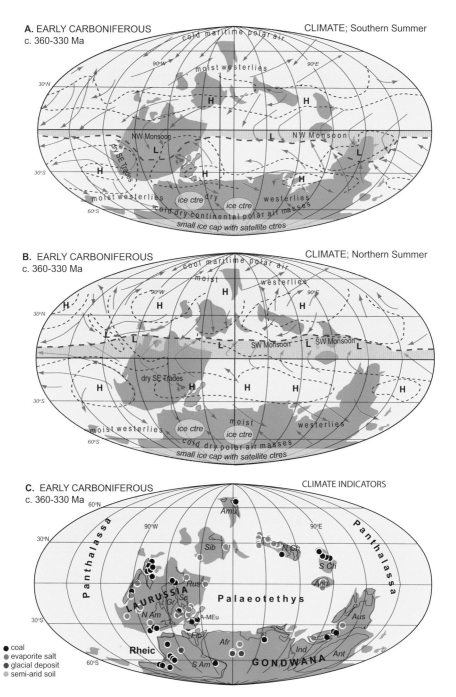

Figure 8.3 **A–B** Reconstructed Early Carboniferous world climate. **A, B** Summer and winter climatic regimes respectively. Reconstructions inspired by the general circulation and distribution of modern world climatic belts. **C** Hydro-climatic indicators largely after Torsvik and Cocks (2017) using the prior work of Scotese (2000) as detailed in the captions to Figure 7.2.

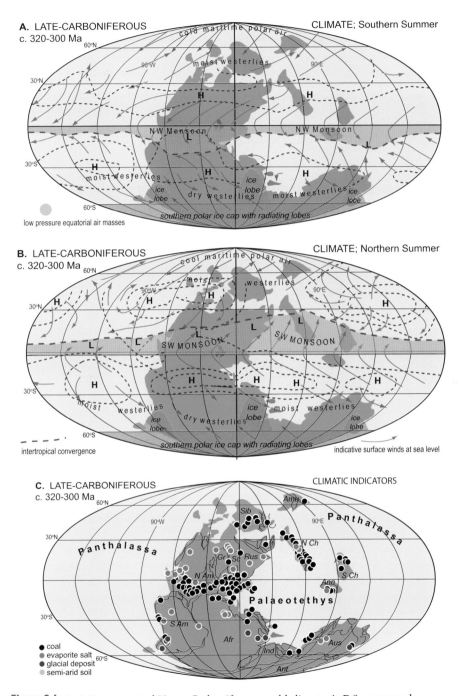

Figure 8.4 **A**–**B** Reconstructed Upper Carboniferous world climate. **A, B** Summer and winter climatic regimes respectively. Reconstructions inspired by the general circulation and distribution of modern world climatic belts. **C** Hydro-climatic indicators largely after Torsvik and Cocks (2017) using the prior work of Scotese (2000) as detailed in the captions to Figure 7.2.

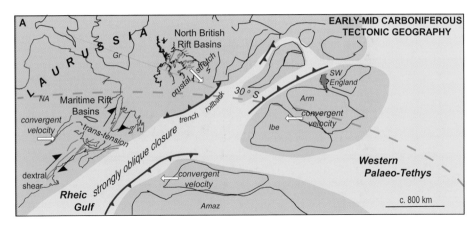

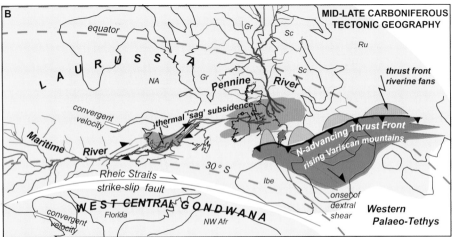

Figure 8.5 **A, B** Chief elements of the Carboniferous Ultra-Variscides within the broader tectonic geography of the Carboniferous. Redrawn and crafted from material in Hibbard and Waldron (2009), Waldron *et al.* (2015), Torsvik and Cocks (2017) and numerous workers on European Variscan tectonics and palaeogeography. Reconstruction **B** for Mid-Late Carboniferous times just predates the beginning of the 'hard collision' that caused Appalachian mountain building. The Pennine River drained Caledonian, Proterozoic and Archaean sourcelands along the Greenland–Scandinavia gap to empty a mighty discharge into the ever-wetlands of Ultra-Variscan sedimentary basins. A similarly large system flowed from a Northern Appalachian hinterland into the Maritime Canadian rift to the north-east (palaeocurrent data from Gibling *et al.*, 1992) with an offshoot into Ireland and the British Isles. Both were fed by monsoonal rainfall.

populations. The climate was hot enough for brine development and the precipitation of evaporite salts, but also periodically wet enough for seasonal river floods in coastal and riverine marshes. Such fluctuations defined a 'schizohaline' environment, analogous to a combination of the modern-day freshwater marshes of southern Iraq fed by Mesopotamian rivers and the saline coastal sabkhas (salt flats) of neighbouring Kuwait. Palaeontologists have surmised that such salinity-stressed environments could have been conducive to natural selection pressure on the prolific and important early amphibians.[3] Algal-rich organic deposits featured in the extensive waters of Lake Cadell in eastern midland Scotland – upon later burial these became the first economically exploited oil shales in the world.

Over many northern, central and southern areas of England, rift basins and their tilted, subsiding ramp-like margins were gradually covered by shallow seas. In these were deposited calcareous sediments. The deepening flanks to some (e.g. Craven Lowlands) featured protruding mound-like reefal build-ups of finely crystalline calcium carbonate. As sea level oscillated, local shorelines shifted back and forth. Coarse-grained detrital sediment was brought into northern rift basins by relatively small rivers and deltas sourced from the north-east. Their muddy plumes of effluent disturbed the delicate balance between calcareous-shelled organisms in clean, clear seawater, giving rise to alternating deposits of shelly carbonate and muddy sediments, with widespread but usually short-lived peat development in deltaic Lepidodendralian forest swamps. Such alternations define the Yoredale Formation of the northern Pennines and equivalent strata in Scotland that hosted the peatlands responsible for the oldest workable coals in northern Europe.

In Middle Carboniferous times the much larger Pennine River (Fig. 8.5B) fed in from its monsoon-drenched catchments. Its delta infilled previously deep-water (several hundred metres or so in some cases) rift basins of the English Midlands, laying down first muddy sediments rich in organic detritus in the foetid depths (the future Bowland Group). These were succeeded by the coarse sandstone deposits of river deltas and their fronting submarine channels, which were to become the iconic Millstone Grit.

For the next thirteen or so million years of Late Carboniferous time, the Ultra-Variscan foreland and the still-rotating North China terrane were the loci of astonishing levels of biological productivity and carbon burial. At the same time, and by no coincidence, the southern polar ice sheets reached their greatest extent. Sea-level oscillations up to seventy metres or so in amplitude periodically sent marine waters with their free-swimming marine faunas far into the interior of equatorial Pangea, flooding over low-lying and gentle-gradient forest swamps. This extraordinary interlinked, mainly depositional, landscape extended over scores of degrees longitude in foreland basins from New Mexico eastwards across the Mid-West to the Allegheny Mountains and on to the former rift provinces of Maritime Canada. As time went on some of the now regionally subsiding Ultra-Variscide basins morphed into foreland basins (for example, South Wales), weighted down by advancing thrust sheets as the Variscan mountain chain grew and propagated northwards (Fig. 8.5B). It is only an apparent paradox that Ultra-Variscan subsidence could co-exist adjacent to the Variscide mountains that were under vice-like compression. This is because Variscan mountain-building was accompanied by crustal-scale telescopic shortening, including prodigious overfolds (nappes) in the Massif Central and elsewhere. For example, the distance today from London south to the Montagne Noir (some 1100 km) was very much greater, perhaps by >200–300%, in early Carboniferous times.

A Modern Plate-Tectonic Analogue

It is a feature of areas undergoing modern continent-to-continent collision that the strains involved cause complex patterns of crustal tectonics. These are best illustrated from the pattern of active deformation seen today in the Middle East and Eastern Mediterranean as the African/Arabian continent collides with Eurasia, particularly along the course of the North Anatolian Fault from the Black Sea across the northern Aegean to the Gulf of Corinth in Central Greece (Fig. 8.6A). The maps show numerous similarities with the situation described for Carboniferous tectonic evolution of the wider North Atlantic area, witness the various tectonic regimes transposed onto the map forming Figure 8.6B.

Sourcelands: To See a World in a Grain of Zircon

The reconstructed Pennine River system of Figure 8.6B tapped a huge area of old continental crust in its drainage hinterland. The evidence for this comes from the mineral composition of Carboniferous sandstones and the evidence from their current structures. Such techniques were originally used by H.C. Sorby and A. Gilligan to infer a probable northern source extending perhaps to Scandinavia for sedimentary detritus for much of the Millstone Grit (often rich in feldspar) and the Coal Measures.[4] Numerous studies since the 1920s have confirmed these views, though with some reservations. For example, a southern source has been indicated for quartz-rich sandstones from a landmass that stretched over much of Central Wales, the South Midlands, East Anglia and across the southern North Sea into Belgium.[5] Most exciting, several sandstone horizons in the Pennine Coal Measures not only show current flow from west to east, but also contain abundant dense minerals (up to thirty per cent of some populations) of a most unusual and distinctive type.[6] These are known as chrome spinels: dense mineral oxides derived from rocks of upper mantle origin. Such rocks of Early Palaeozoic age outcrop extensively today in Newfoundland, north-west Ireland (e.g. Mayo) and Ayrshire (south of Girvan, around Ballantrae). Together with the presence of abundant reworked Devonian spores and geochemical evidence these discoveries point to an eroding drainage catchment feeding eastwards that may have covered a significant portion of the north-west Atlantic between Maritime Canada and north-western Ireland (Fig. 8.5B).

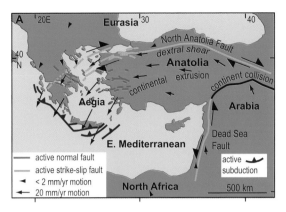

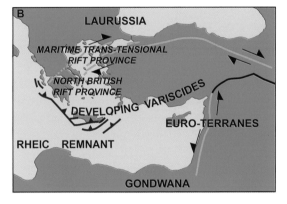

Figure 8.6 **A–B**: **A** Modern plate-tectonic analogue for the Carboniferous rift provinces of Maritime Canada and North Britain/Ireland from the Aegean and neighbouring areas. The present day motion of crustal blocks is simplified from satellite geodesy platform GPS data on plate surface velocities in the region by McCluskey *et al.* (2000). **B** An early attempt at such an analogue was by Maynard *et al.* (1997).

From the mid-1980s, hints as to the age structure of the Pennine River's geological hinterlands came from determination of uranium–lead isotopic ratios in sand-sized sedimentary grains of uranium-bearing minerals such as zircon and monazite.[7] But there was a problem. At the time of the first research it was only just possible at one or two advanced labs in the world to analyse single crystal grains. Instead, multi-grain samples sorted into populations by colour and shape had to be used. A northern source was certainly implied, but more distant and much older (including Archaean rocks >2500 Ma old, probably from northern Greenland) than previously imagined by Gilligan, and seemingly not including any Proterozoic-age (540–2500 Ma) rocks. By the later 1980s, single zircon and monazite grains had begun to be routinely analysed, the results confirming the major facts revealed in the multigrain studies but shedding no further light on the 'Proterozoic Deficit'.[8] More recent work in an era of advanced chemical preparation and analytical techniques has enabled analysis of hundreds of random individual zircon grains.[9] These have upheld the pioneering conclusions outlined above, but with one huge caveat: there was no 'Proterozoic Deficit' from the northern catchments – precise dates from that era occur throughout Pennine Carboniferous strata.

Polar Perspectives

The first field evidence for south polar ice during the Late Devonian to Early Permian time interval (383–290 Ma) came from observations published by three young officers in the first bulletin of the Geological Survey of India in 1859, just two years after the Indian 'Mutiny' had interrupted their lives.[10] It was just sixteen years after the Swiss Louis Agassiz had presented in English his revolutionary glacial theory for the existence of a Pleistocene Ice Age. Brothers W.T. and H.F. Blanford and their colleague W. Theobald Jr., all in their twenties, carefully assessed the evidence for the glacial origin of Late Carboniferous/Early Permian rocks they had been mapping in a fruitless search for coal in the district of Cuttack, south-west of Calcutta in north-east India. Their evidence included an illustrated example of what we now call 'dropstones', released from melting icebergs and impacting the Palaeotethyan sea bed. The trio carefully referenced Lyell's account of melting ground ice presented in the eighth edition (1853) of his *Principles of Geology* and proposed a similar origin for the Late Carboniferous Talchir Boulder Bed that lay below coal- and plant-bearing Early Permian strata.

Over the following eighty years, wider fieldwork in central and southern Africa, India, Australia and South America provided further evidence: most spectacular were ice-scratched rock pavements and ground moraine deposits full of far-travelled glacial erratics. It was clear to Alfred Wegener, marshalling his facts concerning continental drift just before the First World War, that:

> The riddle of the Permo-Carboniferous glacial period now finds an extremely impressive solution in the displacement theory [continental drift]: directly those parts of the earth which bear these traces of ice-action are concentrically crowded together around South Africa, then the whole area formerly covered with ice becomes of no greater extent than that of the Pleistocene glaciation on the northern hemisphere…[11]

Further, given the stratigraphic arrangement of the glacial deposits, being older than the plant-bearing strata with the form *Glossopteris* in South America, Africa and India but younger, of Permian age, in Australia, it seemed probable that west to east south polar wandering had also occurred. Even by 1937, the year of publication of A.L. Du Toit's *Our Wandering Continents* – a lawyerly, field-led evisceration of the considerable, mostly American, opposition to Wegener's theory – it was obvious that there was no single stratum of glaciated sediment in Gondwana. Du Toit could rightly insist that:

> Each year makes it clearer that this great 'Ice Age'…was not a single episode, but a complex series of glaciations with milder interludes, embracing in all a very considerable period of time.'[12]

He listed eleven regional occurences dated by fossils of Early Carboniferous to Late Carboniferous age, pointing out evidence for multiple glaciations in some, and from others a regional radial ice outflow from Gondwana deduced from the orientation of ice-scratch marks on underlying rock surfaces. A dependable modern synthesis of subsequent developments is presented by Fielding *et al.* (2008).

Coda: 'Far-Field' Effects of Glaciation

In the early 1930s, impressed by the field evidence for the reality of the Gondwanan ice cap, it seemed certain to geologists H.R. Wanless and F.P. Shepard, working on the Late Carboniferous coal-bearing deposits in the Illinois Basin (similar in most respects to the Pennine Yoredale Formation noted above), that the waxing and waning of the ice could have influenced global sea level, and hence the advance and retreat of equatorial Carboniferous coastal deltas.[13] This was later christened a eustatic or worldwide mechanism, a 'far-field' consequence of glaciation. What is more, the to-and-fro shift of coastlines due to the change would have been synchronous and periodic the world over, juxtaposing marine sediments with terrestrial sediments containing peat accumulations. By the late-1980s it was widely realized that the eustatic mechanism was correct in its essentials, since it had a richer predictive basis than other theories and, decisively, was approximately periodical according to orbital time frames for climate change.[14]

9

Carboniferous Equatorial Swamp Forests

Last night as I lay dreaming
There came a dream so fair,
I stood mid ancient Gymnosperms
Beside the Ginkgo rare.

I saw the Medullosae
With multipartite fronds,
And watched the sunset rosy
Through Calamites wands.

Oh Cryptograms, Pteridosperms
And Sphenophyllum cones,
Why did ye ever fossilize
To Palaeozoic stones?

A clever and charming Carboniferous palaeobotanical ditty
attributed by W.G. Chaloner (2005) to Marie Stopes.

Background: Modern Wetlands as Peat 'Factories'

Wetlands are poorly drained, often transitional environments between better-drained land and open, shallow-water bodies that may be saline, brackish or fresh. Their bottom waters and sedimentary substrates are often poorly aerated so that colonizing life forms must tolerate low-oxygen conditions. Various names for wetland resonate from dictionary pages – mire, bog, fen, marsh, swamp – and from the vernacular tradition of our familiar native dialects and languages – carr, moss, waste, flow (North British); cors, ffen, llaid (Cymru); portach, caorán, corcah, seascann (Gàidhlig).

Wetland habitats can be classified a little more specifically.[1] Marshes and fens are freshwater, mostly founded on mineral sediment or on thin peats resting on such materials. They have permanent or seasonal inundations of standing or slow-moving

Figure 9.1 The winter-flooded liminal environments of a temperate maritime valley carr and fen wetland with brushwood stands (mostly willow and alder; note the fern epiphytes rooted in moss), specialized clumps of tufted sedges and (distant, brownish) dense reed-beds of *Phragmites*. These record the gradual Holocene infill of a valley lake behind the Long Strand at Ownahincha, Rosscarbery, West Cork (51.335449, 8.584789). The combination of brushwood carr and reed made the best-burning peats in the Holocene coastal riverine fens of both Ireland and Britain, in particular those of the Norfolk Broads and the Somerset Levels. Source: Author.

flood water, generally nutrient-rich and of neutral acidity, supporting reed (*Phragmites*) beds and forest stands (Fig. 9.1). Swamp forests often fringe fens and marshes and have aerated surface layers of neutral acidity above oxygen-poor peaty or mineral-rich substrates. Bogs comprise water-saturated sphagnum moss over a peaty substrate. They are isolated from incursions of nutrient-rich groundwaters, depending solely upon nutrient-poor and slightly acidic rainwater for replenishment.

The chief feature of wetlands that concerns us here is their ability to prevent the oxidation of plant remains. Instead there is degradation to vegetable peat – a watery peat 'factory'. The high preservation potential of wetland peat makes it one of nature's chief vehicles for carbon storage. A homely human analogue is the garden composting stall. By way of contrast, well-drained forest floors receive seasonal organic litter, leaves, fallen branches and trunks together with roots, but these are consumed by life forms (chiefly

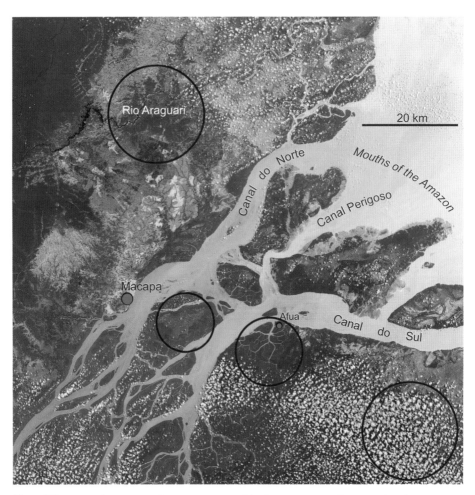

Figure 9.2 This satellite image of 22 August 2017 of the lower Amazon illustrates that the preservation of wetland forest swamps depends upon the interplay between two factors: erosion and sediment deposition by channelized areas of active river and tidal flow (pale yellow brown colours) and areas of backswamp (uninterrupted green) whose interior parts are completely free from the influx of sediment-depositing floodwater. It is in areas like the latter around the Rio Araguan that thick peats can accumulate over the millennia; that is, until the main river channel happens to migrate into the area to cover it with alluvial sediment. Similar closed basins feature on the island east of Macapa and over the large area of the Ilha de Marajo south of the Canal do Sol. By way of contrast the area south of Afua is drained by a network of shifting estuarine channels that limit the lateral extent of forest swamp vegetation. The cumulus 'Popcorn' clouds visible in parts of the image are a common occurrence during the Amazon's later dry season, formed by condensed water vapour released by transpiring plants and trees, and convected upwards during the sunny day (*see* text section 'Did Late Carboniferous Equatorial Swamp Forests Influence Climate?'). Extensive human forest clearance is visible in the upper-left section of the image; the large brown areas on the basin margins are where vegetation has been cleared away. Source: Wikimedia Commons/ European Space Agency; File: Amazon River ESA387332.jpg.

fungi and insects), by oxidation and by periodic wildfire burn-off. Peat is transformed by richly diversified and specialized bacteria. These obtain energy from organic degradation in zero or low-level oxygen conditions, making use of their specialized metabolisms to chemically reduce nitrates, sulphates and ferric iron. The key thing is that the organic carbon-forming tissues of plants and trees are largely preserved. Wood partly decays to humic acids and other mobile organic complexes, but the majority of carbon fixed by photosynthesis in life stays *in situ* after death.

Tropical riverine and deltaic reed fens, marshes and swamp forests are a large category of modern wetlands analogous to Carboniferous swamps. Some are sediment-rich; others are isolated from the sources of sediment supply and entirely comprise organic detritus (Fig. 9.2). Inland examples like the Okavango of Botswana are rich in endangered habitats and many are still little explored, witness the recent discovery of the largest area of swamp forest in the tropics, the Cuvette Centrale of the central Congo Basin in Zaire.[2] Here the Holocene-active peatlands occupy inter-channel basins isolated from mineral sediment influxes and whose water is provided by direct rainfall. The variety of living plant and tree species, and therefore of insect and animal life, is rich under such conditions – organic productivity per unit area and the rate of formation of peat are very much greater than in temperate climes. As a consequence, post-glacial tropical peat accumulations may be extremely thick, sometimes over ten metres accumulating over a few thousand years.

Generally, wetland vegetative cover is sensitive in its species make-up to the degree of soil wetness. Subtle gradients of elevation and relief provide completely different substrate conditions for growth. Natural levees, for example, are well-drained raised banks formed by rapid sediment deposition dumped during flooding when the highest floodwaters top the levee crests. They comprise coarser sediment than found in neighbouring watery environments, and during most of the year are exposed to the atmosphere, providing better-drained conditions suitable for certain species of deciduous tree colonization, e.g. the ubiquitous willows. Other vegetation, like horsetails (*Equisetum*), sedges (including hot climate papyrus), rushes and reeds (*Phragmites*) are adapted to more watery substrates and prefer to colonize the tops of sediment bars and the shallow peripheries of standing and slow-moving water bodies. Papyrus itself can form immense floating bogs.[3]

The Big Squelch

As we have seen, the 'tree-like' Lepidodendrales emerged in latest Devonian to Early Carboniferous times (365–340 Ma) with their widespread adaptations to wetland conditions. They were to dominate such environments until the last third of Carboniferous times in Europe and North America. A unique location in Maritime Canada exposes an Early Carboniferous *Lagerstätte* dominated by early forms.[4] They grew several metres to tens of metres high, and had already developed specialized cell activity promoting the growth of secondary lignin-rich (woody) strengthening tissues, especially bark. Power-

Figure 9.3 The most long-lived and abundant of the Lepidodendrales, the form *Lepidodendron*, underwent explosive evolution and areal expansion in Late Carboniferous swamp forests – microscopic examination shows that they commonly make up 75% by volume of Late Carboniferous bituminous coals. The sketch is superimposed on the preserved root structures featured in Figure 9.4B. The twin-dividing end-branches to canopy crowns bore narrow needle-like leaves and copious male and female organs (the microspores are named *Lycospora*). The branchless trunks were adorned with spiralling diamond-shaped scars of former stem leaf and crown branches. Sources: coloured and recrafted after Oakley and Muir-Wood (1962) and Natural History Museum (1964).

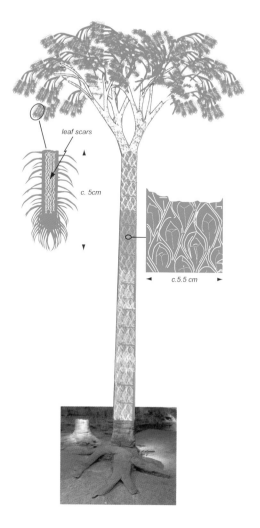

ful roots provided the physical support needed for gigantism of trunk and crown, seen most spectacularly in their Late Carboniferous descendants, most abundantly the form *Lepidodendron* (Figs 9.3, 9.4). These periodically lost their scaly tree leaves and crown branches – fossil specimens feature both attached and spirally arranged leaf cushions with the scars left from vascular tissue. The trunk base is seen to divide downwards into four branches that further divide over several metres into successive increasingly horizontal, double-branching root appendages – known by the separate name, *Stigmaria*.

Preserved *in situ* vertical trunks of Lepidodendrales such as those in Figure 9.4 are not uncommon in the Coal Measures, known to miners as 'coal-pipes'. They created an extra (and perhaps unexpected) hazard to coal miners. Here is the original account by Charles Lyell in his digest of the vast *Principles of Geology* (1830–34) – his influential semi-popular *Elements of Geology* (1838):

Figure 9.4 **A–B** A preserved *Lepidodendron* forest stand. **A** The central Glasgow forest *Lagerstätte* at 'Fossil Grove' in Victoria Park (55.523263, 4.201617) was discovered in the 1880s at a horizon within the later part of the Early Carboniferous Limestone Coal Group. It originally comprised a dozen or more stumps of *Lepidodendron* preserved as sandstone casts up to half a metre high, the roots protruding from a pale-grey soil deposit, known as fireclay from its refractory properties when made into bricks. **B** The *Stigmaria* rooting system in this fine specimen is seen to perfection housed under protective cover today in the Victoria Park Museum. It acted as both anchorage and baffling filter for water and nutrients to be taken up by trunk vascular cells to the crown. The roots were anchored in water-saturated peat rather than detrital mineral sediment. The reasons for the destruction of the Glasgow forest prior to its burial and the infill of its trunks by riverine sands is not known – maybe a monsoonal storm blow? Source: https://www.theglasgowstory.com/image/?inum=TGSE00534

These 'coal-pipes' are much dreaded by our miners, for almost every year in the Bristol, Newcastle and other coal-fields, they are the cause of fatal accidents. Each cylindrical cast of a tree, formed of solid sandstone, and increasing gradually in size towards its base, and being without branches, has its whole weight thrown downwards, and receives no support from the coating of friable coal which has replaced the bark. As soon, therefore, as the cohesion of this external layer is overcome, the heavy column falls suddenly in a perpendicular or oblique direction from the roof of the gallery whence coal has been extracted, wounding and killing the workman who stands below. It is strange to reflect how many thousands of these trees fell originally in their native forests in obedience to the law of gravity; and how the few which continued to stand erect, obeying after myriads of ages, the same force, are cast down to immolate their human victims.[5]

It would be misleading to imagine Late Carboniferous forest swamps as dank and gloomy places beneath their stands of immense *Lepidodendron*.[6] The evidence of their branchless trunks, abundant fern and other undergrowth and the spine-like nature of the crown foliage all point to rather more sunlit scenes below their canopies (that is, when monsoon rain clouds were absent). The needle-like leaves would have whistled and moaned in the wind like modern pines. Light competition must have become an issue as latitudinal shift and concomitant climate change caused lowland wetland habitats to spread widely in the Middle and Late Carboniferous. It enabled natural selection of specialized cells that could quickly grow trunk bark that thickens to up to fifty per cent of lower trunk volume in large adult trees. The apparent lack of secondary cell tissue to transport back down to the roots photosynthetic food manufactured in the crown is seen as an issue by some palaeobotanists. It has been suggested[6] that a 'diffuse' local photosynthesis took place involving trunk leaf cushions and root spines under shallow water.

Tall Lepidodendrales undoubtedly maximized spore dispersal by the enhanced velocity and turbulence of wind currents experienced at the ≤ 50 m altitude of the forest crowns. Additionally, the upward-thinning bark would have enabled the whole tree to flex like a skyscraper during tropical storms above the increasingly immobile trunk and roots lower in the forest's boundary layer. The lack of trunk branches would also have lessened air resistance, the additional strength provided by thick bark augmenting biomechanical resistance to the increased wind drag expected at the high oxygen levels[7] postulated for the Late Carboniferous atmosphere. In modern forests, blowdowns during gales are nucleated along valleyed margins and through wind-gaps created previously by the spreading action of leaping wildfire on tree crown foliage. Such fires must have been common under tropical storm tracks acting on wispy crowns in the highly oxygenated Late Carboniferous atmosphere – witness the abundant fossil charcoal at certain levels in some coal seams.[8]

Although the Lepidodendrales dominated Late Carboniferous forest swamps by their sheer numbers and size, wider forest communities were considerably more diverse. Large trees like the ancestral conifers, the Cordiatales, colonized drier habitats on channel levees and in bordering higher ground where their >40 metre canopies would have provided shelter from erosion for bedrock.[9] They even colonized arid Late Carboniferous hinterlands bordering saline coastal salt flats (sabkha).[10] Opportunistic and pioneering underfloras of ferns, seed ferns, tree ferns and horsetails also existed – notably the arboreal *Calamites* genus that grew up to ten metres high. Recent discoveries point towards them as primary aquatic colonizers of shallow subaqueous sediment bars at the mouths of deltaic channels.[11] Wider detailed palaeo-ecological studies reveal that the distribution of fossil plant species was variable, with a subtle zonation of species that adjusted to local relief, drainage, soil type, shade, and habitat disturbance by both freshwater and saline floods, gravitational bank collapse, water stress and wildfire.[12]

It has not been definitely established that Carboniferous equatorial environments featured bogs sourced by direct rainfall, although some researchers note that certain fern species may have been able to form such ombotrophic systems.[6] Certainly there would have been more than sufficient rainfall for the purpose. At any rate, preservation of the dominant Lepidodendrales peats owed as much to the lack of oxidizing conditions in Carboniferous wetlands as to the lack of fungae capable of degrading their lignin-rich bark and other secondary tissues. Such organisms evolved later, especially lignin-loving fungi and related forms.[13]

Did Late Carboniferous Equatorial Swamp Forests Influence Climate?

Recent research on the Amazonian rainforest has led to headlines in the semi-popular science press such as the caption above a photo of rainforest that proclaims 'These trees in the Amazon make their own rain.'[14] The authors of the original study[15] point out that although it is well established that transpiration (the water vapour produced by plants during photosynthesis) contributes to rainfall over Amazonia, it remains unclear whether this actually helps to drive the seasonal cycle of rainfall or merely responds to the weather physics involved; i.e. the forest might be either passive or active in the process. As it turns out, the clever use of satellite data determined that the timing of the seasonal cycle is independent of the physical weather. It owes its initiation to pre-monsoonal peak forest transpiration creating its own penumbra of vapour. This causes increased shallow-level convection that warms, moistens and destabilizes the atmosphere during the early stage of the dry-to-wet season transformation. This rain-inducing process induces the 'popcorn cloud' arrays seen in Figure 9.2, and typically occurs two to three months before the arrival of the intertropical convergence in the region and the onset of true monsoon. Elimination of forest habitat by natural climate change or by human destruction would therefore eliminate this mechanism, extend the dry season, and enhance regional vulnerability to drought.

A further, deeper question arises. How does the rainforest induce this added transpirational rainfall? It seems that here again a key role is played by the trees themselves. An important study involved careful collection and analysis of fine particles that were brought into air masses from trade winds journeying over 1000 kilometres of pristine forest to the Amazonian Tall Tower Observatory.[16] The tiny (sub-micron) secondary aerosol particles comprised organic material around nuclei of c.400 nanometre potassium-rich salt directly emitted by biota in the rainforest, especially biogenic processes active during the formation of fungal spores. In the closing words of the authors:

> Our findings support the hypothesis that the Amazonian rainforest ecosystem can be regarded as a biogeochemical reactor in which the formation of clouds and precipitation in the atmosphere are triggered by particles emitted from the biosphere...In view of the large impact of tropical rainforests on biogeochemistry and climate, the biological activity and diversity of particle-emitting organisms seem likely to play important roles in Earth history and future global change.[16]

Consider further the behaviour of equatorial southern Amazonian forests during Pleistocene glacial periods, as tracked by the analysis of stable isotopes in neighbouring cave stalagmites. Far from shrinking into isolated refugia as once thought, and despite a five degree Celsius reduction in mean annual temperatures and an almost fifty per cent reduction in rainfall, the tropical forest cover there barely changed during glacial intervals, with no sign of a change to savannah vegetation.[17] Such isotopic evidence indicates that water availability was not an issue because of lower transpiration levels during the cooler glacial climates. Though lowered, the level of transpiration must still have been sufficient to facilitate shallow-level convection.

Perhaps pre-monsoon transpirational convection, salt nuclei from fungi for aerosol seeding, and temperature control on transpiration levels led to the continued dominance of Lepidodendrales forest swamps during repeated Late Carboniferous climatic excursions from interglacial to glacial and back again?

Wetland Successions, Goniatites and Changing Sea Levels

Like most sediment, peat accumulates upwards. In most cases the accumulation is more or less *in situ,* the remains are as they died – toppled trunks with their roots traces; fallen branches; layers of seasonal leaf fall with pollen and spores settling into still water. Because of this, peat and its immediate substrate constitutes that most valuable of archives, a series of undispersed snapshots of successive plant communities perhaps existing in the mire for hundreds or thousands of years. Peat is therefore the ultimate *Lagerstätte* and has a stratigraphic context within the detrital sedimentary layers that bound it above and below.

A detrital sedimentary succession that gradually coarsens upwards is the simplest case in point (Fig. 9.5). The succession may be from dark-coloured, fine-grained muds or mudstones deposited towards the centre of a lake or a coastal delta-front bay. These pass upwards into siltstones, then sandstones with ripple-like forms, bearing witness to the passage of stronger water currents (both underflows and wave-formed) bringing in coarser detrital sediment. Wider examination of contiguous sections may enable an interpretation involving the infill of a lake or a coastal bay from one side or other by shallow flood-relief channels. Repeated successions record eventual reversion to lake- or bay-like conditions induced by relative sea- or lake-level rise, perhaps over hundreds or thousands of years – the buried peats suffering compaction and water expulsion as they transform into more solid and carbon-rich states. In some areas traces of thicker sandstone strata cut across and truncate the infilling successions – sedimentary bar deposits of migrating deltaic channels, as at the top of the succession in Figure 9.5.

In Middle and Late Carboniferous successions the most exotic and exciting strata, to the geological imagination, apart from the coals themselves, are black, carbon-rich, freely splittable mudrocks usually just a few decimetres to a metre or so thick, often closely overlying coal seams. These black mudrocks define 'marine bands', for they bear fossilized remains of a myriad of sea-dwelling creatures. Most notable are the aforementioned pelagic goniatites, along with other free-swimmers like pectenid bivalves (think small, thin-shelled scallops – a tiny mouthful). Goniatite fossils were

Figure 9.5 View of a perfectly formed, 8 m thick, coarsening-upwards stratal succession of late Lower Carboniferous age exposed along the coast of East Fife in midland Scotland (*c*.56.125741, 2.424428; Fielding & Frank 2015). The gradual predominance of thicker sandstone interbeds (the pale ledges standing proud from the outcrop) in relation to the darker more-easily abraded mudrock layers record the gradual infill of a coastal delta front bay by sedimentation from outflowing freshwater discharges. Towards the top of the succession the sharp base of a thick, coarse sandstone forms a prominent vertical clifflet. This sandstone erodes partly into the coarsening-upwards succession and may be interpreted as the deposit of a deltaic channel – the water body fronting the delta had by this time been infilled with sediment and had given way to a terrestrial, sometimes peat-forming deltaic environment. Source: Professor Chris Fielding, University of Nebraska.

long regarded with interest for their obvious role as harbingers of marine flooding of the forest swamps, but it took a lifelong effort by amateur geologist W.S. Bisat to untangle their evolution and use for correlation of Carboniferous strata. He published his great opus in 1924,[18] still the longest paper ever published by the Yorkshire Geological Society, and beautifully illustrated by professionally taken black and white photographs (Fig. 9.6). These clearly show the intricate traces of the 'sutures' – casts of internal flotation-cavity walls within the inner coiled shell – used by Bisat to trace their evolution, the rapidly evolving forms providing firm correlation over regions and continents. The form *Cravenoceras*, for example, defining the boundary between the Early and Middle Carboniferous, is named from North Yorkshire's Craven Lowlands – it is staggeringly cosmopolitan, found in marine strata all the way from Arkansas to Australia via Ireland and Kazakhstan. Bisat modestly wrote nearly one hundred years ago:

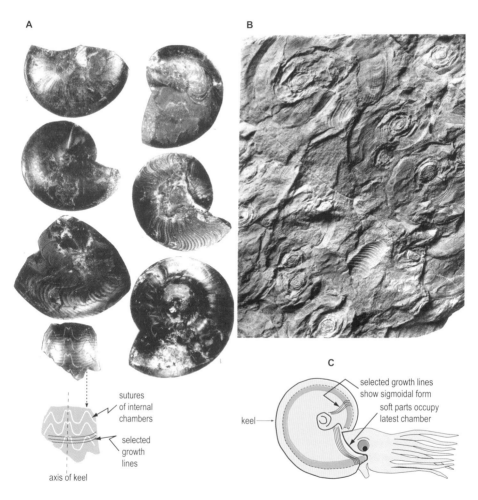

Figure 9.6 **A–C** Late Carboniferous marine band fossils. **A** An example of W.S. Bisat's original 1924 plates of beautifully preserved and photographed fossil goniatite specimens (objects are 1–2 cm diameter) from Middle Carboniferous marine bands of the South Pennines. They are preserved unflattened by compaction, mineralized by early post-mortem calcium and iron carbonates or by iron sulphide (pyrite). Several different genera are represented, notably *Homoceras*, as in the bottom two examples. Source: Bisat, 1924, Plate I (image inverted here). **B** The usual appearance of fossil goniatites, *Homoceras spp.* in this case, from Congleton Edge Quarry, Cheshire. They are preserved as compression casts in black fissile mudrock ('paper shales') of a marine band, originally a fetid, stinking, black mud. Source: Bisat, 1924, Plate VIII. **C** A reconstruction (roughly true scale) of the well-known and distinctive genus *Reticuloceras* as it might have appeared in life, happily surfing the world's oceans in Middle Carboniferous times. Source: redrawn and coloured after Ramsbottom, 1978.

> The faunal zones [of goniatites]…are put forward in the hope that they will
> be of use in correlation. They have been fairly well tested in Lancashire,
> Yorkshire and Derbyshire, and it will be interesting to see to what extent they
> can be traced in other Provinces.

It is now clear that Bisat's goniatites bear silent but incontrovertible witness to the aforementioned far-field effect of changes acting on the south polar Gondwanan ice cap, causing it to wax and wane over intervals of tens to hundreds of thousands of years. During waning times it has been calculated that the retreating ice released enough water into the world ocean to raise global sea level by up to seventy metres.[19] This could have happened quickly by geological standards, in just a few thousand years at rates of several millimetres a year, by analogy with the termination of the Pleistocene glaciations.

The effect of such rapid rises would have been catastrophic to Carboniferous equatorial swamp forests (Fig. 9.7). Swamp and river waters on the gently seawards-sloping coastal plains would have first turned brackish, then saline as their deltaic parts were pushed far back into the riverine hinterlands by the marine flooding – the goniatites propelling themselves ever-shorewards to feed on small crustacean prey in the surface waters. Over vast tracts of now-abandoned swamp forests in the deepening sea, aerated bottom muds and silts were deposited. At the climax of marine flooding (highstand) the deepest water became still and stagnant – tidal mixing being generally minimal.[20] Carcasses of goniatite and other creatures were buried in droves as they sank into fetid

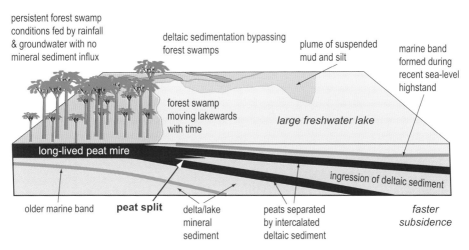

Figure 9.7 An environmental sketch and stratigraphic panel (front view) to show the interrelations between peat accumulation, deltaic sediment influx, marine flooding, and subsidence. A consequence is peat-splitting, whereby a thick peat passes laterally into two or more thinner ones, a feature seen to perfection in the South Staffordshire coalfield as individual seams are traced from south to north (*see* Fig. 19.2B). Source: adapted from Hains and Horton (1969, Fig. 8) and other sources cited therein.

bottom-waters where all oxygen had been utilized over long periods of very low sediment deposition rates.

As polar glaciers grew once more in response to lessened solar radiation, the ocean water volume shrank, sea level fell a little more slowly than it rose, and the great equatorial deltas pushed back in, sedimenting as they went, reclaiming their former domains. And so it went on, with sea-level yo-yoing in over eighty repetitions from Middle to Late Carboniferous times. And all the time the great coal basins, from Iowa via Appalachia to the southern North Sea and Donetsk, were subsiding and preserving most of their deposits in thick successions that reached up to three kilometres by the end-Carboniferous.

Coda: Where Did All the Swamp Forests Go?

A parochial answer for eastern equatorial Pangea relies on sedimentary portents of the coming Variscan tectonic storm. Above the last coal-bearing strata, long-lived forest swamp habitats became rarer and marine incursions ceased altogether after *c*.310 Ma. The latest Carboniferous successions in Britain comprise sometimes very thick, brown-red strata – the Warwickshire Group – a product of deposition on well-drained and oxidized riverine plains.[21] Modern analogues are the vast alluvial cones that abut the Indo-Gangetic plain and sourced from the Himalayas to the north. Calcareous soils provide evidence of seasonal precipitation rather than the leached iron/aluminium-rich soils indicative of the 'everwet' climates of the Coal Measures proper. Referring back briefly to Figure 8.5B, where had the Pennine monsoon-fed drainage gone? And the west-sourced streams of detritus from Newfoundland and north-west Ireland? Why had all connections with the Palaeotethys Ocean been cut?

The answer to these questions is a combined tectonic and climatic one, involving mountain building in the core Variscan regions of Iberia, Armorica, the Massif Central and Middle Europe. We must imagine a rising Pan-Variscan frontal range of mountainous relief cutting off marine connections and bearing the brunt of the northern summer monsoon rains. This put the shrinking depositional plains of the Ultra-Variscides and their hinterlands to the north into a 'rain-shadow' with often prolonged moisture-deficit, the riverine fans of the Warwickshire Group fed by rainfall from the winter monsoon over the rising northern slopes of the Variscides.

By way of contrast, such a scenario did not arise in western equatorial Pangea until very close to the end of Carboniferous times. Here, forest swamps and marine bands persisted in the sedimentary basins of the Mid-Western states of Illinois and Iowa. For most of Late Carboniferous time, periodic far-field changes to sea level, climate and areal extent of habitat had no effect on the dominance of Lepidodendrales in these swamplands.[22] Clearly, enough swamp habitats remained intact to enable periodic re-establishment of the flora during lowstands.

Such conditions ceased abruptly in the late Upper Carboniferous around 307 Ma when the Lepidodendrales race of forest giants (the Ents of the Carboniferous world) became geographically extinct (the genus *Sigillaria* excepted) in western equatorial Pangea.[23] This event coincided with a massive melt-out of Gondwanan ice that raised global sea levels to an unprecedented degree, followed by an equally impressive sea-level fall. The supposition is that such extreme marine excursions more or less eliminated swamp forest refugia over the extended riverine hinterlands. For the remainder of Carboniferous time the swamp biome here was dominated by tree ferns and other arborescent forms. Further eastwards in the Variscides of Ukraine, the Massif Central and Cantabria, intermontane basins hosted larger refugia where Lepidodendrales survived with great vigour into the very latest Carboniferous. Remarkably, some of the same race of survivors flourished in the extensive forest swamps of the Cathaysian archipelagos and island terranes of East Asia into the Late Permian, *c*.255 Ma, a remarkable tribute to an order of trees that existed for more than a hundred million years of geological time.

10

Carbon Accumulations and Mineral Additions

Full fathom five thy father lies;
Of his bones are coral made;
Those are pearl that were his eyes;
Nothing of him that doth fade
But doth suffer a sea-change
Into something rich and strange…

Second verse of 'Ariel's Song' from Act I Scene ii of Shakespeare's *The Tempest* (1610)

Diagenesis - The Effects of Sedimentary Burial

Deposited sediment undergoes physical and chemical changes as it is progressively buried by later sedimentation. Coal, coal balls, clay ironstone and oil shale are all thought to have formed during burial – part of the general process known as sediment *diagenesis* (Greek root: 'continuing origin'). The progressive burial causes mechanical compaction and expulsion of water – as in a loaded tofu slab prepared for a stir-fry. Sediment loading is self-generated and gravitational – it is known as autocompaction. The process is rapid at first, slowing down with depth and time as the clay particles close up to make the water expulsion process progressively more difficult and slower.[1] The same process occurs when humans drain wetland peats by drainage cuts and pumping – the former peat surface falls rapidly at first and then ever more slowly as the peat shrinks and its once open, water-saturated, porous framework collapses (Fig. 10.1). All the squeezed-out water has to go somewhere, the direction of travel determined by the local pressure gradient. At depth in the crust the migrating water warms up and is better able to transport dissolved constituents scavenged from the ambient sediment that it passes through.

Figure 10.1 *Black Fen Fantasy* by Joy Lawlor (20.6 x 29.5 cm; watercolour). The artist is pictured viewing the celebrated iron post which replaced an original oak post whose top was formerly at ground level in Holme Fen, Cambridgeshire (52.492541, –0.232709). She is imagining some of the succession of ghost-like visitors over the last 172 years, each standing at appropriate levels for the time and dressed in a contemporary way – early-Victorian, late-Victorian, 1920s Flapper. The background story is that the fen was formerly part of the wider Whittlesea Mere (*see* Rotherham, 2013). In 1848 the original post was driven down through 6.9 metres of Holocene peat to earlier post-glacial Holocene clays and trimmed to ground level before drainage ('reclamation') and compaction began. Since then around four metres of compaction over Holme fen has occurred due to peat shrinkage during dewatering. The five steel bands below the top of the post record the successively lowered ground surface in 1860, 1870, 1875, 1892 and 1932. the shrinkage rate decreasing with time from around 120 millimetres per year in the first 12 years, 89 over the next ten years, 30 over the next 5 years to around 9 in the last 80 years. The general area is currently the lowest in Britain, 2.75 m below mean sea level. Reproduced by kind permission of the artist.

Under Pressure, Warm, No Oxygen

To change one word of Miranda's speech in Shakepeare's *The Tempest*, we are now in a 'brave new world that has such beings in it…'. In our case not an island paradise but smelly muds with bacteria as the beings. By their mediation of chemical reactions they leaven organic-rich sediment, stewing it up and changing original organic tissues. For us it is not so much Ariel's bones and eyes that undergo a 'sea-change', but organic molecules in organic-rich sediments – lignin, protein, carbohydrate and fats. All are altered by bacterial mediation and react with dissolved chemicals to form new by-products – coal, liquid and gaseous hydrocarbon, carbon dioxide, sulphur dioxide, hydrogen sulphide and associated mineral precipitates of calcium, magnesium and iron. Diagenesis is the universal route to the formation of natural materials – things that are truly 'rich and strange'.

Why should burial make such changes come about in organic-rich sediments? First there is the steady downwards increase of temperature from the Earth's surface that persists for scores of kilometres at a normal rate of between 20 and 30 degrees centigrade per kilometre – directly felt if you have visited or worked in a deep mine. This gradient of temperature, the geothermal gradient, exists because of the conduction of heat from about 100 kilometres depth at the base of the lithosphere where the temperature is maintained at a steady 1200 or so degrees centigrade. In the present context the elevated temperatures at depths of up to a few kilometres allow chemical reactions to proceed much more quickly according to well-founded chemical laws, allowing easier breakdown of organic compounds and the liberation of crude oil at temperatures of *c.* ≤150 degrees centigrade.

Second, it is impossible for oxygen in seawater or freshwater to diffuse very far below the sediment/water interface at the sea or lake bottom, even in sandier, more porous sediments. It is often limited by the depth, a few decimetres at most, that can be reached by burrowing organisms (molluscs, crustacea, worms) that make their homes in sediment and into which they draw down ambient oxygenated bottom water. Between oxygenated surface waters and such dwelling places, the 'pore water' trapped between grains of deposited sediment is in a state of permanent reduction – a hidden anaerobic world free of oxygen. Chemical conditions are mediated by many different strains of bacteria inhabiting in unimaginable numbers, all doing different things in their reactions with the ions and molecules present in sediment pore waters and adsorbed onto sedimentary particles. Such reactions enable the aforementioned mineral precipitates to crystallize out of solution as solid deposits, either locally by aqueous percolation and diffusion or after long-distance travel (advection).

Flowers of the Forest - Coal from Particular Peats

In pristine terrestrial vegetation, the building blocks of cellulose and lignin are the predominant constituents. Both are altered and broken down by the process known as humification. Cellulose is cell-wall carbohydrate with carbon, oxygen and hydrogen atoms in chainlike and interwoven strands of joined-up glucose molecules manufactured by photosynthesis. It is relatively easily broken down during burial and decay. Indeed, its position high on the food scale for ruminant beasts and vegetarian humans depends on specialized gut bacteria breaking it down quickly and easily (also gassily!). Lignin (Latin root, *lignare*; wood; Greek root, *riza*, rhizome) is the great plant tissue stiffener. It links cellulose strands and is quite another matter to break down. Chemically it is a polymer made up of phenol groups. As in the more familiar carbolic acid, it has hydroxyl groups attached to a ring-like structure. In the Carboniferous, lignin made up the thick external bark and roots of the aforementioned Archaeopterids and Lepidodendrales, also the secondary wood of the Cordaites.

Raw peat has already begun to be affected by the humification process. Vegetative cellulose and lignin are de-photosynthesized by conversion into organic humic and fulvic acids and their residues. It marks the beginning of coalification in the coal series that leads with increased burial, compaction and subsurface temperature to lignite (aka brown coal), crudely laminated bituminous coal (including house coal) and, eventually, at the highest temperatures (150–200 degrees centigrade), anthracite.[2] The process is accompanied by a definite increase in carbon content, from around 55% in peat, 65–79 % in lignite, 80–90% in bituminous coal, finally up to >90% in anthracite. At the same time oxygen and water are driven out as cellulose and lignin tissues alter, yet hydrogen is still present in methane gas even in the highest grade bituminous coals. This methane is available during coalification to be given off as a migrant species into any adjacent porous rock – our 'natural' gas.

Marie Stopes christened the common constituents of bituminous coal in a pioneering paper of 1919 using field specimens she collected from a West Midlands colliery. These were:

Fusain: equivalent of mineral charcoal via the French term from the Latin *fusus*

Durain: equivalent of 'dull' hard coal

Clarain: equivalent of 'bright' or clear (glance) coal

Vitrain; as clarain but with a conchoidal fracture and brilliant appearance

She made the typically pithy comment:

These names, I am fully aware, do not represent chemical entities (with the possible exception of vitrain), but they do represent tangible entities of the same useful order as 'jet', 'granite', or 'cheese'.[3]

Nowadays, adaptations of Stopes' names have become uglier, as in inertinite (fusain), huminite (clarain), vitrinite; with liptinite coals formed from lipid-rich (aka fat- or oil-rich) spores and algae.

Reviewing the constituents of coal sixteen years later, Stopes revealed that she had long wanted her beloved subject of coal microscopy to have the same status as that of rock microscopy, with its emphasis on the nature and abundance of constituent minerals.

It may have been a response to an arrogant male geological colleague, who pointed out this deficiency, that in 1935 she proposed the term *maceral* for coalified plant fragments, her elegant transformation of 'mineral'. She wrote vividly:

> The concept behind the word '*macerals*' is that the complex of biological units represented by a forest tree which crashed into a watery swamp and there partly decomposed and was macerated in the process of coal formation, did not in that process become uniform throughout but still retains delimited regions optically differing under the microscope, which may or may not have different chemical formulae and properties. These organic units, composing the coal mass I propose to call macerals, and they are the descriptive equivalent of the inorganic units composing most rock masses and universally called minerals.[3]

Coalification paths define what is known as the 'rank' of a given coal. This can be measured under the microscope from fragments of 'woody' or 'barky' material preserved as huminite and vitrinite macerals. Huminite fragments (roughly Stopes' clarain) still have unfilled cellular structure and show a relatively low degree of light reflectance on the microscope stage. They have a lower rank than vitrinite, whose cell spaces are largely infilled by humic breakdown products like bitumens and whose reflectance values are therefore greater. Since the process of infill is gradual and progressive, it is possible to give a percentage rank to the light-reflectance of the coal in question. Knowledge of the subsurface temperatures experienced by certain coals buried to known depths (the best data coming from the Western Canadian foreland basin) then enables an estimate of the highest temperatures experienced by other coals with similar reflectance values.[2] Coal rank in Carboniferous sedimentary basins is seen to vary with burial depth, past temperature gradients and the intensity of rock deformation during subsequent tectonics.

During the course of the early Industrial Revolution it became apparent that there were a wide range of bituminous coals with specific uses. Soft bituminous/low volatile coal would cake well and also retain its strength (essential for the furnace), enabling coke manufacture under conditions of slow-burn away from the open atmosphere, first in soil-covered heaps and later in coke ovens. House coal and smithy/foundry coals were of moderate volatile content and needed to burn strongly and brightly with only moderate caking in the grate. The spread of gas lighting sourced from municipal gasworks required high-volatile lean-bituminous coals that generated plenty of methane. As steam power gradually rose to ascendancy in all manners of transport (steam locomotives, marine screw-driven vessels, etc.) following Watt's various inventions, the higher ranks of low-ash, low-volatile (dry), high-bituminous to low-anthracites became immensely sought after for steam-raising. Later on, as turbine-driven power grids covered entire nations, the common middle and low grades of bituminous coals once used exclusively as house coals found uses in the great power stations built in the early to mid twentieth century. The drive against urban pollution after the 1950s saw a huge rise in demand for anthracite as 'smokeless fuel' in urban centres – though of course it still gave off its invisible carbon dioxide.

Iron and Bacteria - Ironstones, Marine Bands and Sulphurous Coals

In one of the many geological and geographical coincidences explored in this book, the periodic spread of Late Carboniferous riverine and deltaic swamps to form vast equatorial wetlands with their peaty foundations also gave rise to the precipitation of trillions of tonnes of low-grade iron ore in contiguous deposits. This iron mineralization came about as part of what microbiologists Dorion Sagan and Lynn Margulis vividly call a 'Garden of Microbial Delights'.[4] The story starts in lakes and lagoons in front of advancing deltas whose discharging freshwater was laden with suspended mud particles. When seen from an imaginary Carboniferous space shuttle, it would have appeared as ever-shifting, yellow-brown threaded veins running through the otherwise clear blue freshwaters – much as the Amazon or Yellow Rivers appear today during the wet season. Since the warm Carboniferous monsoon-fed rains were efficacious in chemical rock weathering in the mountainous hinterlands, the discharge also contained plenty of dissolved reduced (ferrous) iron. Like the Yellow River today, the particulate clays and silts suspended in the floodwaters also contained particles and bore surface smears of oxidized (ferric) iron compounds – insoluble yellow-reddish-brown oxides and hydroxides making up the minerals limonite and goethite – the common ochres used by artists. As the sediment-laden plumes decelerated, depositing the iron-bearing particulates and diffusing the dissolved fractions, the sediment soon found itself in a completely new chemical environment.

In this hidden microbial garden, ferric iron, unstable in the absence of oxygen, became the focus of attention for legions of the specially adapted bacteria *Geobacter metallireducens*. Over previous billions of years (and still today) these had evolved the ability to reduce ferric iron to ferrous iron. This process facilitated the digestion of organic matter in the deposited sediment, and as a byproduct producing bicarbonate ions (rich in the light carbon stable isotope). The process involves ferric iron accepting an electron produced during bacterial digestion. Such bacteria are truly remarkable, for it seems that once they have used up a particular tranche of ferric iron they 'sniff' out further sites nearby and move to them – in other words they are *chemotactic* life forms, attracted by chemical gradients. In the words of the discoverers of this process from the University of Massachusetts:

> *G. metallireducens* senses when soluble electron acceptors are depleted and then synthesizes the appropriate appendages [flagella – whip-like cellular propellers] to permit it to search for, and establish contact with, insoluble [ferric iron or manganese] oxide. This approach to the use of an insoluble electron acceptor may explain why *Geobacter* species predominate over other oxide-reducing microorganisms in a wide variety of sedimentary environments.[5]

As dissolved ferrous iron concentration reaches above five per cent or so of calcium concentration, iron can be precipitated as a carbonate, the iron ore mineral siderite. Pure siderite has 48% iron in its makeup, but the presence of clay-grade particles forming its matrix reduces this to a value of around 30% or so in most raw-picked ores – hence the name 'clay ironstones'. These retain the lighter carbon isotope signatures of their parent carbon. During subsequent deeper burial, iron reduction is accompanied by other bacteria that control the process of anaerobic fermentation (as in brewing beer or fermenting wine) by reducing remaining organic matter. The combined liberation of methane and bicarbonate, the latter with a heavier carbon stable isotope signature, dramatically favours renewed siderite precipitation.

In these ways, conditions favourable for iron precipitation enabled siderite to precipitate as microcrystalline clumps when concentrations of iron and bicarbonate were patchy in the local sediment – the iron and bicarbonate ions diffusing towards adjacent areas of higher concentration. Here the iron carbonate grew into ellipsoidal masses (nodules) of centimetric to decimetric dimensions within the mud-rich host sediment.[6] The occurrence of fossilized molluscs and other creatures in the core of such nodules (as in the famous Late Carboniferous Mazon Creek *Lagerstätte* of Illinois, *see* Fig. 10.2) indicate that their rotting organic soft-parts helped to enhance local gradients of pore water chemistry that encouraged the initial nucleation of iron precipitate. The rarer presence of persistent siderite layers points to conditions of more general precipitation, with their bedded nature suggesting they formed in depth zones below the sediment – water interface along distinct geochemical gradients.

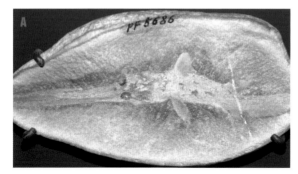

Figure 10.2 **A–B** Decimetric-scale ellipsoidal siderite nodules (clay ironstone) from the famous Mazon Creek *Lagerstätte*, Illinois. **A** This example has been split open to show an exquisite uncompacted fossil specimen of the baby shark *Bandringa* in the nodule's inner core. Note the surround of a ghostly trace around the deceased creature of the earliest siderite to precipitate. **B** A more oxidized nodule (the ochreous colour is a limonitic crust) split to show a leaf impression of the common Carboniferous seed fern *Neuropteris*. Sources: Wikimedia Commons. A File: *Bandringa rayi* fossil shark Mazon Creek lagerstätten.jpg; author: James St John. B File: Macroneuropteris2.jpg; author: Rutgers University, C. Zambell.

So-called 'blackband' ironstones are special. They occur above certain coal seams in the transition to overlying mudrocks of lake origins. They are especially common in the Limestone Coal Group of the Central Coalfield of midland Scotland and the Upper Pennine Coal Measures of the North Staffordshire coalfield. They occur as thin (<1 metre) layers containing distinctive siderite laminations along with coaly (hence the name blackband) and calcareous shell layers. Their occurrence adjacent to workable coals (they could be mined together), the general absence of muddy silicate mineral particles, their coaly intercalations and calcareous layers, meant they were cheaper to mine, calcine (the coaly laminae doing much of the work), smelt and flux than the normal run of clay ironstones. Their relatively high manganese content may also have helped in the Bessemer and Open Hearth steel smelting processes (*see* Chapter 20). Mining of such ironstone in North Staffordshire continued very late, until the 1950s, long after its extinction everywhere else in the face of competition from the opencast iron ores of the Midland Jurassic scarplands.

E.L. Boardman's work[7] establishes a likely origin for the blackband ironstones as having formed in shallow depressions upon partly flooded bogland peat, in small lakes and ponds. Groundwater driven from adjacent, more-elevated alluvial fringes contained ferrous iron which oxidized in the standing water to precipitate as ochres. Perhaps it was a seasonal process, with plant-rich surface runoff from the peat basins providing the next layer of carbonaceous materials, and so on over perhaps hundreds of years. Once buried below the sediment–water interface above the impermeable forest peat, legions of *G. metallireducens* got to work and converted the ochre to siderite. Boardman saw such conditions as analogous to those of Holocene and modern post-glacial peatlands bordering alluvial valleys. Here the ferrous iron brought in by groundwaters precipitates as ochres to accumulate as 'bog iron ore', but in this case rarely going all the way to a siderite state in the absence of standing freshwater rich in bicarbonate. Such ores were much used over northern and central Europe for bloomery smelting in the early Iron Age.[8]

The mode of formation of both blackband ironstone and the commoner clay ironstone in thin seams (exceptionally up to a metre thick, as in the famous Tankersley Ironstone of the West Yorkshire Middle Coal Measures[9]) posed physical difficulties for miners working by the longwall method, as witnessed by this testimony from Govan, near Glasgow, in 1841:

> The labour of ironstone miners is often worse than that of colliers. I have seen them at work in a space of from 22 inches to two foot high, where even seated a man could not keep his neck straight…Two men take between them 14 yards of the band of stone, and make their own walls of the roof which comes down when the stone is extracted, leaving a road six feet wide to every space of 14 yards.[10]

Coal Ball Games

In 1907 Marie Stopes and D.M.S. Watson of the University of Manchester read a paper on 'coal balls' to the Royal Society of London.[11] It still gives intellectual pleasure to the modern reader. Watson's contribution was part of an undergraduate project supervised by Stopes (Fig. 10.2), then a young lecturer, the first female to be such a thing at Manchester. Watson later became prominent in the field of fossil vertebrate morphology and evolution, but should also be remembered for being the father of Janet Vida Watson, the first female President of the Geological Society of London, herself an astute, modest and inspiring all-round geologist who shone light on the previously dark worlds of Archaean and Proterozoic Scotland.

In their introduction, Stopes and Watson define coal balls and their interest in them. They are:

> calcareous masses which occur actually embedded in the coal of some districts [and] have proved of vital importance to students of botanical anatomy and phylogeny [evolution], because they contain plant tissues so perfectly petrified that the complete structure of the plants can be discovered by microscopic investigation…

The pair wanted to know why coal balls gave such unique preservation of plant detail when most other Carboniferous plant fossils were carbonized and flattened impressions in mudrocks, the flora collapsing as their muddy graves were compacted and lost water. They determined that coal ball plant remains, by way of contrast, must have formed *in situ* but *post mortem* as rigid mineralized growths, since they resisted the later four-fold compaction of the peat to coal (Fig. 10.3) that they observed in north Lancashire coal seams. The calcareous nature of the coal balls was explained by recourse to late-nineteenth century marine biochemical results from the epoch-making *Challenger* oceanographic expedition of 1873–1877. This had established that the action of bacteria could reduce the sulphate ions abundantly present in seawater to hydrogen sulphide. The strain responsible was later named as *Desulfovibrio vulgaris*, providing sulphur that enabled precipitation of iron monosulphide (eventually to become iron pyrites) and of calcium carbonates rich in magnesium.

The authors proposed that seawater had percolated into *in situ* peat by the aforementioned periodic marine flooding events, which also brought in the fossil goniatite creatures found in marine bands. Downward percolation and lateral invasion by flooding seawater mingling into the peat substrate enabled precipitation of calcite, dolomite and microscopic aggregates of pyrite crystals as a consequence of the bacteria. It also explained why sulphur-rich coals (the 'stinking' coals of many coalfields) usually underlie marine bands – they were contaminated by infiltration of the iron sulphide mineralization. We now know that the most important factor limiting pyrite formation is the availability of organic matter that can be metabolized by the bacteria. Since

Figure 10.3 **A** Marie Stopes photographed in the early twentieth century in her laboratory at the University of Manchester examining a thin-section slide under a light-transmitting microscope. A paraffin oil lamp provides the light source as concentrated through a water-filled glass sphere onto the transmitting mirror seen at the base of the microscope. **B** A glass-mounted thin section of a coal ball originally belonging to Stopes (perhaps with her writing) of a cut and ground piece of almost perfectly circular and therefore uncompacted *Stigmaria* root (the greyish 2 cm diameter central mass with radiating structure) from within the outer calcitic part of the coal ball (*see* Gelsthorpe, 2007). Note the root's thick prismatic tissue layer, surrounded by brownish epidermal cells outside an only partially mineralized central cavity. The enigmatic fine-laminated structure to the coal ball's outer calcitic framework is suggestive of periodic lamina-on-lamina precipitation, perhaps bacterially mediated? Source: both images courtesy of Kate Sherburn, ©Manchester Museum, The University of Manchester.

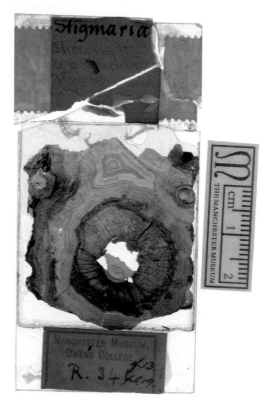

both ferrous iron and sulphate are usually present in abundance in marine sediment porewaters, the greater the amount of organic matter present, the greater the amount of pyrite produced.[12]

The First Retort

The twentieth century was the 'Age of Hydrocarbon', when the whole world and its dog sought personal, petrol-driven automotive transport, and oil and natural gas gradually replaced coal in industrial, heating and power generating plants. The first successful oil well in 1859 went down to around twenty-five metres below surface, its liquid crude slurping out eventually into a waiting washtub. The well was courtesy of pioneer driller 'Colonel' Edwin Laurentine Drake in Pennsylvania, USA. He had the bright idea of drilling coaxially (i.e. within stationary outer piping) to prevent hole collapse. Almost a decade before, Scottish miners in West Lothian had provided James 'Paraffin' Young with a solid hydrocarbon source hewed from a stratum of cannel coal. This provided the world's first significant industrial-scale production of hydrocarbon using a retorting and cracking process patented by Young in 1851. The same patented process was later used to crack Pennsylvanian crude oil – volumes were subsequently measured in barrels rather than washtubs.

Cannel is a word of north British provenance deriving from dialect pronunciation of 'candle' (the 'd' elided) both in Lancashire and in Scotland, where it was also known as 'parrot'-coal and 'boghead' coal, the latter after the Boghead locality in West Lothian where it was originally mined. 'Cannel' first appears in print by John Leland (as canel), the noted 'King's Antiquary' (himself from a Lancashire family) in 1538. It refers to a particularly light but hard carbon-rich rock – flammable and bright-burning – smokeless like a good candle. The Wigan area of South Lancashire was famous for its polished, ebony-like carved ornaments, which Celia Fiennes so much admired and purchased. It has little in common with the layered (bright/dull/sooty/charcoal) maceral content of Marie Stopes' bituminous coal, having originated as a subaqueous deposit of spores and algae rather than as a swamp forest *Lagerstätte*. It records the persistent accumulation and preservation in a shallow, isolated body of water of wind-borne organic detritus derived from nearby forest canopies and from *in situ* seasonal algal 'blooms'. We noted earlier that several thick Middle Devonian coals from Siberia were of this type.

Oil shale is loosely defined as retortable carbon-rich mudrock containing at least five per cent liptinite – waxy, fine-grained, hydrocarbon-bearing detritus (spores, cysts, cuticles, resins and planktonic algal colonies). Such rocks are not that common in the stratigraphic record because of the rather special environmental conditions required to concentrate the liptinite. This kind of scenario arose in midland Scotland during Early Carboniferous times when, in what is now West Lothian, a large lake existed for a few million years around 335 Ma. It was named Lake Cadell by sedimentologist J.T. Greensmith, honouring the influential and innovative nineteenth century Scottish geologist Henry Moubray Cadell, great-great-grandson of William Cadell, co-founder of the Carron Ironworks (*see* Chapter 24). A dozen or so individual oil shales are known from the West Lothian basin, comprising about three per cent of the total sediment thickness that accumulated in Lake Cadell.[13] They are laminated, dark grey to black mudrocks and silty mudrocks ranging from a few centimetres to many metres in thickness, with 1–5 m being fairly typical. Fossils include plant stems, fish fragments, abundant ostracods and algal remains. The latter include the wide-ranging planktonic green alga *Botryococcus braunii* – still-extant, thriving and used in modern-day bioculture production. Chemical analysis of the shales records total organic carbon ranging from four to twenty-nine per cent, similar to other oil-prone petroleum source rocks.[14]

It was in Lake Cadell, fed by runoff, periodically isolated from detrital sedimentation and under a seasonally humid monsoonal climate, that high shoreline levels were periodically attained. These occurred when waxing monsoonal conditions every few tens of thousands of years (tuned to Earth's orbital fluctuations) caused freshwater inflow from runoff to greatly exceed evaporation. In the deepest parts of the lake permanent stratification was set up involving a stagnant bottom reservoir (cool, dense, gloomy, oxygen-poor) isolated from overlying (shallow, warm, sunlit) waters. The latter featured high organic productivity by *B. braunii* as surface colonies held together by an organic matrix rich in hydrocarbon oils. Today such algae bloom spectacularly when the water column is enriched in dissolved phosphate or when higher alkalinity is produced by influxes of wood ash rich in potassium after wildfire episodes. The blooms release free fatty acids that are toxic to other micro-organisms and fish. After such blooms in the Early Carboniferous, the algal colonies would have sunk to the lake floor, accumulating over time as thick organic oozes. Enhanced productivity in this case could also have been provided by the influx of discharges rich in phosphorus from tropically weathered basaltic lavas to the west in midland Scotland, the Clyde Plateau Lavas.

The Last Resort

The prospect of a new hydrocarbon industry – based on the unconventional gas and oil production method infamously known as fracking – had for many years haunted citizens in large tracts of rural and suburban Britain. Luckily, as of November 2019 the reality of that nightmare finally disappeared, though in these unsettled times, easily made political promises are too often just as easily broken on some pretext or other. The background to this is that after the general hype and scare-stories based on real or imagined events in Appalachian shale oil/gas development there came the publication by the British Geological Survey of a prospective fracking map that finally gave some definite geographic limits to the extent of the issue for the British public.[15] It depicted three regionally frackable source rocks in: eastern midland Scotland – mostly oil-prone, from the West Lothian oil shales; northern England from the Irish Sea to North Yorkshire – gas-prone Middle Carboniferous Bowland Shales; south-east England centring on the Weald of Sussex – oil-prone Late Jurassic Kimmeridge Clay.

For some citizens, the drilling of many closely spaced fracked production wells was an anathema, not only from an understandable NIMBY point of view, but also because present and past governments have paid lip-surface to creating a low-carbon energy mix. Other objectors, perhaps more geologically in the know and sensitive to the landscape, pointed out that the blanket of potential shale gas source rocks lies beneath surface landscapes of great variety and distinctiveness, depending as they do on Mesozoic and Late Palaeozoic surface geology, the effects of glaciation and thousands of years of human settlement. These comprise the densely populated coastal plains and hillscapes over the oil shales under Fife and Lothian; fells, moorlands, dales, vales, chalkland downs, coastal wetlands and karst over Bowland Shale targets in northern England (e.g. Kirby Misperton in the idyllic, prehistory-rich Vale of Pickering, North Yorkshire); the unique High Wealden Cretaceous landscapes over the south-east England oil prospects.

In all these places licensing to drill exploration wells had already taken place in what can only be regarded as an environmental scandal involving both British and Scottish parliaments. Kirby Misperton was likely to have been one of the first areas for test-fracking on what looked like the largest and most likely target of all to produce gas. But exploitation in such a precious rural and prehistoric landscape seemed nugatory given the massive unused renewable resources of wave-, tide- and wave-generating power around our coasts, not to mention solar. We shall see what happens.

Vein Mineralization and Metal Ores in Carboniferous strata

In addition to interbedded coal, iron ores and oil shales, Carboniferous strata host a great deal of metal mineralization in veins that cross-cut the (chiefly limestone) strata involved. These include vein minerals of iron oxide (hematite), lead sulphide (galena), zinc sulphide (zinc blende/sphalerite) and copper-iron sulphide (chalcopyrite). They were accompanied by non-metallic minerals worthy of mining in their own right in later times such as calcium fluoride (fluorite/fluorspar) and calcium sulphate (barite/barytes). The chief ore fields involving Carboniferous strata were those of the Forest of Dean (iron), Mendip (zinc, lead), Halkyn Mountain in Flintshire (lead and zinc), Peak District (lead), West Cumbria (iron) and the North Pennines (lead, zinc). Cornwall and Devon metal mineralization is hosted in mostly Devonian strata and is of Early to Mid Permian age and not considered further in this book.

The many uses of iron are well known. Lead was much sought after first by the Romans for roofing, cisterns and baths, but also for its modest silver content. Zinc became more sought after in the sixteenth to eighteenth centuries with the rising popularity of brass alloy (a zinc/copper mixture) for ecclesiastical (rococo ornamental work and inscriptional tablets), musical (wind instruments), domestic and industrial uses. Further increases in zinc production followed the invention in the nineteenth century of the hot-dip galvanizing process whereby a zinc-coated film is laid upon corrugated 'iron' or wire (a mild steel), the oxidized zinc carbonate giving it a weather-resistant protective sheath.

Amongst the non-metallic minerals present in many of the lead/zinc orefields, barite has a high density, about one-third greater than common silica. It was once discarded onto the waste tips of lead and zinc workings, which were subsequently reworked. It finds many modern uses in medicine (barium 'meal' x-ray and scanner tracers), paint fillers and, finely powdered, in drilling fluid suspensions where its high density helps to counteract adverse subsurface pressure. Fluorite was also a discarded accessory in both the Peak District and North Pennine orefields, but found major uses as a flux in steel-making and as the source for fluoride chemicals, including the powerful hydrofluoric acid. Both barite and fluorite outlasted exploitation of base metal ores where present in the major mining districts: they were the last-mined minerals at the eventual extinction of deep-shaft metal mining in the mid- to late-twentieth century. Several of the mining districts are featured in the relevant chapters to Part Four of this book.

11

Tectonic Inversion: Preservation of Coal Basins

I will here give a very brief sketch of the geological structure of these
mountains: first, of the Peuquenes, or western line…Even at the very crest…
at a height of 13,210 feet, and above it, the black clay-slate contained numerous
marine remains…It is an old story, but not the less wonderful, to hear of
shells, which formerly were crawling about at the bottom of the sea, being
now elevated nearly fourteen thousand feet above its level…These great
piles of strata have been penetrated, upheaved, and overturned, in the most
extraordinary manner, by masses of injected rock, equalling mountains in size.

The strong geological side to Charles Darwin, writing about the inversion of former Pacific
sea-floor sediment and its metamorphosis into the Andean Western Cordilllera of Chile
(1840, p.390).

The Pangea Supercontinent Assembled

By end-Carboniferous to Early Permian times (*c*.300 Ma) the long independence of
Gondwana and Laurussia as separate continents was at an end – both had performed
the finale of their fifty million year sliding and spinning double-act (Fig. 11.1). The final
contractions of the Variscan mountain chain with its core of thickened crust occurred
in central, southern and north-west Europe – the terranes ploughing northwards into
the coal basins of the Ultra-Variscides. As in the Andes described above by Darwin,
thorough-going intense deformation associated with thrust faults led to overfolded
strata, development of rock cleavage, and, in places, even the folding of that cleavage. The
thrust-faulted northern limit to such extreme deformation across south-west England
and southern Ireland marks the course of what is known as the 'Variscan Front', the
martial analogue giving a sense of the crustal mayhem involved.

In what was to become western Pangea, the right-handed sliding motion along the
southern margin of Laurussia, from modern-day Pennsylvania to Tennessee, ceased

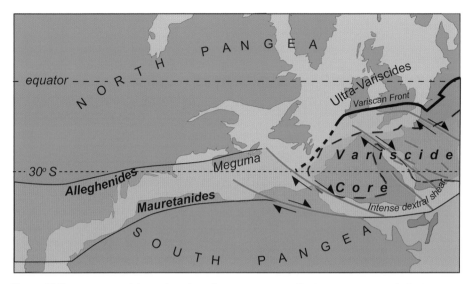

Figure 11.1 The extent of the end-Carboniferous Variscan–Alleghenian mountain belt along central tropical Pangea. Red lines are major 'strike-slip' faults with the arrows indicating sense of shear along the fault trends – dextral (to the right) in all cases. Source: recrafted and coloured after Torsvik and Cocks (2017) with fault data from Zeigler (1989) and other works.

as the two continental masses docked their leading edges into head-on collisional mode. South America, north-west Africa and intervening Florida with parts of coastal Alabama, Georgia and the Carolinas were Gondwanan fugitives stitching themselves onto the southern Laurussia continental margin (Fig. 11.1). All trace of the Rheic seaway was now finally obliterated along the line of the Alleghenian fold-and-thrust mountain belt in the Appalachians and its Oachita continuation into Arkansas and Texas. Opposite was the mirror-image Mauretanian fold-and-thrust belt of north-west Africa.

'Alarums and Excursions': Coal Basin Inversion

Inversion is the term that describes the change from basin subsidence and sediment deposition to uplift and erosion caused by a change of tectonic regime – from crustal stretching and subsidence to compression and uplift. Across the Ultra-Variscides up to seven kilometres of Late Devonian to Late Carboniferous sediment had accumulated in places. The reader will recall that these deposits were arranged in two contrasting layers. Below were discontinuous Early Carboniferous rift sediments bounded by normal faults. Above was a more continuous and uniform arrangement of Middle to Late Carboniferous strata (Fig. 11.2) arranged around a 'bulls-eye' of greatest subsidence in south-east Lancashire (Fig. 11.3) where the strata were thickest (up to 1500 m). Such multi-layered strata reacted quite differently to the extreme compressional stresses that now moved through them from the south.

Inversion-related uplift and erosion: former rift-phase normal fault locked: out-of-basin fold & thrust belt develops

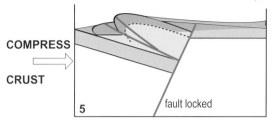

COMPRESS

CRUST

5   fault locked

Inversion-related uplift and erosion: former rift basin fault reversed to form growth fold

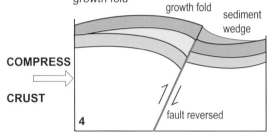

growth fold   sediment wedge

COMPRESS

CRUST

4   fault reversed

Post-rift ('sag') subsidence & sedimentation

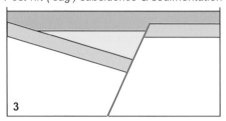

3

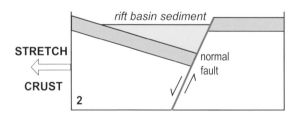

rift basin sediment

STRETCH

CRUST

normal fault

2

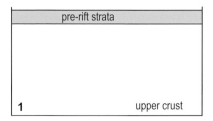

pre-rift strata

1   upper crust

[Left] Figure 11.2 Diagrammatic sections through the upper crust (top ten kilometres or so) to show the effect of inversion tectonics upon rift basins. **1** to **3** are sequential stages in rift evolution, while **4** and **5** are alternative scenarios for inversion tectonics caused by subsequent crustal compression. Source: author as inspired by Bayona & Lawton (2003).

[Opposite] Figure 11.3 A–B: A Early Carboniferous rift basins. **B** Map to show the 'bulls-eye' nature of the South Pennine trends in Late Carboniferous Coal Measures thickness due to thermal subsidence (it ignores the rift trends completely) together with end-Carboniferous inversion structures and coalfields mostly in synclinal downfolds adjacent to regional axes of uplift. Source: coloured and recrafted after Corfield *at al.* (1996) with additional information on stratal thickness from other published sources after Leeder (1982).

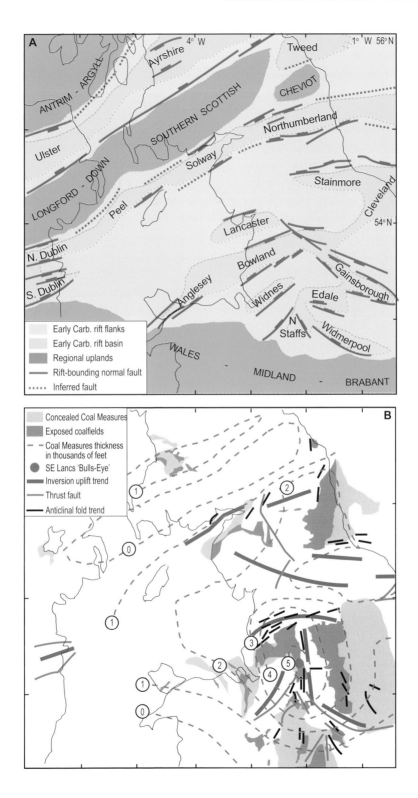

Variscan tectonics rejuvenated and reversed the Early Carboniferous rift-bounding normal faults. They morphed into thrust faults that pushed folds outwards and upwards, piling up fold-and-thrust sediments like crumpled snow gathered in front of a moving snow plough. The net result was the formation of a mosaic of multi-sized folds and thrust faults across the entire former rift province.[1] The anticlinal upfolds formed elevated scarp ridges and summits whose resistance to erosion was much diminished by the presence of structural cracks (joints) formed during the folding process that caused them to be deeply dissected over the first few million years of the succeeding Permian period. Soon, the entire Late Carboniferous coal-bearing stratal succession was removed from the crests of these structures, remaining only in the flanking synclinal downfolds that formed the nuclei of almost all the numerous future British coalfields.

Fifteen million or so years into the Permian, most if not all of the remaining coal basins were preserved under a veneer of desert sediments, and subsequently by younger Mesozoic and Cenozoic strata. These were deposited in new rift and thermal sag basins as the North Sea and Irish Sea developed in response to stresses generated by the sequential break-up of Pangea, notably the northward propagation of Atlantic sea-floor spreading. Under these additional strata, up to four kilometres in the southern North Sea, further coalification processes ensued, including the emission and migration of vast amounts of methane gas. Some of the gas was efficiently trapped in porous Permian desert sandstones and, less abundantly, in Carboniferous sandstones associated with the degassing coals themselves. Late in the Cretaceous Period stresses of compressional deformation were transmitted northwards by the beginnings of Alpine mountain-building. This caused many of these Mesozoic basins, together with the deeply buried Carboniferous, to be themselves inverted – gently folded and uplifted.

Final Inversion in the West

A final spectacular event affected the distribution of coal-bearing strata – an episode of uplift and erosion connected to rifting of the northernmost Atlantic continents at the beginning of the Cenozoic era, around 60 Ma.[2] To best understand the extent of this radical change in geology, we need look no further than the geological map of onshore Ireland, which once contained voluminous coal-bearing Late Carboniferous and marine Mesozoic strata. The tiny remnant outliers of Coal Measures rocks in Tyrone, Tipperary and Kilkenny and of Cretaceous chalks in Antrim contrast with the extensive outcrops of the eastern United Kingdom. The uplift is thought to have been caused by the deep generation and movement of partially molten rock during oceanic rifting, from Rockall in the south to the Faroes and Iceland, taking in the seaboard of Ulster and the Inner Hebrides. The result was not only an active volcanic province of huge extent (the North Atlantic Volcanic Province – think Giant's Causeway and the Black Cuillins of Skye) but an even wider uplift covering the whole of onshore Ireland and west Britain. The cover of remnant Late Palaeozoic and of more widespread Mesozoic strata were raised into uplands and exposed to merciless erosion by plentiful orographic rainfall from

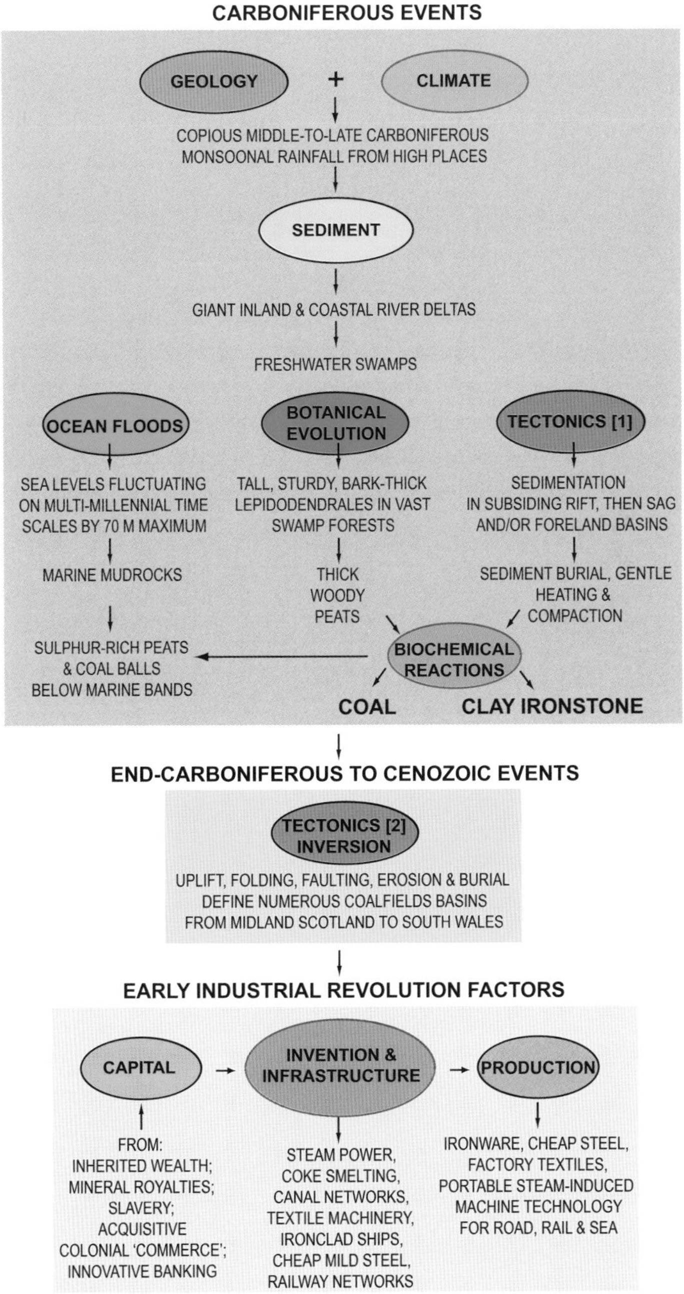

CARBONIFEROUS EVENTS

Figure 11.4 Flow diagram indicative of Carboniferous geological and other events that led to the early Industrial Revolution taking place in Britain. Source: Author.

abnormally warm westerlies for up to ten million years or so. The eroded sediment was dumped into the surrounding marine basins of offshore Ireland, Wales and Scotland. What was left after the firestorm of extrusive basaltic volcanism were pitiful remnants of outcropping Carboniferous and Mesozoic strata whose areal extent was further reduced by the ravages of Pleistocene glaciations. Even today there is a residual excess of topography up to a kilometre or so above thinned crust (down to 20–25 km in places) of much of western Scotland and north-west Ireland. This has been attributed to ongoing uplift from still hotter than normal mantle.[2]

Coda

Since most of Britain and Ireland were covered at one time by Coal Measures strata, it is clear that the surviving measures owe their existence to a long series of geological 'accidents of preservation'. Figure 11.4 is a flow diagram that includes attempts to summarize the various stages in the evolution of present-day coal basins by such events.

PART 3

Legacies: Carbon Cycling, Chimneys and Creativity

In the 1980s full industrial consummation came to the hydrocarbon-poor but coal-rich populous nations of south and east Asia. Today their smoggy palls from unprecedented levels of coal-based emissions echo the coal-burning smogs of Victorian London, threatening not only the health of their billions of citizens, but also to change planetary climate irrevocably. The link between the carbon cycle, fossil fuel burning, atmospheric composition and surface temperatures was unknown to the Industrial Revolution's first pioneers. It seems nugatory to denote our modern atmospheric crisis as a consequence of any longer-term Anthropocene effect – rather it reflects a 'hand-me-down' from our mid-Georgian ancestors' resolute pursuit of coal-based energy. The reasons why the Industrial Revolution started in Britain rather than elsewhere in the early eighteenth century are still mulled over by economic historians, but it seems clear that the long-drawn-out process had a good deal of serendipity to it. It centred initially around the need to solve three essentially different practical conundrums: first, the shortage and growing expense of charcoal for iron smelting; second, drainage of deep mines; third, provision of fuel needed to heat the houses and fuel the traditional trades of Europe's biggest city during the cold winters of the 'Little Ice Age'. Coal became the common factor, which together with textile machinery inventions that cut the cost of mass-produced quality yarn, saw the world in pursuit of Lancashire cottons and West Yorkshire worsteds. By the late 1870s, great nation states had emerged or were emerging from civil war and 'unification' (Germany, USA, Italy) and because of free trade were able to obtain the most modern of available technologies to form, join and keenly compete in the industrial 'club'. Yet by 1912 only two per cent of coal cut in Germany was by machine compared to eight per cent in Britain. Industrialization and the 'Machine Age' also brought urbanization used elsewhere. The vast social and cultural changes that ensued were reflected in a new era of artistic and literary creativity and, eventually, the political and personal freedoms that developed in largely-urbanized parliamentary democracies, mostly in Europe – mostly conjoined in peaceful economic union.

12

Atmospheres, Global Carbon Cycling and Glaciations

The force of vapour [steam] is another fertile source of moving power;
but even in this case it cannot be maintained that power is created. Water
is converted into elastic vapour by the combustion of fuel. The chemical
changes which thus take place are constantly increasing the atmosphere by
large quantities of carbonic acid [carbon dioxide] and other gases noxious
to animal life. The means by which nature decomposes these elements, or
reconverts them into solid form [carbon], are not sufficiently known.

From Charles Babbage *On the Economy of Machinery and Manufactures* (1835)

A Fact and an Historical Perspective

The fact is that there has been a steady rise, also an increasing rate of rise, in the concentration of atmospheric carbon dioxide recorded at the Mauna Loa observatory in Hawaii (Fig. 12.1) over the past sixty years, from 315 to 411 parts per million, a mean increase of around thirty per cent. The probability is that this is due to the effects of the post-1940s global industrial and transport revolutions with their exponential expansion of fossil fuel burning.

The earliest and most prescient remark concerning the effects of machine- and human-generated carbon dioxide came from polymath Charles Babbage in 1835, reproduced above.[1] He warned (in a book extolling the advantages of industrialization) that there may be, as we would put it today, unintended consequences to widespread coal burning. A brilliant reply to this cautious request for further research came ten years later from another fertile mind, that of a young French geochemist, Jacques-Joseph Ébelmen, born in Baume in eastern France (although we don't know whether he was aware of Babbage's treatise). Originally a mining engineer, his early work on the chemical assaying of industrial rocks and minerals led to improvements in Sèvres porcelain manufacture. He was fêted by Michael Faraday at the 1851 Great Exhibition in London

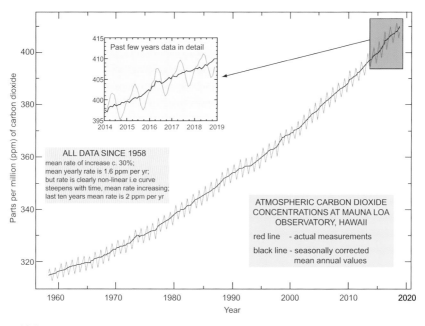

Figure 12.1 The sixty year record of atmospheric carbon dioxide at the Hawaii Observatory. Redrawn and coloured after: www.esrl.noaa.gov/gmd/ccgg/trends/full.html

where he was an official French delegate. The two men talked together, probably mostly about the chemical analysis of ancient and modern pottery glazes, a subject of great mutual interest. Faraday invited him to one of his Royal Institution evening lectures, but if they ever did discuss Ébelmen's conclusions concerning the nature of chemical weathering by the atmosphere, there is no hint that Faraday ever returned to the topic or referred to it. Tragically Ébelmen died the next year at only thirty-seven years of age.

In his paper of 1845 Ébelmen had listed (Table 12.1) the chief natural and human-induced chemical reactions that might change the proportions of carbon dioxide and oxygen in the atmosphere.[2] The results were published in the scientific journal of the French Bureau of Mines and virtually lost to science, never cited nor even acknowledged more than just once until the mid-1980s, by which time Ébelmen's whole scheme had been re-invented by Nobel Laureate Harold Urey and called the Carbon Cycle.[3] One is tempted to suggest that Carbon Cycling should be known as Ébelmen Cycling in his memory (one thinks of the Krebs Cycle in biochemistry, for example).

Ébelmen's trump card was his deep understanding of the nature of the chemical weathering of rocks. He identified a major 'sink' (a depository) for atmospheric and soil carbon dioxide, pointing out first that limestone weathering by natural carbonic acid in the atmosphere has no effect whatsoever on the carbon cycle. This is because once drained into the oceans, reactions that precipitate calcium carbonate give off carbon dioxide in exactly the same proportions as that used up during the original weathering. He highlighted that

1. Causes which tend to raise the proportion of carbon dioxide contained in air
    A. Without diminution of the proportion of oxygen
        i.) Emission of gases associated with volcanic orifices
    B. With diminution of the proportion of oxygen
        i.) Destruction of organic matter contained in sedimentary rocks (oil shales, lignites, coal)
        ii.) Weathering of iron pyrites and some iron minerals

2. Causes which tend to diminish the proportion of carbon dioxide
    A. With the liberation of oxygen
        i.) Formation of iron pyrites
        ii.) Formation of mineralized combustibles (coal,etc) and the conservation (burial) of organic debris
    B. With or without the absorption of oxygen
        i.) Weathering of silicates of igneous rocks

3. Causes which tend to raise the proportion of oxygen contained in the air
    A. Without diminishing the proportion of carbon dioxide
        i.) none
    B. In lowering the proportion of carbon dioxide
        i.) Formation of mineralized combustibles (coaI ,etc)
        ii.) Formation of iron pyrites

4. Causes that produce a diminution of oxygen contained in the air
    A. With formation of carbon dioxide
        i.) The destruction (oxidation) of organic matter contained in the ground
        ii.) The weathering of iron pyrites and some iron minerals
    B. With the absorption of carbon dioxide
        i.) Weathering of silicates in igneous rock

Table 12.1 J-J Ébelmen's list of controls on the *relative* proportions of oxygen and carbon dioxide. The 'iron pyrites' mentioned is the common disulphide of iron, known as 'fool's gold' in visible crystalline form and as disseminated microcrystals in muddy sediments (see Chapter 9 for some interesting details concerning its bacterial origins). Ébelmen views the sources of carbon dioxide as coming from: emissions during volcanic eruptions, the destruction by combustion of sedimentary carbon and hydrocarbon and the oxidation of soil organic matter. Regarding sinks for the gas, these are: chemical reactions involved in the weathering of mineral silicates in igneous rocks, photosynthesis and the preservation by burial of plant materials (by what we identify as tectonic subsidence).

the situation is quite different for the weathering of calcium-bearing silicate minerals in rocks like common granite and basalt, which dominate the Earth's crust. Here, for two molecules of carbon dioxide used up in weathering only one is given off by subsequent reactions back to calcium carbonate. The overall carbon dioxide of the atmosphere is thus reduced by this sink effect. However, it is important to realize that neither Ébelmen nor anyone else at the time had knowledge of the rates of such weathering reactions – a critical point, given what we now know of the extreme magnitude of human CO_2 emissions compared with their natural reduction by weathering; almost one hundred times as much.

On another tack, Michael Faraday's successor as Director of the Royal Institution, John Tyndall, was also a hugely talented experimental physicist. He had closely studied Joseph Fourier's work on the diffusion of heat by conduction, in particular the idea that the Earth's atmosphere might be instrumental in keeping incoming solar heat bound closely to the planetary surface. He decided to experimentally test the ability of various atmospheric gases to absorb heat energy using a novel instrument of his own invention. He identified water vapour, carbon dioxide and hydrocarbon as the chief absorbers. These were all polyatomic, i.e. comprising more than one type of atom, unlike, say, oxygen or nitrogen. He wrote concisely and authoritatively in 1859:

> The bearing of this experiment upon the action of planetary atmospheres is obvious... the atmosphere admits of the entrance of the solar heat, but checks its exit; and the result is a tendency to accumulate heat at the surface of the planet...gases absorb radiant heat of different qualities in different degrees.[4]

Neither Tyndall nor anyone else at the time considered that human activity could influence the amount of insulating gases like carbon dioxide in the atmosphere. Yet, perversely, the cooling effect of a reduction in carbon dioxide concentration was topical. Only eighteen years before, Louis Agassiz had staggered the scientific world by proposing the existence of past ice ages, and by this time in the late 1850s the geological evidence for this theory was widely accepted.

It was left to the future Swedish Nobel prize winner, Svante Arrhenius, to have a further crack at exploring the possibility of a carbon dioxide effect arising from industrial emissions. In December 1895 he read a landmark paper to the Swedish Academy of Sciences making use of Tyndall's and later researchers' experimental results. He calculated the timing and effect of changes in carbon dioxide concentrations on the atmospheric balance that might cause global cooling or warming. An extract from this lecture was published in English the next year.[5] Strangely, he never considered a human (industrial) source for carbon dioxide, rather he was concerned with changing carbon dioxide concentrations in general, from natural causes not specified, and how they might be responsible. As he writes, rather wearily, in his summary discussion:

> I should certainly not have undertaken these tedious calculations if an extraordinary interest had not been connected with them...on the probable causes of the Ice Age; and...that there exists as yet no satisfactory hypothesis that could explain how the climatic conditions for an ice age could be realized in so short a time as that which has elapsed from the days of the glacial epoch.

As it turns out, it seems he was misled by the faulty conclusions of a close colleague. In his summary discussion, Arrhenius provides lengthy quotes translated from one Professor Arvid Högbom who, clearly ignorant of the rates of the reactions inherent in weathering, had concluded that the:

> quantity of carbonic acid, which is supplied to the atmosphere chiefly by modern industry, may be regarded as completely compensating the quantity of carbonic acid that is consumed in the formation of limestone (or other mineral carbonates) by the weathering or decomposition of silicates.

And that:

> the most important of all the processes by means of which carbonic acid has been removed from the atmosphere in all times, namely the chemical weathering of siliceous minerals, is of the same order of magnitude as a process of contrary effect, which is caused by the industrial development of our time, and which must be conceived as being of a temporary nature.

In other words there was nothing for us humans to worry about. But, vitally, Högbom failed to mention that the chemical weathering of silicate minerals is a very, very slow process. Any buffering effect by weathering cannot begin to influence the volumes of human-induced outpourings of carbon dioxide on the timescales involved. The length of time of this buffering 'lag' is likely to be many hundreds to thousands of years or more. Högbom considered as wholly random the ups and downs of carbon dioxide concentrations that might have occurred in the geological past – due to variations in volcanic eruption intensities. Arrhenius and others by this time were calling the atmospheric warming due to carbon dioxide the 'hot-house effect' (aka 'greenhouse effect').

Global Cycling and the Carbon Dioxide Crisis

The cycling of the solid Earth we recognize today comes courtesy of plate tectonics. The necessity for such cycling was originally a thought experiment in the fertile mind of James Hutton. By 1795, in old age, his vision of a forever recycling Earth had been given unequivocal support from geological field evidence – in the form of unconformities. He now saw the planet in terms of an equilibrium achieved by 'circulation':

> We are thus led to see a circulation in the matter of this globe, and a system of beautiful economy in the works of nature. This earth, like the body of an animal, is wasted at the same time as it is repaired. It has a state of growth and augmentation; it has another state, which is of diminution and decay…and the operations by which this world is thus constantly renewed, are as evident to the scientific eye, as are those by which it is necessarily destroyed.[6]

Hutton's elegant prose and the idea of 'circulation' were taken up by Charles Lyell in his epoch-making three-volume *Principles of Geology* (1830–34). He lent it his own terminology, contrasting in Volume One what he termed the restorative or 'reproductive' processes of the planet with those of its reduction by 'destructive' forces. Charles Darwin carried Volume One onto the Beagle in late 1831, and there can be little doubt that Lyell's eloquence guided his subsequent observations and thoughts on the nature of uplift and subsidence during earthquakes, and of the slower oceanic subsidence that caused coral atolls to form. Yet such grand projects and whole-earth views appealed little to the majority of thoroughly practical geologists of the early-nineteenth to mid-twentieth

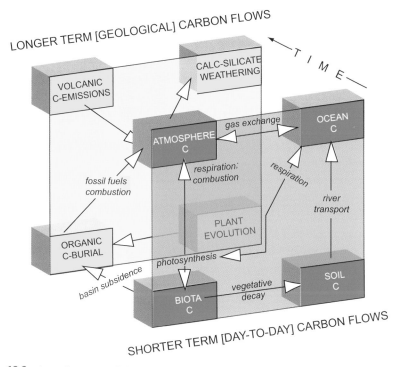

Figure 12.2 Three-dimensional depiction of short- and long-term carbon cycles and flows. The carbon cycle is envisaged diagrammatically as a group of interconnecting boxes representing the major reservoirs of solid, dissolved and gaseous carbon. The arrows with words represent the flow of carbon from one box to another. The total amounts of carbon in these boxes varies – the size of boxes is not proportional to the actual amounts of carbon dioxide. Incoming arrows to a box represent additions (sinks) of carbon dioxide; outgoing arrows from a box represent subtractions (sources). Source: after one-dimensional versions by Berner (2004).

centuries. In Thomas Browne's words of 1646, perhaps they mistrusted 'playing much upon the simile or illustrative argumentation…'.

At the same time as plate-tectonic concepts were evolving, there were attempts from the early 1970s to revisit Hutton's vision of global cycling by examining the chemical elements involved in Earth surface and atmospheric reactions. These involved carbon, oxygen, nitrogen and sulphur, in what is known as biogeochemical cycling.[7] This describes gaseous and aqueous reactions during volcanic eruptions, chemical weathering and photosynthesis. Such approaches bore fruit when applied to sedimentary cycling, and laid the basis for 'hindcast' modelling (predicting the consequences of what is already there and known), in particular that of the geological carbon cycle (Fig. 12.2) deduced by R.A. Berner and co-workers.[8]

The concepts of biogeochemical cycling led to a number of questions concerning the nature of the young Earth (i.e. pre-2500 Ma), notably the timing and mechanism for the appearance of an oxygenated atmosphere. That an atmosphere had evolved from reducing to oxidizing was ascribed to the cumulative effects of early plant photosynthesis and carbon burial. These ideas and lateral thinking from research on planetary atmospheres led to the concept of Earth as a self-regulating entity, the state of Gaia, as proposed in a widely read book by James Lovelock in 1979.[9] When first proposed, this scheme operated without reference to, and entirely independently of plate tectonics – which seemed a fatal omission as one eagerly read through the book in the early 1980s.

Tectonically Controlled Weathering Changes Climate

The essential point is that sedimentation and tectonic subsidence will preserve carbon until it is released by tectonic uplift and erosion. Therefore elemental burial must be seen in the time context of plate-tectonic recycling, i.e. over tens to hundreds of million years. It can be argued that this combination of tectonic and biogeochemical cycling is the great fulfilment of the Huttonian circulation scheme. It came about with the realization of the implications of Ébelemen-type carbon sequestration by silicate mineral weathering in the newly formed Alpine–Himalayan mountain belt around 40–45 Ma. These new mountains, stretching from the Pyrenees to Sichuan, were full of pristine silicate-bearing rock just waiting to slowly draw down carbon dioxide from the atmosphere as they were weathered by warm monsoon rains. This realization led Maureen Raymo and William Ruddiman in the 1990s to propose and proselytize it as the cause of the undeniable global cooling seen since the mid-Cenozoic – ending in the Pleistocene Ice Ages.[10]

In support of this theory, enhanced silicate weathering had entirely independent consequences. The young crustal rocks produced by melting and fusion during mountain building had relatively high concentrations of the strontium-87, the radiogenic isotope (derived from radioactive rubidium-87) of the element, compared to stable strontium-86. Over Mid-Cenozoic to Holocene times the oceans received increasing amounts of radiogenic strontium-87 relative to strontium-86: i.e. the mean ratio of strontium 87/86 increased. It was recorded exactly in fossilized calcitic and phosphatic shell-forming organisms collected from deep-sea cores. We therefore see a global outcome – silicate weathering joining plate tectonics as a major environmental control on climate. This realization opened up the search for other such linkages in the geological past. Given the efficacy of the greenhouse effect and the knowledge that the geological carbon cycle interacts with tectonics over long timescales, the causes of past global climate changes need to be carefully assessed, something we briefly turn to.

It has been suggested that Gondwanan south polar glaciations became possible as Middle and Upper Devonian forests drew down carbon dioxide.[11] The timing of this – after Acadian mountain building – implies that silicate weathering may also have played a role. At the same time colonization of fluvio-deltaic habitats by Lepidodendrales swamp forests was providing steadily increasing abstraction. That this had begun before the traditional view of the onset of the main Gonwanan glaciation in early Middle Carboniferous times is reinforced by both biogeochemical stable isotope trends and the recent discovery of (presumed) Early Carboniferous ice-stream erosional valleys and glacial deposits in Niger and Sudan.[12] These were at palaeolatitudes of around sixty degrees south, with trajectories from southern polar ice caps northwards to the southern margins of Palaeotethys, where they doubtless debouched fleets of icebergs.

For the later Carboniferous period, the further geographic expansion of the forest swamp ecosystem must clearly have played a major role in carbon dioxide removal – a one-way arrow of ever-increasing photosynthetic magnitude and efficiency. Add to this the growth of the main Appalachian–Variscide mountain chain featured in Part 2 of this book, then we have a perfect storm of causes and effects to choose from. Indeed, as I began to finally revise this section I became satisfyingly aware of recent detailed and precise strontium-87/86 measurements from Carboniferous microfossils that entirely support application of the Raymo–Ruddiman hypothesis to Variscan tectonics, erosion and carbon dioxide sequestration.[13]

In the light of the above remarks it is worth quoting here the succinct words and footnote of J.M. Robinson from a classic paper published almost thirty years ago concerning the consequences of Carboniferous carbon burial, particularly for the Gaian hypothesis:

> the emerging picture is inconsistent with a Gaian biosphere. The Palaeozoic biosphere, led by the evolution of the plant kingdom, appears, not as a stabilizing influence, but as a source of major instability...the hypothetical Gaian regulatory system...cannot be imposed on a planet where crustal recycling introduces 100 m.y. lags into the cycles of biologically important elements.[14]

Robinson adds in a footnote:

> I see the Gaia hypothesis (the assertion that the biosphere controls the atmosphere and the Earth system to its own advantage) as the functional equivalent of the Turing Machine in computer science, or Maxwell's Demon in physics...as a hypothetical, half whimsical construct, useful for extending the range of scientific imagination.

Quite!

Broken Cycle Needs Mending

We have no direct evidence of the composition of the Carboniferous atmosphere. To obtain it we would need to sample preserved air for comparative analysis – which we cannot do. There are certain indirect geochemical and biophysical estimates for carbon dioxide levels – from fossil calcium carbonate soils[15] and fossil leaf pore (stomatal) dimensions and density.[16] These yield results indicating generally reduced levels of atmospheric carbon dioxide for the period compared to the preceding late-Devonian and later Permian times. But the methodologies are indirect, involve several assumptions, and have high statistical errors.

There is one other indirect line of evidence, also with high errors. We can back-calculate the amount of carbon that was sequestrated over the main thirteen or so million year interlude of coal formation and examine its likely consequences to atmospheric composition to order of magnitude accuracy.[17] The equatorial Euro-American Carboniferous swamp province alone had an area of at least 3.23 million square kilometres (we are ignoring the largely Permian Indian and North China coals). A conservative estimate would involve upper and lower limits of 500 metres to one kilometre for the thickness of coal-bearing sedimentary strata originally present before end-Carboniferous inversion and erosion. The typical proportion of coal present in those strata is between one and three per cent, representing accumulations of anything between five to thirty metres of solid carbon in swamp forest peat (originally around four times as thick) over thirteen or so million years. The total mass of solid carbon stored by burial in this sedimentary succession would have been in the range 22,000 to 131,000 gigatonnes (Gt), with a mean sequestration rate by sedimentation and subsidence of 0.00167 to 0.01 Gt per year.

Such average estimates for sequestration are obviously misleading, since coal-forming peats were not deposited for all of Late Carboniferous time. The simplest arithmetical assumption would be to assume that the time under peat deposition at any one location would have been proportional to the percentage of coal now observed in the entire coal-bearing sequence. The one to three per cent limits noted above would require total durations of peat production in the range 0.13–0.39 Myr and potential sequestration rates during these intervals varying from 0.17–1.0 Gt per year. A more sophisticated take would recognize that peat sequestration would be at maximum (say, 100%) during glacial lowstand and rising sea-level conditions when deltas were at their greatest areal extent, and at minimum (say 0%) during interglacial highstands when they were largely flooded by marine waters (although not all coal basins had marine connections). This would require marked unsteadiness in sequestration over the longer term orbital periodicities of 100–400 kyr, perhaps yo-yoing by fifty per cent or more.

How do these figures chime with the likely magnitudes of the remaining parts of the Late Carboniferous carbon cycle? We have no independent way of knowing what the sequestration rates were for carbon dioxide via silicate weathering at that time, neither for the provision of 'new' carbon dioxide from contemporary volcanic eruptions. Our only solace is to turn to modern values in a blatantly uniformitarian gambit that one feels Charles Lyell might have approved of, though others might not. Luckily, accurate calculations for the relevant fluxes are essential to efforts to geo-engineer the current planetary surface out of global warming. They provide us with some reasoned estimates for their likely magnitude.[18] Thus the mean consumption flux of carbon dioxide in rock weathering is computed at around 0.25 Gt per year, with modern volcanic emissions providing 'new' carbon dioxide at a similar rate, between the limits of 0.18–0.44 Gt per year with a preferred mean estimate of around 0.26 Gt per year. By way of comparison, current anthropogenic CO_2 emissions are 35 Gt per year, dwarfing this estimate.[19]

Since these weathering inputs and volcanic outputs roughly cancel each other out, it seems fair to conclude that plant carbon sequestration must have been unsustainable without comparable massively (and unlikely) extra carbon dioxide fluxes from volcanic and metamorphic degassing. Atmospheric drawdown would have been rapid over the short timescales appropriate to peat formation. The process would have encouraged polar cooling, glaciation and sea-level fall as interesting positive feedbacks. The weathered Variscan highlands with their pioneer forests at lower altitudes supplied runoff and sediment that filled the later Carboniferous coal basins of Euro-America: little wonder that the combined Raymo–Ruddiman effect and plant carbon sequestration by tectonic subsidence led to the great Late Palaeozoic (post-370 Ma) cooling event, with periodic widespread Gondwanan glacial activity over >100 Ma, well into the Permian period.

Coda: Enhanced Oxygen in the Carboniferous Atmosphere?

R.A. Berner's modelling of the imbalance of the Late Carboniferous carbon cycle led to the calculation of an atmosphere containing up to fifty per cent greater oxygen than present-day levels – far more oxygen was being produced by plants than being used up *post mortem* in the oxidation of that same plant biomass and of iron sulphides.[8] A prior idea due to A.J. Watson and co-workers[20] was that any 'hyperoxic' conditions that have evolved on Earth would increase flammability of vegetation, by analogy with experiments on cardboard flammability. Wildfire from lightning strikes would then reduce plant cover proportionally, creating a stasis in oxygen concentration around modern levels – evidence for a Gaian self-regulatory mechanism. Unfortunately cardboard is not remotely like tree lignin in its flammability, and there is no evidence for universal or even regional conflagration. We have known since Marie Stopes' pioneering 1935 observations (Chapter 11, note 3) that there are (highly) variable amounts of particulate charcoal present in Carboniferous bituminous coal (her fusain), proving that wildfires did occur,[21] but there are no statistically sound estimates (i.e. with error margins) of their magnitude, frequency or extent.

Hyperoxia would have led to an atmosphere around fourteen per cent denser, since oxygen is denser than nitrogen. Testable consequences reviewed by Robert Dudley[22] are that increased oxygen partial pressure would have increased diffusion rates in insect trachea. This enabled the gigantism observed in the fossil record of walking insects like millepedes and of giant dragonflies, whose flight efficiency was increased due to higher aerodynamic lift produced by stronger wing muscles beating in the denser atmosphere. We also know that wind drag is directly proportional to air density, so that for given wind speeds there would have been a higher potential for severe monsoonal tropical storms to bend and break the trunks of the tallest trees in equatorial Lepidodendrales forests, adding to the accumulation rate of their peaty substrates.

13

Britain: First 'Chimney of the World'

I used to know the landmarks on this route
The industries of Britain left and right.
Once I'd know exactly where we were
From the shapes of spoil heaps and from winding gear
Spinning their spokes and winching down a shift
Miles deep into this sealed and filled-in shaft...

Tony Harrison from *Cremation eclogue* in *Under the Clock* (2005)

Pioneer of Pollution: London's 'Great Stinking Fogs'

The 'London Trade' in coal from Tyneside noted in Part 1 lasted for at least six hundred years. The following is the terse counter-command of July 1264 from Henry III's court at Windsor Castle to the Sheriffs of London:

> to purvey for the King in the City of London without delay and without fail a boatload of sea-coal and four millstones for the King's mills at Windsor Castle and convey them thither by water for delivery to the constable of the castle.[1]

Henry was a prisoner of Simon de Montfort at the time, just before the epochal Battle of Evesham; hence the note has unexplored historical significance. It also reminds us that the Thames was navigable up to Windsor. The requested 'boatload' could have comprised several score tonnes of coal, originally shipped from Tyneside in a clinker-built cog vessel, subsequently offloaded into barges on the fringes of the inner Thames estuary. We are left with the somehow very un-medieval image of the great Castle's domestic and kitchen fireplaces belching coal smoke out of roof apertures during the winter of 1264–65, with Henry and son Edward comfortable inside, de Montfort's 'Second Baronial Rebellion' finally defeated.

As emphasized by Peter Brimblecombe in his classic account (1987; 2011) the smell and smoke from coal burning had always provoked strong reactions in London. By the sixteenth century the limestone stonework of old St Paul's Cathedral was severely soot-coated and, worse, corroded – due, as we now know, to acid dissolution. Many ineffectual efforts were made down the years to keep the commercial burning of coal out of the main residential districts and restrict it to the peripheries. As the city started to grow outside its medieval walls, the problems increased. Little was done to alleviate the situation, for politicians knew that it was coal that kept the wealthier urban population warm, cheaply, in winter. I repeat Brimblecombe's quote from the ever-observant John Evelyn in this 1659 rant from Restoration London against the ugliness of unplanned medieval London before the Great Fire, with its polluted and disgusting atmosphere:

> such a cloud of sea-coal, as if there be some resemblance of hell upon earth,
> it is in this volcano in a foggy day: this pestilent smoak, which corrodes the
> very yron [iron], and spoils all the moveables [garments, laundry], leaving
> a soot on all things that it lights: and so fatally seizing on the lungs of the
> inhabitants, that cough and consumption spare no man.[2]

In 1661 Evelyn added that in his opinion the true origins of the pollution were not from kitchen or domestic fires:

> but from some few particular Tunnels and Issues, belonging only to *Brewers,*
> *Diers, Limeburners, Salt* and *Sope-Boylers*, and some other Trades, *One* of
> whose *Spiracles* alone, does manifestly infect the *Aer*, more than all the
> Chimnies of *London* put together besides.[3]

Limeburners in particular provided the means of making mortar in the burgeoning city with all the rebuilding after the Great Fire. They brought their cheap raw chalk from the Downs into the City to meet the more expensive coal at urban limekilns. Elsewhere it was the other way round, with coastal limestone outcrops feeding local kilns to which the coal was brought by ship, and from which the lime was taken out for wider export (Fig. 13.1).

Louis Simond, writing in early March 1810, echoes the earlier sentiments in his fine prose concerning the London winter days of his visit:

> the smoke of fossil coals forms an atmosphere, perceivable for many miles,
> like a great round cloud attached to the earth. In the town itself, when the
> weather is cloudy and foggy, which is frequently the case in winter, this
> smoke increases the general dingy hue, and terminates the length of every
> street with a fixed grey mist, receding as you advance. But when some rays
> of sun happens to fall on this artificial atmosphere, its impure mass assumes
> immediately a pale orange tint, similar to the effect of Claude Lorraine glasses

Figure 13.1 Unique late-eighteenth-century multiple limekiln stacks forming a clustered assemblage whose raw materials were dropped in from a common roof: Beadnell, Northumberland. Local limestone from the foreshore was burnt with local coals interbedded within the Yoredale succession and augmented by imported coal shipped in from the main Northumberland coalfield to the south. Photo: Author.

[small convex mirrors used by artists to see gradations of natural colour tone] – a mild golden hue, quite beautiful. The air in the meantime is loaded with small flakes of smoke, in sublimation – a sort of flower of soot, so light as to float without falling. This black snow sticks to your clothes and linen, or lights on your face. You just feel something on your nose, or your cheek – the finger is applied mechanically, and fixes it into a black patch![4]

By 1818 even the fine white Portland stone of the new St Paul's was blackened. John Keats poked fun at efforts to discern its source and to seek a remedy in this fizzy letter to the young daughters of a family friend in Teignmouth. 'My dear Girls', he writes:

I promised to send you all the news. Harkee! The whole city corporation with a deputation from the Fire offices are now engaged at the London Coffee house in secret conclave concerning Saint Paul's Cathedral its being washed clean. Many interesting speeches have been demosthenized in said Coffee house as to the Cause of the black appearance of the said Cathedral.[5]

The strongest corrosive effects of burning certain sea coal arose from sulphurous constituents in the form of mineral pyrite. On burning it oxidizes to sulphur dioxide gas, which combines with water droplets to form very dilute, but still potent, sulphuric acid. This reacts with limestone, and together with deposited soot produces a flaky, black, insoluble gypsum crust. The added effects of coal smoke formed the 'Great Stinking Fogs' – the cutting out of winter sunlight and a consequent increase in the occurrence of rickets amongst the young. There is no better description of urban fogs along the Thames estuary than that provided in the mid-nineteenth century by Charles Dickens in 'Bleak House' (1852–3). There was:

fog everywhere, fog up the river where it flows among green aits [heights] and meadows – fog down the river, where it rolls defiled among the tiers of shipping and the waterside pollutions of a great (and dirty) city. Fog on the Essex marshes, fog on the Kentish heights. Fog creeping into the cabooses of collier-brigs; fog lying out on the yards, and hovering in the rigging of great ships; fog drooping on the gun-whales of barges and small boats.

There were in fact near-smokeless coals available at the time to alleviate such specific ills: hard west-Glamorgan and Pembrokeshire anthracite and cannel coals from several Midlands and Lancashire coalfields. Yet Tyneside coal was cheap, and as Defoe well knew, the 'London Trade' followed a long-established sea route. The trade also raised considerable taxation revenues.

Levels of urban pollution were enhanced in the mid-nineteenth century by coking plants supplying piped methane 'town gas' for heating, cooking and lighting, the latter seen as one of the wonders of the world on its first appearance in Victorian London. Steam coals powered the giant locomotives of the rail networks, their summer journeys marked by spark-induced fire scars along many a desiccated embankment and cutting. By the 1880s, lasting to the early 1950s, a stasis prevailed; but a tipping point was eventually reached in urban pollution, for in addition to untrammelled domestic and industrial use, in the 1930s great coal-fired power stations like Battersea and several others were built close to central London. Also, after the Second World War, legions of motorized commuter transports replaced electric trams. Frequent late-autumn smogs hung over the coal basins of the Midlands, the North, midland Scotland and, especially, the London Basin. Here the topography of the low-lying Thames valley between ridges of surrounding chalk downlands caused temperature inversions during conditions of high pressure. Dense cold air lay motionless in the topographic low, trapping the particulates and noxious gases at ground level for days or weeks on end.

Such 'pea-souper' smogs culminated in the high-pressure weather of early December 1952 that caused thousands of 'extra' deaths, hastening the much-needed Clean Air Act of 1956. To the disgrace of the Churchill Conservative administration, it had disfavoured legislation prohibiting freedom to burn as one pleased. It was prodded into action by outraged MPs in a Private Member's Bill. The Act included the compulsory introduction of smokeless fuels in designated areas, the establishment of smoke control areas, the prohibition of 'dark' heavily particulate smoke from chimneys and furnaces, the monitoring of such particulates and a minimum height for chimneys. Yet no attempt was made in the Act to discourage the combustion of sulphurous coals; no mention made of sulphur dioxide as a specific coal-based pollutant or to the nitrogen oxides known to be present in the exhaust gases of diesel engines.

Also, unknown at the time, for a hundred and more years the prevailing westerlies over Britain had carried tele-pollution as 'acid rain' far over the unspoilt Scandinavian mountain alpine tundra, disturbing whole ecologies and decreasing the pH of already slightly acidic alpine soils and lake waters dangerously further. Present-day traffic-borne diesel hydrocarbon emissions have similarly become tele-pollutants in their own right, depositing noxious gases rich in nitrogen dioxide and particulates with little regard for *in situ* human health in the capital city – mayoral laissez-faire of a decade ago is on a par with the ineffectual opposition to the more visible smogs of yesteryear.

Causes of the Industrial Revolution

'Inventions enabled industrialization when cheap energy vied with high material and labour costs'. I paraphrase the viewpoint championed by economic historian Robert Allen in his dissection of the causes of the Industrial Revolution.[6] Continuing in Allen's vein, why did the British invent machines to do the work previously done by hand or by animals, while other nations invented the Jacquard loom, the cuckoo clock, comfortable chairs, delicious chocolates, the guillotine and the hot-air balloon? It is not that the British at this time were particularly industrially inventive human beings, for this insults the record, scope and breadth of north European and other nations whose secular judicial codes enabled freedom of thought to do scientific experiments, and for engineers to try out new machines and ideas to solve practical problems. Remember that Newcomen borrowed key concepts for his engine from the Frenchman Papin. Also, though we rightly praise and admire James Brindley's skills as a canal engineer, it had all been done before and on a vaster scale by Pierre-Paul Riquet along the Canal de Midi.

No, the answer is simple – by the time of the mid-eighteenth century, Britain had developed a unique and burgeoning interlinked coal and iron industry, thanks to its geological history and to the voracious appetite for coal fires from Londoners in the harsh winters of the 'Little Ice Age'. The very real problems that arose with mine drainage and ventilation threatened production levels and the profits of mine leaseholders and landed gentry who owned the mineral rights and shared in these profits. Newcomen's pumping technology was highly focused and novel – it was the result of the inventor's own problem-solving enterprise. The mine drainage problem in the metal mines of his native south-west had stimulated in his inventive mind the idea that a mechanical means was possible, using thermal energy from burning coal to cause the up-and-down motion of the common parish-pump handle. Later engineers (Smeaton, Watt, Trevithick) with their own ad hoc additions made the basic steam pump more and more efficient, so that it used less and less coal to raise a given amount of steam. The result summarized by Robert Allen is incontrovertible – that by the mid-nineteenth century the British consumption of coal in industrial plants had decreased by twentyfold in terms of mass per power output: an amazing achievement.

A similar case for discoveries made by problem-solving (as opposed to labour-saving) as the backbone of the Industrial Revolution can be made for Darby's success in identifying and 'charking' certain local coals from his own back yard in Coalbrookdale (minimal transport costs) and using them as a cheaper and more robust substitute for charcoal in iron smelting. This again was a highly-focused endeavour made possible by repeated experiments with different raw materials to obtain a usable product. Coke still needed labour to produce it – mining and transporting the coal, manning the coke tips (and, later, ovens), loading the wagons and so on – and it may be that the effect upon employment was neutral at the end of the day.

From the geological perspective, although coal did outcrop widely in northern France, north Germany and Poland, it was not needed there (wood was plentiful) at the critical time of the early to mid-eighteenth century It was negligent of economic historian E.J. Hobsbawm to write:

> If Britain's ample reserves of coal explain her priority [as the locus for the Industrial Revolution], then we may well wonder why her comparatively scant natural supplies of most other raw materials (for example iron ore) did not hamper her just as much, or alternatively why the great Silesian coalfields did not produce an equally early industrial start.[7]

There was never an iron ore shortage; it was just that the native Coal Measures clay ironstone strata were low-grade, a fact assuaged by close proximity to coal seams. All that changed by the mid-nineteenth century as the vast and near-surface reserves of Jurassic sedimentary iron ore in Alsace-Lorraine and Luxembourg and of the Ruhr and Silesia's own clay ironstone could be smelted using the basic-lined Bessemer and Open Hearth processes. Geological knowledge also increased the efficacy and economics of deep-shaft mining. Steam-powered rail and ship transport came to the fore everywhere.

In other developing industries of the eighteenth century, chiefly textiles, Allen's main thesis that fossil capital and invented machines combined to replace expensive labour is undoubtedly the correct one, though the income and wages of hand-loom weavers and agricultural labourers would hardly have seemed 'high' to themselves at the time. Here, in the cotton industry of Lancashire in particular, machine-led production using the water frame and spinning-jenny enabled fantastic increases in productivity. Yet, as stressed by Andreas Malm,[8] steam power was subordinate to water power in operating these machines in many, perhaps the majority, of Pennine and Lanarkshire mills. It remained so across Britain until the early nineteenth century, when portable and super-efficient steam engines providing rotative power (courtesy of James Watt) took over almost everywhere.

The 'Republic of Letters' and the Industrial Revolution

Alternative views as to the main causes for the beginning of the Industrial Revolution in eighteenth-century Britain have also been proposed. One is the superficially appealing idea that inventors and engineers made use of the scientific advances of the preceding century, notably in Newtonian physics and in the experimental sciences stimulated by leading lights of the Royal Society. Allen has crisply dismissed such links, pointing out that Newton's discoveries and theories led to as many invented machines (excepting telescopes) as did those of Galileo Galilei in Pisa – i.e. none. Also, we should add that even brief familiarity with the advances made by eighteenth-century French and German scientists and mathematicians reveals they were on a par with, or even more impressive than, those of the Royal Society – but they too did not lead to a French or German Industrial Revolution at the critical time.

Quite simply, as Allen stresses, the main applications of science to engineering during the early Industrial Revolution came from Greek (Aristotelian) roots, updated by Torricelli's demonstration of the vacuum and the existence of atmospheric pressure, chiefly in the field of hydrostatics (as we noted in Chapter 2, George Sinclair's mathematical speciality). It was the common-sense application of such imperfectly understood phenomena that mattered – like that of the condensation of steam to form a cooling vacuum (Newcomen) or the use of pressured steam (Trevithick). Explanations for such observations would have to wait until the advent of kinetic theory and the formulation of the laws of thermodynamics by William Thomson (Lord Kelvin) one hundred years later.

On a similar cultural-scientific tack, economic historian Joel Mokyr has played up the role of international Enlightenment culture – the 'Republic of Letters' – in Scotland and England as encouraging to inventors and engineers.[9] As we have seen, this mostly comprised Edinburgh and Birmingham *savants*, perhaps with around three dozen individuals (men) in total at both centres (see Part 1). Many had supreme all-round abilities – such polymaths would hardly recognize our modern-day descriptions of them: Black and Priestley as chemists, Wedgwood as potter, Watt as power engineer, Boulton as industrialist, Darwin as poet and botanist, Hutton and Whitehurst as geologists, Keir as glass-maker, and so on. Their frequent correspondence from one to another and between the two groups (especially the close relations between Hutton, Black, Watt, Wedgwood and Boulton) are crammed with brilliant observations, witty asides, cultural insights, original poetry and helpful suggestions concerning all manner of scientific and engineering topics.[10] For example: Darwin designed canal machinery; Wedgwood made hundreds of experiments on mineral-chemical glazes and invented a specialized kiln thermometer, leading him to be elected as a Fellow of the Royal Society; Hutton wrote extensively on agricultural science and the philosophy of scientific logic; Watt improved ocular surveying apparatus. Black, perhaps the most brilliant and original of them all, discovered the latent heat of steam, invented the eighteenth century's best precision balance, established that animal respiration and microbial fermentation both gave off carbon dioxide – his own self-discovered 'fixed air' – and discovered calcium carbonate precipitation.

There was one field in which the British *savants* excelled equally – the material world of geology and minerals. 'Mineral mania' was at its height in mid- to late-Georgian Britain, exemplified by the breadth of knowledge and extensive collections of that stupendous woman, Georgiana Cavendish, Duchess of Devonshire.[11] Such geological items excited the natural curiosity of the men of midland Scotland and the English Midlands as they personally collected specimens from the rich stratal repositories of the Lothian Coal Measures, South Lanarkshire lead mines, Derbyshire Carboniferous Limestone, Wrekin quartzite outcrops and Cornish mines. They gained knowledge of mineral chemical compositions, properties and potential practical uses, such as in the manufactures of glass and glazes (Keir, Wedgwood), lime and cement (Black), sal-ammoniac (Hutton), barium carbonate, barium sulphate and kaolinized feldspar in stoneware pottery (Watt, Wedgwood, Withering) and so on. We may contrast the practical aspect of the British mineralogists with the 'collectors' of Europe, such as Goethe who, although recorded as having large mineral and rock collections (the iron hydroxide, goethite is named after him), seems to have done little of practical use with them.

Each group in Edinburgh and Birmingham had its own resident geological genius: James Hutton in Edinburgh was supreme, but the older man, John Whitehurst, in Birmingham was also a serious and thoughtful geologist. Both undertook laborious fieldwork, Hutton more widely and with more guile. They both went on to write influential tomes –Whitehurst's *An Inquiry into the Original State and Formation of the Earth* (1778) and Hutton's *Theory of the Earth* (1788). The two productions differ markedly in scope and logic. Whitehurst was ever-anxious, and perhaps over-anxious, to preserve the legacy of God and the Flood. He based his theory of pre-Flood Earth on the gradual emergence of a fluid, then partly solid planet from the void with one universal 'revolution' whose earthquakes and volcanic eruptions upset global stratal patterns before the subsequent Creation and Flood. His ideas have much similarity with those of Buffon, whose *Les Époques de la Nature* was also published in 1778, but which referred back to essentials published over the previous decades. As we saw in Chapter 12, Hutton's was the most general and particular dissection of Earth history that recognized the necessity of cyclic 'revolutions' – from mountain erosion to renewed mountain building. Over the centuries this has proven to be the most rational and testable of theories.

However, it would be negligent to assume that all the driven men of Birmingham and Edinburgh were equally at home with all manner of natural things. Here is an undated letter from Wedgwood to Thomas Bentley beseeching for some guidance on the origin of coal:

> I should be glad to know from some of you Gentlemen learned in Natural History and Philosophy the most probable theory to account for these vegetables (as they once were) forming part of a stratum, which dips into the Earth to our knowledge 60 or 100 yards deep, and for ought we know to the Centre! These various strata, the Coals included, seem from various circumstances to have been in a Liquid state, and to have travelled along what was then the surface of the Earth; something like the Lava from Mount Vesuvius…But I have done. I have got beyond my depth…I must bid adieu to them for the present, and attend to what better suits my capacity. The forming of a Jug or Teapot.[12]

Wedgwood was an honest man!

While we can agree with many of the different strands of Mokyr's argument concerning the general intellectual milieu of eighteenth-century Britain, we can also add the rapid development of the infant disciplines of pure and applied geology to the mix. Yet their influence on the course of the early Industrial Revolution were nugatory. Neither Newcomen nor Darby could remotely be described as paid-up contributors to Mokyr's *Culture of Growth* – one can search in vain for a single mention of mine drainage, charcoal or coke in his index. Yet they provided the true foundations to the Industrial Revolution and its subsequent adoption all the way to modern East and South Asia.

Social Consequences of Industrial Revolution

The factory system of manufacturing led to the concentration of labour and the replacement of hand-made goods by machines. Accompanying urbanization spread across Britain from the mid-eighteenth century as the chief externally recognizable social consequence of the Industrial Revolution. Whence came the growth of industry, and settlement in high places such as the moorlands of the north and south Pennines and South Wales, where abundant water power, coal, iron and base metal minerals compensated for the lack of other considerations like weather, pre-existing amenities, fertile hinterlands and so on. It is with this history of new settlement in mind that we can appreciate the general sentiments of archaeologist Jacquetta Hawkes in the 1940s on her particular journey out from central Leeds to the south-west into the surrounding moor-top suburbs and the neighbouring mill town of Batley. She writes:

> As mills, factories, foundries and kilns multiplied, the little streets of the workers' houses spread their lines over hills that belonged to wild birds and mountain sheep, and up valleys where there was nothing busier than a rushing beck…The results were grim, but sometimes and particularly in the Pennine towns they had their own grandeur. Where houses and factories are still built from the local rocks and where straight streets climb uncompromisingly up hillsides, their roofs stepping up and up against the sky, they have a geometric beauty that is harsh but true, while the texture of smoke-blackened lime or sandstone can be curiously soft and rich, like the wings of some of our sombre night-flying moths. Nor do such cities ever quite lose the modelling of their natural foundations. On my first visit to the industrial north I rode on the top of a tram all the way from Leeds to Batley and all the way I rode through urban streets. In the last daylight it seemed a melancholy and formless jumble of brick and stone, but as darkness closed and a few smoky stars soothed and extended my thoughts, the lamps going up in innumerable little houses restored the contours of hill and dale in shimmering lines of light.[13]

In other coal- and iron-rich coastal and deep-water estuarine locations like Clydeside, Tyneside, Wearside, West Cumbria and Teesside, new population centres grew up to exploit cheap energy and steel production for shipbuilding, locomotive manufacture, and transport infrastructure like steel bridges, as in Middlesbrough. The water-powered mills of Lancashire fuelled the initial eighteenth-century growth of Manchester and its dozens of satellite textile towns with further growth as cheap coal transported by canal encouraged the spread of portable steam power provided by James Watt's rotary engines to drive mill machinery. Urban terraced 'back-to-backs' and new street networks snaked out from valley bottoms along and over steep slopes. As expansion continued, new settlements nucleated at canal nodes, and by the 1840s along burgeoning national and suburban railway routes and, later, along tramways and around suburban underground stations. Yet in the midst of all this untrammelled and often cramped and squalid urban expansion there were notable efforts made by philanthropic-driven employers to ameliorate industrial housing, notably in the 'model industrial villages' of Saltaire, New Lanark and, later, Port Sunlight, to name but a few.[14]

A corollary of the Industrial Revolution in England was a near-parallel Agricultural Revolution from the late seventeenth to eighteenth centuries, the industrial changes lagging the agricultural by several decades in England, more in Wales and Scotland. Increased agricultural production was also enhanced by mechanization and machinery. This marked the onset of rural labour-shedding during the manpower and food crises of the long (22 years) Napoleonic wars and their aftermath. Initially, horse-powered machines were invented for drilling and threshing. Technically advanced one-man/ two-horse ploughs were rapid and deep-turning – James Hutton took one from Norfolk (along with a skilled horse ploughman) for use in his Berwickshire farm improvement initiatives of 1754–64.

Much of the new agricultural machinery replaced labour-intensive winter work such as hand-threshing, with the result that rural labourers on eastern and southern arable farms were deprived of such work. As a consequence they were seasonally 'put on the parish', i.e. liable for relief from parish overseers under the Poor Laws. Numbers rose alarmingly after the wars from 1816 onwards, generating sporadic rural unrest, machine-breaking and incendiarism.[15] Rural depopulation by both internal and external emigration was the eventual result, reinforcing the trend to urbanism across the breadth of the lands. In Scotland the Highland Gaels suffered the dire consequences of societal breakdown and clearances that carried on into the century after the vicious aftermath to the failure of the second Jacobite rebellion. Together with victims of Southern Scottish economic clearances[16] and the arrival of hundreds of thousands of starving Irish in the 1840s, they joined a proletarian workforce that satisfied the rising economies of midland Scotland's burgeoning coal, iron and steel industries.

All-in-all there was a complete reorganization of British society based upon mechanization and the centripetal concentration of people in new cities and towns. The inexorable growth of London as the world's greatest financial and mercantile city has already been mentioned. Urbanization also encouraged the emergence by the mid-nineteenth century of education by self-study in municipal libraries (helped later by Andrew Carnegie), polytechnic colleges, mechanical institutes and a huge number of scientific societies across the land. Proud industrial cities and towns built their beautiful civic halls in a variety of architectural styles – the mid-Victorians had an extraordinarily eclectic 'jackdaw' tendency in architecture. The many outbreaks of cholera led to an increasing awareness that municipal water supplies needed to be secure and untainted at source, leading to the construction of numerous major reservoirs in both adjoining and distant ultra-urban upland catchments.

Educational provision by central government and local authorities improved through various education initiatives. First was Forster's Act of 1870, which required universal education from the ages of five to thirteen. Provision of secondary and higher education was established by Butler's wartime act of 1944, a companion to the Beveridge social provision and health reforms that followed peacetime in 1945. Ordinary families were now able to send their children into secondary then (for just a few) into higher education, giving them the chance to escape the necessity of following their fathers down the local pit or their mothers to the local mill. As put by Dave McDougal, a volunteer and ex-mining engineer at the National Mining Museum of Scotland, a Prestonpans man from a family with a long tradition of mining at Preston Grange: 'For the first time we had a choice...'

From Opposition to Organization

Though the spread of industry in the eighteenth century was remorseless, there were often checks. Landowners could refuse to lease or sell land, though in upland Coal Measures basins it generally had low rental value – in most cases money talked, and industry spread willy-nilly. Into such new developments migratory labour entered freely with none of the restraints that the Poor Laws exerted elsewhere, for instance in southern and eastern England. Mobile labour might have been expected to lead to wage competition, but the early organization of proprietors into informal local or regional cartels stymied this possibility. Nevertheless, as documented in Scotland by historian Robert Duncan,[17] there were countless local struggles for increased wages in pits, factories and mills throughout the eighteenth century, exacerbated during times of food (chiefly bread) shortages brought on by poor harvests.

Workers could not organize into co-operative unions after passage into law of the Combination Acts of 1799–1800, but even when repealed in 1825, strikes, picketing and collective bargaining were still illegal. Given this background, it was no surprise that direct violent action against machines or persons by discontented and frustrated labour, and by those dispossessed manual workers made redundant by machines, broke out at intervals. This was most noticeable in the early decades of the nineteenth century by the Luddite activities of weavers, croppers and finishers in the specialized, labour-intensive wool textile trades. Add to this the rise of radical republican movements ('Jacobins', à la Revolutionary France) then the course of economic history towards complete industrialization seemed in danger of stalling. Historian E.P. Thompson has written extensively on the (mostly) English experience and summarizes:

However different their judgements of value, conservative, radical and socialist observers suggested the same equation: steam power and the cotton-mill = new working class. The physical instruments of production were seen as giving rise in a direct and more-or-less compulsive way to new social relationships, institutions and cultural modes. At the same time the history of popular agitation during the period 1811–1850 appears to confirm this picture. It is as if the English [British!] nation entered a crucible in the 1790s and emerged after the [Napoleonic] Wars in a different form….It is, perhaps, the scale and intensity of this multiform popular agitation which has, more than anything else, given rise (among contemporary observers and historians alike) to the sense of some catastrophic change.[17]

Yet that catastrophe, i.e. bloody revolution, did not happen, much to the lasting regret of Engels and Marx and their once-numerous acolytes. There came a Second Reform Act in 1867, a Trade Union Act in 1871 that established basic union rights in law, and a Third Reform Act of 1884. Workers organized into regional associations, then national unions. Collective bargaining became the norm, with memorable national strikes such as the great miner's strike of 1912 and the general strike of 1926. The advent of universal suffrage had enabled workers in industrial constituencies to lobby parliamentary candidates to support their causes. A party to represent labour arose in 1893 – Keir Hardie's Independent Labour Party, dedicated to achieving political power by peaceful means. Unions of workers and operatives and the principles of democratic socialism led to collectivization and eventually, after two world wars and the attainment of universal suffrage, the National Health Service and widespread nationalization of three key industries: mining, railway transport, iron and steel production.

Human Consequences of the Industrial Revolution

To begin with, consider the appalling record of casualties in the early mining industry due to explosions of methane gas, such as at Felling Colliery, near Gateshead that killed ninety-one souls, including sixteen boys aged twelve or under (one only eight years old):

About half past eleven o'clock in the morning of the 25th May, 1812, the neighbouring villages were alarmed by a tremendous explosion in this colliery. The subterraneous fire broke forth with two heavy discharges…a slight trembling, as from an earthquake, was felt for about half a mile around the workings; and the noise of the explosion, though dull, was heard to three or four miles distance and much resembled an unsteady fire of infantry. Immense quantities of dust and small coal accompanied these blasts…it caused a darkness like that of early twilight, and covered the roads so thickly, that the foot-steps of passengers were strongly imprinted in it.[18]

The colliery had two years previously just finished mining the High Main seam and had begun to exploit the underlying Low Main, Bensham and other 'sour' (i.e. sulphurous) seams down to a depth of around 250 metres. That Tyneside coals were notoriously gassy is borne out by the observations made just one year earlier by Louis Simond (Chapter 6).

It was the Felling disaster that led George Stephenson to design his safety lamp, the Geordie Lamp and, after later explosions, to Humphrey Davy designing his own version, with the help of the young and eager Michael Faraday.[19] In the widely adopted Davy lamp, the flame was surrounded and protected by closely knit iron gauze that could, nevertheless, freely admit methane (fire damp), which then burned harmlessly inside the lamp. The height of the luminous cone of burning methane above the flame gave some idea of its concentration in the mine's atmosphere.

Industrial workers gradually gained much-needed protection from exploitation through many government acts set in motion by devoted reformers. The Great Reform Bill passed in 1832 did nothing for the representation of the common people, but in that *annus mirabilis* of 1833 not only was slavery itself abolished throughout British overseas territories, but the first effective and generally operative Factory Act was passed. It followed a Royal Commission whose findings of the appalling treatment of child workers in factories horrified the nation. Here are Lord Shaftesbury's observations from Manchester:

> Thirty-five thousand children, under 13 years of age, many not exceeding 5 or 6, are working, at times, for 14 or 15 hours a day, and also, but not in these works, during the night![20]

Leeds-based doctor Charles Thackrah (see the quotation heading Chapter 20) provided medical evidence for Sir Michael Sadler concerning the results of the exposure of child workers in Leeds mills (in this case Marshall's great flax mills) from an early age. Here are the words of a modern biographer:

> Of Marshall's workforce of 1079, Thackrah could find only nine people over 50 years old and 22 who had reached the age of 40. For Thackrah, however, the worst aspect of the factory age was the exploitation of children, who, from the age of seven, were expected to work in a dusty atmosphere from half past six in the morning until eight at night, six days a week, for a pittance. Michael Sadler quoted from Thackrah's book in the House of Commons as he struggled to persuade Parliament to limit the hours of labour for such operatives: 'No man of humanity can reflect without distress on the state of thousands of children, many from six to seven years of age, roused from their beds at an early hour, hurried to the mills, and kept there, with the interval of only 40 minutes, till a late hour at night; kept, moreover, in an atmosphere impure, not only as the air of a town, not only as defective in ventilation, but as loaded with noxious dust.'[21]

And, distressingly so, on and on.

Yet nothing was done about the dire conditions of mine-working children for another nine years. It took a tragic accident at the Huskar Colliery at Silkstone, Barnsley in which twenty-six children, including eleven girls, aged between seven and seventeen were drowned in a ventilation adit as they tried to exit the colliery after the winding engine had become inoperative during a torrential rainstorm. A subsequent and famous Royal Commission report published in 1842 revealed the wretched lot of the tens of thousands of such child labourers in mines, with graphic images of working conditions and witness statements containing details of long solitary hours spent underground in dark, often wet passages performing repetitive tasks or back-breaking manual labour. Lord Ashley, later the seventh Earl of Shaftesbury, sponsored a Mines and Collieries Bill later in 1842 that prohibited all underground work for females and for boys under ten. Subsequent legislation sought to reduce the accident rate in mines by appointing mine inspectors under the remit of the Home Office and by setting stricter standards of underground practice. Yet by 1872 there were still over one thousand lives lost yearly in mining accidents alone. In 1881 a Mines Regulation Act was passed that required the Home Secretary to hold inquiries into the causes of all mine accidents.

The health problems induced by mining, chiefly due to the accumulation of particulate matter in the lungs, were eventually recognized as occupational hazards whose diagnosis established that the person so-injured was entitled to compensatory payments. It was not always thus. For many years (since Edwardian times) it was held by many in the mining business that it was impossible to tell whether lung damage was due to miner's heavy tobacco smoking or to coal and rock dust inhalation. It was only after the First World War (in Cardiff) and following nationalization in 1947, that serious epidemiological research on the problem was carried out in centres for occupational medicine (most importantly in Edinburgh).[22] Tens of thousands of miners were regularly monitered by radiographs, and, *post mortem*, by lung tissue histology. Together with measurements of fine-particle dust concentrations in hundreds of mines, it was established statistically that fibrosis due to both pneumoconiosis and silicosis were indeed products of passive dust inhalation underground. The latter was more deadly because of the abrasive nature of the tiny quartz dust particles present – inhaled when miners operating mechanized cutters sawed partly through bedrock rather than just pure coal.

Unforseen Consequences: The Road to British Mining Extinction[23]

Victorian and later politicians and economists were obsessed with the likely magnitude of British coal reserves and the likely date of their inevitable exhaustion. Here are a couple of examples:

> There is still a quantity of coal in store in England and Wales, sufficient to afford a supply of 60 million tons for about 1000 years.
>
> Edward Hull, 1861. *The Coal-Fields of Great Britain: Their History, Structure, and Duration. With Notices of the Coal-Fields of Other Parts of the World*. London: E. Stanford

This island is made mainly of coal and surrounded by fish. Only an organizing genius could produce a shortage of coal and fish at the same time.

Aneurin Bevan, MP for Ebbw Vale, House of Commons speech, mid-1940s

My former boss at the University of Leeds, the canny and garrulous Welshman, Howel Francis, was fond of recounting that in his days in the 1970s, as the most senior officer in the Geological Survey in Northern Britain, rarely a year would go by without a request from some Westminster politician or civil servant for a definite re-statement of coal reserves. Howel would raise his eyebrows and go away to look up the original estimate first dreamt up by a Victorian predecessor (probably the Edward Hull of the aforementioned quote) that had many noughts in it. He would subtract subsequent production (known exactly) and issue a precise estimate of the remaining tonnage and the likely time of exhaustion of reserves. Given its apparent super-abundance, no-one thought to ask whether British deep-mined coal could continue to compete in an open market in an age when hydrocarbon was rampant and sea-going transport by super-freighter would eventually enable massive cheap coal imports across the world's oceans.

Following the post-war era of cheap oil tankered in from the Middle East, the mid-1960s signified the 'all-change' for the industrial and domestic habit use of coal in Britain. This was thanks to yet another geological provender – massive offshore hydrocarbon resources on the British continental shelf. In the autumn of 1965 a British Petroleum drilling rig in the shallow waters off the East Yorkshire coast struck natural gas in what was to become, by 1967, the West Sole gasfield. Scores more, some giants, were subsequently discovered. Then, decisively (from 1975), oil fields were discovered and developed in the Central and Northern North Sea. Over the next fifty years the North Sea became by far the largest offshore oil and gas industry in the world, with Aberdeen as the new Houston and Scotland as the new Texas. Successive governments of both political persuasions blew much of the tax revenue from this huge domestic resource, failing to create any lasting national wealth fund and using the royalties to fund dole money for the masses of unemployed their own economic policies had simultaneously created. Offshore production immediately brought a decrease in coal use. Eventually, in the absence of any accompanying coherent national energy policy, environmental conservation or mitigation strategy, came the extinction of the entire deep-mined coal industry.

That process took almost fifty years as once-dominant coal-burning 'smokestack' industries and domestic users switched to hydrocarbon, or the industries were swept away in governmental economic cullings. But coal hung on, burnt as essential metallurgical coke in steel-making (Fig. 13.2). More importantly, coal-fired power stations sourced around seventy-five per cent of the national power grid system that had developed since the Second World War. Nevertheless, many older mines of the nationalized Coal Board became uneconomic according to their own internal standards, and in terms of the price of coal on the world market. It was now make or break time for a large chunk of the industry in ageing pits. The deciding question was whether market rules should govern, leading to the immediate closure of uneconomic pits, or whether the government (and/ or the EU) should step in with aid. The latter course (followed in the past decade in Germany) would have allowed both external subsidy (taxpayers' money) and internal transfer (super-pit profits supporting old pits' losses) to keep mining going in managed decline, enabling worker relocation, retraining and community reconstruction before eventual closure.[24]

It didn't work out the second way. A vengeful political backlash occurred after the successful 1972 and 1974 miner's strikes over pay. The incoming 1979 Thatcher administration had set up thorough contingency plans for action to counter such strikes. In the subsequent ill-timed and divisive (over balloting) strike over mine closures, the striking miners were led to ignominious defeat by Arthur Scargill. Closures were pressed on with, so by the time of deep-mining privatization in 1994, only fifteen deep mines remained out of the pre-strike total of one hundred and seventy. With the closure of Kellingley in the Selby coalfield in 2015 the deep-mine industry was extinct.

Down to the present day there exists a legacy of divided families and communities from the consequences of mine closures. Unemployment, reduced employment opportunities and social deprivation raise their ugly faces. Consider the following 2016 statement from the head of the Coalfield Regeneration Trust:

> There are 5,500,000 people living in Britain's former mining communities…
> The greatest challenge is the fact that there are only 50 jobs per 100 working age people in the coalfields. When you compare this figure to London (79), and the South-East (68), it doesn't take much to recognize that there is a major problem here.

> Coalfields Regeneration Trust Supplement: *New Statesman*, 2016

Today, under landscaped spoil tips and verdant countryside are thousands of kilo-metres of collapsed and flooded underground passages in hundreds of abandoned mines. At the surface huge and ghostly concrete power station chimneys and cooling towers still stand awaiting explosive demolition. Former miners' terraced dwellings and younger new-build 60s–80s estates, sometimes on the outskirts of medieval villages, are now stranded and seemingly out of place in their surrounding, apparently timeless, rural landscapes.

Yet, lest it be forgotten, mining was (and is, elsewhere) a skilled and communal occupation with an emphasis on safety-first from the miners themselves. In past days of muscle power, both man and beast put their shoulders to the task. Even in the late 1960s, a scene such as the following described by Henry Tegner at Seaton Burn, Northumberland on the A1 trunk road north of Newcastle has the ability to shock, and at the same time to celebrate, both of the creatures involved – the miner and the pit pony:

> Early one morning on the way to Newcastle upon Tyne, in the near-darkness
> of a late winter dawn, the traffic was brought to a halt by a uniformed figure

Figure 13.2 The Port Talbot Banksy: *Season's Greetings.* A typically wry image from Banksy commenting on both the local and wider scandal of urban air quality in today's Britain – an innocent child welcomes the arrival of winter snow, which turns out to be a fallout of sedimenting ash from a burning pollution source, i.e. the nearby steelworks. Source: Wikimedia Commons File: Seasons Greetings, Banksy (3). Jpg. Author: Fruit Monkey.

within the limits of Seaton Burn RDC. Through the quick flicker of the windscreen wipers, the man who brought the car to a standstill looked like a sort of moon man, only he was pitch black instead of white. His skull hat was shining in the steady sleet, his face was coal-stained with great white orbs around the eyes. His clothes were dark in colour whilst his feet were shod in a pair of massive steel-tipped boots. The procession the pitman was guarding now began to file across the A1. Each animal had a man at its head all dressed exactly like the traffic controller. The creatures they led were a quite astonishing sight. Little fat dwarf quadrupeds each shaven to the skin, their manes hogged and their tails shorn so that they looked like the trunks of miniature elephants. The pit ponies were all harnessed and their eyes blinkered against the dangers of debris and sudden daylight. The file of men and ponies drifted steadily across the road from pit-head to pit-head. There were sixteen of them, queer creatures from the earth's bowels spewed out across the modern motorized highway like some undulating mystical dragon.[25]

I defy anyone not to feel pity for those ponies, and both pity and admiration for the miners who once cared for them.

Chimneys of the Modern World: von Richthofen's 'Planetary Catastrophe'

The cotton of India is conveyed by British ships round half our planet,
to be woven by British skill in the factories of Lancashire: it is again set
in motion by British capital; and, transported to the very plains whereon
it grew, is repurchased by the lords of the soil which gave it birth, at a
cheaper price than that at which their coarser machinery enables them
to manufacture it themselves

Babbage, C. *On the Economy of Machinery and Manufactures*, p. 4, his section 2 (1835)

General

Andreas Malm's definition of a 'Fossil Economy' involves self-sustaining growth based upon increasing consumption of fossil fuel ('fossil capital') and, inevitably, a parallel growth in carbon dioxide emission.[1] At the same time he recognizes that power as a capital resource can be obtained from sustainable sources. That coal is fossil capital was originally of no consequence whatsoever to society in the way it viewed its trajectory towards industrialization. Few persons joined Charles Babbage in wondering what future effects might accrue with the continued oxidation of fossil carbon reserves. Now, as we have seen, we do know. The onward march of fossil fuel consumption can be compared to a plague-carrying bacillus: vectored by innocent carriers in the name of industrial progress and human emancipation. It carries the global affliction of environmental instability that took two hundred and fifty years to come to realization.

Yet Malm reminds us in a timely fashion that the road taken to fossil capital exploitation was not the only choice available. There was another path, less trodden, but sufficiently economically advantageous in many cases. This was via sustainable water power, which stood in direct competition with coal well into the middle of the nineteenth century in much of Britain's upland textile and metal industries (Fig. 14.1). Nowadays we

Figure 14.1 A–C A legacy of past industrial water power reconnected at the former steel furnace, forge and edge-tool works of Abbeydale Industrial Hamlet, Sheffield, South Yorkshire (53.200244, 1.304344). **A** The large hammer pond dammed in 1777 off the River Sheaf. Water was let into two wheel races through separate sluices visible to the left. **B** The fully restored and working high-breast shot wheel that powered the forge tilt hammers. **C** The preserved overshot wheel to the left drove the blowing cylinders for the furnace and forge, undergoing restoration work in June 2019. The wheel paddles seen in B are visible in the opening by the steps up to the grinding hall on the right. Source: Author.

must add tremendous technological breakthroughs – power storage and transmission, solar panels, lightweight strong materials and advanced design of driving mechanisms by wind and water. The tired old water-wheel can come to symbolize a mighty future of sustainable power from wind, tide, river and sun.

Historical Global Economies

History and archaeology teach us that there were numerous literate, highly advanced economies in the pre-industrial world. 'Advanced' here means tightly administered states or empires whose civil servants transferred handed-down laws and directives in writing from a ruling elite. They went to magistrates and enforcement officers amongst the general populace in organized, efficient, though not necessarily virtuous, ways. Prominent among them were the civilizations of Mesopotamia, Pharaonic Egypt (till Cleopatra), Assyrians, Hittites, Minoans and Myceneans; various Persian, Chinese, Indian, Hellenistic, Roman, Central and South American dynastic states; late-medieval to modern-era empires such as those of Spain, Portugal, Netherlands, Britain, France, Germany and Belgium. In such economies the inward and outwards flow of transported food, geological raw materials (precious metals, ores, building stone, etc.) and finished goods were recorded and taxed appropriately, the management of natural resources similarly. Their general attainment of agglomerated civil, mercantile and creative societies might let us use the word 'civilized' in reference to them, though this description is of necessity amoral – none were democratic or accountable in the modern sense.

By such organizational means, states and empires could survive hard times, though, as history and archaeology teach us, none managed to survive forever the scourges of human life – vicissitudes as various as climate change, volcanic eruption, earthquake, tsunami, famine, epidemic and war. Two such societal extinctions and their subsequent reincarnation in modern times are relevant to the history of exploitation of industrial mineral resources and fossil capital. Both occurred during the history of regions now forming the largest and most populous nation-states of Asia – India and China. Both currently feature in the world 'top-five' of coal producers, coal users and pollutant emitters, China being the undisputed leader in each category.

Pre-Industrial Textiles and Economy of India

India had the misfortune that parts of it, chiefly in what was then the populous province of Bengal, became annexed by the British during the early expansion of what was later to become its Indian empire.[2] During the Mughal hegemony from the sixteenth to the eighteenth century, Bengal had built up hugely productive textile manufactories. These were integrated village-based crafts centred on 'tree-cotton' – its harvesting, separation in roller gins, wheel spinning, hand-loom weaving, dyeing, fixing, designing, patterning and cutting. It was the largest pre-industrial cotton-based economy in the world, estimated to constitute an astonishing twenty-five per cent of early eighteenth-century trade. It supplied dyed and undyed calicos, muslins and other cotton cloths to a world hungry for light and durable fabrics to accompany the north European staples woven from wool, flax and hemp.

There was also an iron and crucible steel industry based on the hematite ironstones mapped by those first young geological surveyors of the Indian Geological Survey noted in Chapter 8. Thomas Oldham, the first Superintendent of the Survey, wrote the following conclusion in its first bulletin of 1859 concerning the area's potential for future iron production using hematite ore smelted with charcoal culled from coppiced hardwoods of the extensive jungle forests:

> But it may, I think, be safely stated that there is no sound reason why iron should not, under proper management and in a well-selected locality in the district, be manufactured at a cost equal to, if not less than, that at which it is produced in England. Putting aside all other considerations, therefore, the cost of freight of English iron, which may be taken to average £1 per ton to Calcutta [Welsh pig-iron cost only 1.7 times as much to produce at the time], will be saved, or will be, in fact, so much more additional profit on the manufactured product, supposing it to command the same price in the market.[3]

Why then, ninety years previous to this, did not independent Bengali India itself develop the first Industrial Revolution? Why did Calcutta not become the mechanized 'Cottonopolis' of the world instead of distant Manchester with no indigenous supplies of raw cotton and no tradition of spinning or weaving it?

The short answer is that in Moghul India there was no shortage of cheap labour. Further, the rural Bengali cotton manufactories and Cuttack steel furnaces suffered the twin fate of technical competition and armed invasion by an aggressive, acquisitive and unforgiving foreign power – the effects of the 1750s annexation of their province by Major-General Robert Clive. The technical competition in textiles came from the mechanized mills of Lancashire and Lanarkshire, which could outprice them. In the same way that home-grown industrialization dealt the British rural woollen and linen hand-loom clothiers a factory-led route to extinction, Bengal's village industry was likewise destroyed by the East India Company's administrators, soldiers, mercantile marine, and by a largely partial British Parliament, many of whose MPs, like Edmund Burke for example, were shareholders. After Clive's victory at the Battle of Plassey in 1757 the Company obtained exclusive taxation rights that virtually eliminated the hand-loom weaving tradition by levying taxes and laws against the export of Bengal-produced textiles (but not its raw cotton; that was wanted for Lancashire). This was in addition to the longstanding British prohibition on the import of cotton cloths – only repealed in the 1770s, by which time the mechanized British mills could compete with imports of hand-woven stuffs from any source.

Meanwhile, rural poverty, famine and epidemic hung their long shadow over the usually watery riverine and deltaic plains of lowland Bengal. An astonishing ten million are estimated to have died in the great famine of 1770 after the monsoon had failed the previous year, and in its aftermath in the next few years. The East India Company was ill-prepared to ameliorate famine from its depleted rice stockpiles after thirteen years of often violent and extortionate taxation practices. Over the next decades the Bengal cotton export trade was wholly replaced by imported cloth from Lancashire, as celebrated by Babbage in the chapter's opening quotation. This trade eventually switched to long-staple cotton grown in slave-worked cotton plantations in the southern parts of the then North American colonies. Such was the moral, historical and economic legacy gifted by Britain to both the nascent American republic and to its own 'Jewel in the Crown' of Empire.

It is a great irony of history that in modern Britain the remnants of its once-proud iron and steel industry are now Indian-owned, smelting imported ores, some from India, with imported coke ditto. At the same time the mills of Lancashire and Yorkshire towns are now defunct or derelict, demolished or converted to other uses, while the largely female labour forces of Bangladesh and West Bengal turn out cheap cotton goods to Primark and the whole world. India now stands fifth in the world league of coal-producers; the ever-expanding Indian steel and other industries make use of fossil capital via colossal coal production from its Gondwanan Permian coalfields.

Industrialization and the Re-Invention of the Coal Economy of China

China today uses far more coal than any other nation, almost half of world production. Yet it is not widely known that the modern (post-1950) state has a startling coal-fuelled legacy from its own proud history as the oldest and most peopled nation in the world. This came about during the remarkable 300-year duration of the Song Dynasty[4], from 960–1279, a period coeval with late-Saxon to Angevin England and Dunkeld Scotland. During this time an agricultural revolution expanded early-crop rice cultivation, generating food surpluses and stocks to tide over lean years, leading to the eventual support of over 100 million citizens. The thorough governing of these people was by monarchical and aristocratic fiat above a large, centralized government bureaucracy. The Song pioneered the development of administrative, fiscal, economic, mercantile, naval, philosophical, artistic, scientific, engineering and technological initiatives and discoveries. These led to an increasingly urbanized and literate population that included a large and learned gentry class, whose rise into the administrative machine was controlled by the attainment of a broad education and a system of national examinations into the civil service. This pattern of development is familiar to us, for it uncannily echoes the trajectory of the civil service within the British nation state and its empire.

The Song also had their own industry, based on the use of water power and on the combustion of its abundant Carboniferous coal in their unique porcelain kilns. Marco Polo, travelling in northern China in the early decades after Kublai Khan's takeover from the Song (around 1270), gives us these observations in this 1903 translation by Henry Yule:

> It is a fact that all over the country of Cathay there is a kind of black stone [anthracite] existing in beds in the mountains, which they dig out and burn like firewood. If you supply the fire with them at night, and see that they are well kindled, you will find them still alight in the morning; and they make such capital fuel that no other is used throughout the country. It is true that they have plenty of wood also, but they do not burn it because those stones burn better and cost less.[5]

Yule was an experienced Colonel of the Royal Engineers with geological field knowledge gained in his long and eventful career. He notes some prescient remarks made in the 1880s by the distinguished oriental explorer and geologist Baron Ferdinand von Richthofen, the coiner of the term Silk Road (*Seidenstraße*), also the uncle of the 'Red Baron' fighter ace of the First World War. He wrote concerning the scope and likely impact of the exploitation of Chinese coal reserves. Yule writes (quotes from von Richthofen in italics):

> There is a great consumption of coal in Northern China, especially in the brick stoves, which are universal, even in poor houses. But the most important coal-fields in relation to the future are those of Shantung, Hunan, Honan, and Shansi. The last is eminently the coal and iron province of China, and its coal-field, as described by Baron Richthofen, combines, in an extraordinary manner, all the advantages that can enhance the value of such a field except (at present) that of facile export; whilst the quantity available is so great that from Southern Shansi alone he estimates the whole world could be supplied, at the present rate of consumption, for several thousand years. *Adits, miles in length, could be driven within the body of the coal....These extraordinary conditions...will eventually give rise to some curious features in mining...if a railroad should ever be built from the plain to this region...branches of it will be constructed within the body of one or other of these beds of anthracite.* Baron Richthofen, in the paper which we quote from, indicates the revolution in the deposit of the world's wealth and power, to which such facts, combined with other characteristics of China, point as probable; *a revolution so vast that its contemplation seems like that of a planetary catastrophe* [author's italics here].

Prescient indeed, written by Yule only eighty years or so before Deng Xiaoping's economic reforms of the 1980s that changed the world economic order. During Song times anthracite (bright, smokeless, with a high calorific value) was mined along adits as a cheaper but better fuel than wood to heat the porcelain kilns. Iron, on the other hand, was smelted with charcoal. Canal transport incorporated pound locks taking the pottery and metals from the sites of blast furnaces, puddling sluices, smelters and kilns to centres of urban population, and for export far afield. Water-wheels were used to power the giant bellows used during furnace and kiln operations, enabling both the high temperatures needed for porcelain and steel production, the latter including specialized forms used in sword blade manufacture. Smelted copper formed the basis of a minted currency.

Questions arise, more so than with India. First, given its massive reserves of coal and technological knowledge, could Song China have initiated an early Industrial Revolution based on steam-powered machinery? This is not a new question. It has been examined by Western and Chinese economic historians. The fact that there was no real mechanized industrialization might partly reflect the dynasty's demise during the northern invasion by Khublai Khan and his Yuan Dynasty successors. Even so, during Song times there was no real economic mover that might have intervened to shift the country to steam power, nor to the use of coke or to the factory system of manufacturing. In reality, the answer hinges on the geology. Judging from accounts like those of von Richthofen, the Shaanxi mines had simple geology and were adit-excavated into hillsides, so that the need for draining deep shaft mines did not arise, in contrast to those in Britain where the geological structure was more complex and the drainage problem acute. Hence no Chinese Newcomen thought out how he or she might mimic the action of the common village hand pump by making use of steam power. That meant, in turn, that there could be no Chinese Watt who could transfer portable coal-sourced piston power into rotative and linear motion to drive pumps, wheels and propellers.

The second question is why does China currently produce so much coal? The answer is that in the early 1980s coal was the country's only source of abundant power capable of dragging the country quickly back from destitution after the ravages of Maoism and the 'Gang of Four'. So, once Deng Xiaoping decided to trust the companies of the western world to deliver factory-based industrialization, indigenous cheap, coal-based electrification and labour opened the floodgates. All the western industrialized nations subsequently opened factories in this new China, shifting production ruthlessly from high wage, high energy cost North America and Europe. All this was knowingly and ruthlessly done by brand-name businesses as they sacked their native workforces and decamped production wholesale to East Asia. They called it, and still do – globalization. It was supported by western politicians as an article of faith. In short, China was industrialized by western (largely USA and EU) migratory capitalism and fuelled by coal-produced power.

Fast-forward now to the present day. Chinese coal production in 2017 is reported as 3.52 billion tonnes, just under half of world production and some five times that of India and the USA, the second and third producing nations.[6] Although coal in China is found in strata of all geological periods from the Carboniferous to the Tertiary, three-quarters of reserves are in the thick bituminous and anthracitic seams of the Late Carboniferous and Early Permian age in the north of the country. None of this is exported, for the country has only small reserves of hydrocarbon (though frackable reserves may soon be accessed) and is by far the world's largest importer of oil. China's moves towards annexation of large chunks of maritime south-east Asia over the past decade must be seen in this light – hydrocarbon fields might well lie in the offshore basins. It also imports some 300 million tonnes of coal per year, presumably of low-sulphur metallurgical-grade products for use in coke-smelting. All these figures may be 'just numbers' for many readers of this book, but the comparative point being made is the enormity of carbon emissions from coal-burning by this one country. For some historical comparison, China's 2017 production is around twelve times that of the total UK production in its peak year of 1913. Further, most of China's electricity (and some 40% of that of the USA and Germany) is coal-produced. There is little chance of the ongoing electric car revolution ameliorating the atmospheric carbon dioxide threat to world climate anytime soon – a depressing but unavoidable fact.

Coda: What's in a Name – 'Anthropocene'?

Homo sapiens has been around for about fifty thousand years. In the last three hundred of these it has managed to invent both a means of altering the ecology and climate of the planetary surface and, more recently, destroying itself with nuclear weapons. The regular annual increase in atmospheric carbon dioxide over the past nearly sixty years is well able to be recorded in the polar ice and cave speleothem deposits laid down over that time interval. If there is to be a post-carbon industrial age to come, with dominant use of renewable power, then the period of time since the early eighteenth century will perhaps deserve a special name. The current proposed favourite is 'Anthropocene'.[7] However, any long-term definition seems to outlaw entirely innocent pre- and early-industrial humans who knew nothing of the carbon dioxide problems that would turn up in the future – it is rather a shotgun term for a species that might just solve its own dilemmas. A shorter definition (I paraphrase several authors) lays the blame where it should be laid, at modern-day industrial nations continuing to practise fossil fuel capitalism. It places continuing blame with the living souls who gainsay the facts and consequences of increased carbon dioxide in the late-Holocene global atmosphere. We live in:

> a period of time when Earth surface and atmospheric processes are strongly influenced by the economic activity of humans exploiting a colossal global fossil fuel economy, its onset conveniently recorded by a spike in radiation fallout from the US Army's Trinity nuclear test of 1945 and recorded worldwide in deposited sediments.

How long this period of time will last, though, and whether our species will recognize its eventual going, are other matters entirely.

15

Industrial Sublime and Other Creative Legacies

History can tell you what happened, but the
arts and music tell you what it felt like.

Presenter and musician Kathryn Tickell on her BBC Radio 3 programme,
Music Planet, 9 November 2019

Industrial Sublime

Consider the following fancies envisaged by Hannah Darby, eighteen-year-old
daughter of Abraham Darby II, in a 1753 letter from Coalbrookdale to her aunt,
Rachel Thompson. She readily expresses her fresh reactions to the sight of her
father's blast furnaces and associated machines working in the rural setting of wider
Coalbrookdale. Hannah is expressing herself in terms of what would come to be
referred to as 'sublime':

> Methinks how delightful it would be to walk with thee into fields and
> woods, then to go into the dale to view the works; the stupendous bellows
> whose alternate roars, like the foaming billows, is awful to hear; the mighty
> Cylinders, the wheels that carry on so many different Branches of the work,
> is curious to observe; the many other things which I cannot enumerate; but
> if thou wilt come, I am sure thou would like it. It's really pleasant about our
> house, and so many comes and goes that we forget it's the Country till we
> look out at the window and see the woodland prospect…[1]

This charming account of rural heavy industry was written four years before
publication of Edmund Burke's *A Philosophical Enquiry into the Origin of our Ideas of the
Sublime and Beautiful.*[2] Although our common modern day-to-day usage of the word
sublime differs somewhat, the Oxford English Dictionary defines a particular usage 'of
things in nature and art' as:

Affecting the mind with a sense of overwhelming grandeur or irresistible power; calculated to inspire awe, deep reverence, or lofty emotion, by reason of its beauty, vastness, or grandeur.[3]

Burke himself originally considered that the central irrational passion felt in the presence of the sublime involved not an appreciation of beauty but of astonishment. He argued that this could be brought on by some combination of terror, obscurity, power, privation, vastness, infinity, magnificence, lighting, shading or sound. In general it was a painful and disjointing experience that was out of the ordinary. He found it sourced in nature by the celestial (starry heavens), meteorological (thunder and lightning, storms) and geological (high or deep rocks, cataracts, wild suggestive mountains). It could also be found, amongst others, in: human constructs including words, e.g. certain verses of the Bible, Virgil's Vulcan's cave in Etna, and in Milton and Shakespeare; paintings of places or landscapes (but only those with a certain 'judicious obscurity'); buildings (cathedrals); monuments (Stonehenge); in the terrifying fire of artillery pieces or the 'shouting of multitudes'. Quite a list: perfectly understandable to the modern mind, though perhaps not to the taste of all.

Burke's view of sublimity would certainly seem to permit the inclusion of the effects of the new industrial machinery that was appearing over Britain and that would continue to evolve for the next two hundred years. There were flights of canal locks reaching up gritstone edges, the light, smoke and sounds of blast furnaces and the creaking, groaning action of Newcomen engines. Yet Burke mentions no contemporary example, despite the fact that by mid-century machines had become common in the chief manufacturing districts – perhaps they were out of his metropolitan and cultural ken or interest.

Almost a century later at the height of industrialization, machines were also anathema to the influential critic John Ruskin. In a memorable essay on the history of the sublime in arts and letters, art historian Alison Smith writes of him and his legacy:

Admittedly, industrialization itself could be seen to offer another possible language of the sublime, in contrast to the grand natural scenes that had inspired the Romantics. However, Ruskin's antipathy to the incursion of industrialization into art – as exemplified by his comment that Turner painted *Rain, Steam and Speed* (1844) 'to show what he could do with so ugly a subject' – informed a general perception that the material conditions of the Industrial Revolution helped bring about a crisis in the sublime by degrading the natural world and proposing that it could be manipulated for utilitarian purpose and gain. The influence of the natural sciences, particularly geology, on painting and the concomitant idea that landscape as a genre should aspire to objectivity, as nature was measurable and capable of being defined in precise analytic terms, further problematized the sublime, undermining academic idealist theory and disavowing the possibility of overwhelming aesthetic experience.[4]

Industrial and mechanical processes clearly gave out more than a whiff of the sublime for many observers, yet it was not until recent times that the singularity of the term 'industrial sublime' could be applied to the work and products of industrial civilization. It is now widely recognized as a formal genre – from images of the industrialized banks of the 1930s Hudson River to twentieth-century power stations in Britain (think Bankside and Battersea), France and the USA. We read of a very early example of the industrial sublime in landscape from another visitor to Coalbrookdale, the agricultural writer and traveller, Arthur Young, who in 1776 'admired' the sublimity of the landscape he found there by directly adapting Burke's insistence that horror had a full and influential hold on sublimity (note also use of 'romantic' – an early example):

> Coalbrookdale itself is a very romantic spot, it is a winding glen between two immense hills which break into various forms, all thickly covered with wood, forming the most beautiful sheets of hanging wood. Indeed too beautiful to be much in unison with the variety of horrors art has spread at the bottom; the noise of the forges, mills, etc, with their vast machinery, the flames bursting from the furnaces with the burning of the coal and the smoak [*sic*] of the lime kilns, are altogether sublime, and would unite well with craggy and bare rocks, like St. Vincent's at Bristol.[5]

The experiences of both Hannah Darby and Arthur Young witness the early promulgation of 'industrial sublime'. Young was from a lowland background (Suffolk) so his vision of 'immense hills' might seem tame to many. Yet at this time the depiction of active volcanoes (natural blast furnaces in the visual sense) had also begun, chiefly through the work of Joseph Wright at Vesuvius and Etna. Their emissions of smoke and flame and thunderous eruptions of ash and lava bombs had sublimity in spades. It is to the original genius of Wright, Turner and other artists that we next turn.

Readings of Three Industrial Sublime Artworks

Joseph Wright's *An Iron Forge* (1772)[6] takes us into the light, heat and (imagined) noise of a mid-Georgian version of the hellish environment of the ancient Greek forgemaster Hephaestus (aka Vulcan in Rome). Here the physical work of converting a white-hot block of bar iron (brought from an unseen furnace) into an iron girder or strut by the action of a water-wheel powered trip hammer becomes a scene of almost domestic gentleness. This is beautifully engendered by the presence of the forgemaster's young wife, three pretty and well-dressed daughters and the family dog – the mother holding the youngest child in her best bonnet to face the viewer, with her oldest sister posed similarly holding the mother's back protectively. Together with the proud and smiling father (his muscular arms folded as if to say 'All mine!') the whole group are centred within an arc of intense light from the block of white-glowing iron held firmly by tongs on the anvil by a booted and aproned forge operative who has his back to us. To the left sits another male facing the forge, his right hand drawn upward to his face, perhaps the forgemasters's older kinsman to judge by the familiar grasp of the second daughter's arm as she rests with her back to the anvil on his (her grandfather?) bended right knee. Wright has mixed up the generations in witness to the scene.

It is then that we see the entire reason for the artist's arrangement set against the sturdy stone wall and under a massive roof beam. For tense dramatic effect he has chosen the very moment when the massive trip-hammer head is pulled almost to its apogee by the creaking upward motion of the rotary drum that bears the trip head. Forging hammers of this kind, where the hammer is supported between the head and its fulcrum (the latter not seen here, but along the shaft, leftwards) are known as 'belly-helve' hammers. In just a few moments the power engendered by the (unseen) external water-wheel (this is too early to be steam-driven) turning the drum axle will let the tonne or so of hammer-iron crash down upon the glowing block. Its stupendous kinetic energy will instantaneously draw out the bar somewhat during the time taken for the next trip head to rotate upwards under it and lift it off the bar, enabling the operative to slide the iron this way or that to achieve the task in hand. We can see the still smouldering remains of fragments of iron scale from the last trip excursion caught in the metal tray that surrounds the anvil base.

So, a subtle but undoubtedly sublime scene (we are scared for the children, astonished by the colours, lighting and forces at work) that gaily mixes industrial machine-energy, human strength and tender domesticity. The artist tells us that the bright radiant light and heat and the kinetic excursions of crashing blows are, in the end, both controllable and comforting.

Philip-James Loutherbourg's *Coalbrookdale by Night* (1801; Fig. 15.1) is a dramatic but also confused image[7] – a mix of Gothic mystery, industrial sublime and symbolism. In the metallic grey half-light of a full moon (not seen) an old mansion with curiously large round towers and tall narrow chimneys sits atop a rocky outcrop silhouetted by a backdrop of large heaps of coal being slow-burnt to coke in the open air. The coals periodically erupt into spectacular volcanic-like blazes that illuminate a group of high-pitched buildings with numerous square-sectioned chimneys, one of which is smoking gently above a visible brightly burning hearth enclosed by a Gothic arch. Could this be a smithy or is it a furnace complex? It is unclear. Indistinct figures are grouped to the left of the arch, one of whom seems to be carting a pile of material (perhaps fuel) towards the hearth. A T-frame to the right may be for winding gear, but the purpose of the adjacent pagoda-like tower escapes one.

Meanwhile in the foreground a team of two magnificent horses are moving powerfully away from the inferno down a wide track, watched by a female figure with a child and ignored by a central figure walking back up the track towards the coke fires. The horses are pulling a cart full of what might be logs, urged on by a busily striding man, whip in hand. A youth sits on the cart's side, motioning with his right hand clasping some bolt-like object. Horse power has given way to steam power. The artist has littered the

Figure 15.1 *Coalbrookdale by Night* (1801) by Philip-James de Loutherbourg (1740–1812). Oil on Canvas. 68x106.5 cm. Science Museum, London. Image source: Wikimedia Commons File: Philipp Jakob Loutherbourg d. J. 002.jpg

Figure 15.2 *Keelmen Heaving in the Coals by Moonlight* (1835) by Joseph Mallord William Turner (1775–1851). Oil on canvas. 92.3x122.8 cm. Acc. No. 1942.9.86 Widener collection National Gallery of Art, Washington D.C., Image File: Joseph Mallord William Turner – Keelmen Heaving in Coals by Moonlight – Google Art project.jpg

roadside with random cast-iron tubular artefacts – piston-like objects. On the far right are two enormous steam-engine cylinders with an elbow pipe in the very foreground. Perhaps he felt the need to show explicitly that Coalbrookdale was a site of industrial manufacture and not just a place of conflagration; but the effect is crude and somewhat wanton.

Joseph Turner's *Keelmen Heaving in the Coals by Moonlight* (1835: Cover image and Fig. 15.2) was based on long-previous outdoor charcoal sketches made *in situ* and preceded by a previous, inferior, painted version. Here we have miraculous displays of chiaroscuro – natural and industrial sublime. It is an early autumn night on the narrow, long (*c*.12 km) east-facing estuary. The full harvest moon is resplendent and beautiful, dominating the scene with a breathtaking tunnel of moonlight along mid-channel. By way of contrast, the right bank quayside is full of orange-coloured, brazier-lit human activity with the silhouette of smoking chimneys behind. We are in the outer Tyne estuary, in calm weather, high cirrus clouds parting to reveal strips of dark and starlit lapis-blue sky above. The foreshortened view trends roughly NE towards the open North Sea and the great defensive fort above 'Tynemouth Barres' (*see* Fig. 1.2) at the estuary entrance (not seen here). One of the two Tynemouth lighthouses is shooting up its firelight in the far distance on the left bank (under the bowsprit of the leftmost ship under sail) – this would be the 'High Light' above the village of North 'Sheels' shown on Figure 1.2.

The full moon is low to the north-east, the scene probably mid-autumn – nearly the last sailing time for the collier fleet down to the Thames estuary, the risk of autumnal and winter north-easterly gales onto the usually lee shore of eastern England becoming ever more certain. The high spring-tide water has begun to gurgle away down-channel, forming parabolic wakes from a floating anchored barrel and buoy in the foreground mid-channel. There is further motion in the scene as a distant vessel just left of centre comes up-channel throwing a huge bow-wave – there is clearly sufficient wind for tacking sail against the beginning ebb tide. Ships from the left bank – they may be ordinary merchantmen from further upstream rather than colliers – are under full sail and are moving out slowly, the foremast sails and jib shivering in light breeze on the gently ebbing tide.

The right bank, apart from one full-rigged departing vessel in the middle distance, is all bustle and hustle as keelmen strive to load their coal into the capacious holds of the as-yet unrigged collier vessels. We see them clearly working their shovels, transferring gleaming black coal from their lighters in the brilliant smoky light of coal-fired braziers. One man on the far lower right is seen in mid-swing pouring his coals into the dark maw of a square funnel that will lead through a porthole directly into the central open hold that all colliers had.

This is Turner manufacturing industrial tension under artificial light at the mercy of natural forces under natural light – the efforts of the keelmen to let the coal fleet gain the ebbing tide. They may have just a few hours left to fill the collier's holds, the tide dropping all the time. A dozen or so years after the painting was exhibited, the ancient race of keelmen (many had Scottish Borders ancestry – see header to Chapter 6) were replaced by the mechanical efforts of railway engines. These brought coal directly along-valley from mines upstream, the colliers loaded by gravity rather than muscle power – via chutes on elevated staithes (*see* Chapter 23).

Romantic Poets and Industrial Sublime

The Romantics came of age with the Industrial Revolution ongoing around them. The older men, William Blake, William Wordsworth and Samuel Coleridge, would have nothing to do with any visual or auditory charms of industrial sublime or, indeed, with rational explanations for any natural phenomena when compared with what they (solipsists all) could sense, feel or imagine themselves.

What kind of mind Coleridge actually had may be judged by his risible reactions to reading Newton's great work *Opticks*. He opined in an 1801 letter to Tom Poole that:

> deep Thinking is attainable only by a man of deep feeling, and that all Truth is a Species of Revelation…I believe the souls of 500 Sir Isaac Newtons would go to the making up of a Shakespeare or a Milton…his whole Theory [of Optics] is, I am persuaded, so exceedingly superficial as without impropriety to be deemed false. Newton was a mere materialist. Mind, in his system is always *passive*, – a lazy *Looker-on* on an external world.[8]

Blake, as we all know from his most familiar poem, intended to strike down 'dark satanic mills' and all, wanting England rebuilt as a new Jerusalem, whatever that meant. Further, this mystical man had no need of the thoroughly practical products being made available by advanced industrial engineering – he was entirely uninterested in measurement or scientific experiment on the natural world. Here he is on the heroes of the Enlightenment in his manuscript poem (none of his work was published in his lifetime), *Newton*:

> Mock on, Mock on Voltaire, Rousseau:
> Mock on, Mock on: 'tis all in vain!
> You throw the sand against the wind,
> And the wind blows it back again.
> And every sand becomes a gem
> Reflected in the beams divine;
> Blown back they blind the mocking Eye,
> But still in Israel's paths they shine.
> The Atoms of Democritus
> And Newton's Particles of light
> Are sands upon the Red sea shore
> Where Israel's tents do shine so bright.

His disdain for science and engineering is illustrated in the following anecdotes in one of the first biographies, by Alexander Gilchrist in 1863. It tells of events in or about 1823:

> Some persons of a scientific turn were once discoursing pompously, and, to him, distastefully, about the incredible distance of the planets, the length of time light takes to travel to the earth, etc., when he burst out: 'It is false. I walked the other evening to the end of the earth, and touched the sky with my finger'…

In society, once, a cultivated stranger, as a mark of polite attention, was showing him the first number of the *Mechanic's Magazine*. 'Ah, sir,' remarked Blake, with bland emphasis, 'these things we artists HATE!'[9]

Wordsworth, ensconced in his rural Lakeland kernel, became dismissive of and disinterested in the consequences of the machine age to his fellow citizens as he himself aged. He expressed distaste for both human lives and industry most frankly in the second sonnet of the long sequence, *The River Duddon* (1806–1820). He addresses the upland sources to the river thus:

> Child of the clouds! Remote from every taint
> Of sordid industry thy lot is cast;
> Thine are the honours of the lofty waste…

Downstream, as Norman Nicholson pointed out in his *On the Dismantling of Millom Ironworks* (1981), there had been a blast furnace in Duddon woods for a hundred years or so, where:

> shallow-draft barges shot their ore
> …for the charcoal-burning furnace
> Sited like a badger's set deep in Duddon woods.[10]

He stresses in the last lines that the de-industrialization that Wordsworth would doubtless have welcomed involved empty landscapes devoid of human life.

The younger Romantics, Percy Shelley and John Keats, grew up with the beginning stirrings and discoveries of modern science all around them, and with a growing sense that humans had the right, willy-nilly, to witness nature, reflect on the rationale behind it, experience natural sublime and enjoy machinery and industrial sublime with their own senses that gained nothing from any deity. This was stated early and most definitely in Shelley's long ode, *Mont Blanc* (1816) where he contrasts the clamour and noisy activity of its bounding Arve valley gorge – a rugged-sided, glacier-fed, mountain-sourced river – with the physical fact of the mountain's continued existence. It seems to have a resilient silent power of solitude and strength whose origins human thought can decipher, feel or make felt as in 'a trance sublime and strange'.

He says that it is not enough (as in Coleridge's *Hymn Before Sun-Rise* of 1798) to celebrate the elements of sublime by calling on the reader to glorify God, neither to repress the malevolence of nature's destructive forces in purely pastoral rhapsodies (as in Wordsworth's *Tintern Abbey* of the same year).[11] What is necessary is a profoundly calm reflection on the highly energetic aspect of nature's ways – by this route comprehension and feeling can interact for the good of both. These sentiments are what makes *Mont Blanc* special – modern nature observed, like the mountain landscapes that Cézanne would paint in Provence eighty years in the future. John Keats was to achieve a similar result with his *To Ailsa Rock* (1818), though in this case with a more-developed and direct geological sense and metaphor.

Three years later we see a more playful and ironic face to Shelley's atheism. It comes in an appreciation of the efforts of his close friend, Henry Raveley (who was to become an architect and civil engineer) in manufacturing a steam cylinder and an air pump for a maritime enterprise he was engaged in (both men were nautical fanatics). In his letter of 12 November 1819 to Shelley, Raveley is describing the actual casting as a sublime experience:

> The fire was lighted in the furnace at nine, and in three hours the metal was fused. At three o'clock it was ready to cast, the fusion being remarkably rapid, owing to the perfection of the furnace. The metal was also heated to an extreme degree, boiling with fury, and seeming to dance with the pleasure of running in its proper form. The plug was struck, and a massy stream of a bluish dazzling whiteness filled the moulds in the twinkling of a shooting star…

In his reply Shelley is in sparky mood as he riffs impressively along cosmic and geological lines:

> Your volcanic description of the birth of the cylinder is very characteristic both of you, and of it. One might imagine God when he made the earth, and saw the granite mountains & flinty promontories flow into their craggy forms, & the splendour of their fusion filling millions of miles of void space, like the tail of a comet, so looking, & so delighting in his work. God sees his machine spinning round the sun, & delights in its success, and has taken out patents to supply all suns in space with same manufacture. Your boat will be to the Ocean of Water, what this earth is to the Ocean of Æther – a prosperous and swift voyager.[12]

Late-Industrial Sublime

Charles Dickens gave a version of the sublime in his description of the menacing coal-induced fogs of London's docklands quoted earlier. Shortly after came the Pre-Raphaelites, again as emphasized by Alison Smith:

The challenges presented by science, religious doubt and positivist philosophy which accompanied the shift to an urban secular society, informed in turn the Pre-Raphaelite aesthetic. Pre-Raphaelite painters set out to valorize the familiar and everyday in a spirit of reaction to the artificiality and elitism of the Romantic sublime, which they felt had descended into pictorial cliché in the work of contemporary academic painters…the Pre-Raphaelite's 'democratic' approach underpinned the novels of George Eliot, who was influenced by developments in landscape sensibility through the work of Ruskin and the cult of detailed naturalism.[4]

Eliot's aims are epitomized, writes Smith, in *Adam Bede* (1859), where the narrator loftily espouses the cause of the Pre-Raphaelites in no uncertain terms:

In this world there are so many of these common coarse people… It is so needful we should remember their existence, else we may happen to leave them quite out of our religion and philosophy and frame lofty theories which only fit a world of extremes. Therefore, let Art always remind us of them; therefore let us always have men ready to give the loving pains of a life to the faithful representing of commonplace things – men who see beauty in these commonplace things, and delight in showing how kindly the light of heaven falls on them. There are few prophets in the world; few sublimely beautiful women; few heroes.[4]

So it was that the Pre-Raphaelites and the followers of Ruskin came to either welcome or eschew industrialization in their various ways. Like William Blake they might reject it as ugly, conformable and repetitious. Or using visual allegory, create carefully crafted scenes of human and animal lives set in conformable landscapes whose dangerous sublimity was sometimes there, sometimes not. Such are the beautifully groomed sheep straying and gambolling over foundered coastal landslips along the Sussex coast in William Holman Hunt's *Our English Coasts*[13], 1852. A later exception to these rather pastoral and posed studies was Eyre Crowe's *The Dinner Hour, Wigan* (1874; Fig. 15.3). He depicts female mill hands on cobbled sandstone setts of open spaces adjacent to the Victoria mill in the South Lancashire cotton town of Wigan. The women are at their lunch break; they still have their work gowns on over their coloured skirts and blouses – most of them wear clogs, just one barefoot. The centrepiece group occupy a low stone-built wall – they have finished eating and drinking, their baskets and billycans closed up. They are in pairs, some sitting chatting, some seemingly ready to return to work. One has a basket over-arm; her friend exchanging confidences is barefoot, looking directly at the artist. Two carefree friends walk nonchalantly left to right across the scene, one arm each around waist and shoulder. To the right, by a gas lamp another is holding up her apron with one hand and waving with the other to someone off-picture. Her neighbour has her legs up on the wall resting, revealing her feet in a pair of red stockings! Behind this central group are further gatherings,

Figure 15.3 *The Dinner Hour, Wigan* (1874) by Eyre Crowe (1824–1910). Oil on Canvas. Object Number 1922.48a Manchester Art Gallery. Source: Wikimedia Commons. Author: Martin of Sheffield.

Figure 15.4 *Opencast Coal Mining* (1943) by G.V. Sutherland (1903–1980). Pencil, chalk, gouache, pen & ink. 23.2x18.5 cm. Location and Source: Private Collection, In Gough, P. *et al.* (2013)

sitting and standing around. A uniformed policeman stands upright with his stick in the distant vacant centre.

Looking up past the frontside of a pub towards the vast brick-built mill complex, the chimneys smoking into a grey, yellow-streaked sky, one gets a strong feeling of the women's relaxed and joyous escape from the management of thousands of looms, threads and bobbins within the mill sheds. Their break will end soon with the blast of a steam-whistle signalling them back inside the noisy and dusty mill floors for the rest of the working day. The artist makes us consider the women as real individuals released temporarily from monotonous toil, none of them seemingly broken by their circumstances – rather their demeanour seems to project great individualism and optimism.

Responding to Crowe's painting when on exhibition in Manchester, Elaine Webster wrote in 2012:

> I grew up in the 50s and 60s in Wallgate and used to walk past Taylor's mill on my way to school up Miry Lane and Wallgate. I turned the corner by the New Star Inn every day, (the replacement of the pub in the picture) and I know exactly where the artist stood. I remember the mill girls coming out in their dinner hour covered in cotton, going to the fish and chip shops and sitting around having a laugh, before later making their mass exodus at five o'clock when the whistles blew. Taylor's was the satanic mill at the end of my street, Mason Street, and I remember seeing it burning from every window during a major fire in the early 60s. The warehouse at the end of Wigan Pier is the building you can just see on the left. This is a fascinating picture on so many levels, and looks stunning when you see it for real in Manchester Art Gallery.[14]

It was left to Modernists and NeoExpressionists to depict industrially ravaged landscapes and mining activities with a renewed sense of the sublime. From the battered First World War battlefields of northern France and Belgium to the Second World War depictions of opencast strip mining, blast furnaces, and the realities of life underground in metal mines, artists like Paul Nash and Graham Sutherland were able to produce visceral images to signify the dismemberment of the land surface by war and mechanized mining activity (Fig. 15.4).

L.S. Lowry and His Bleakness

The twentieth century's most popular 'industrial' artist was W.S. Lowry, born in Stretford and for much of his life a resident of Pendlebury near Salford. His industrial landscapes of red-brick chimneyed mills, terraced housing, pubs and factories are peopled by legions of apparent automatons, their bodies caught bent against rain, cold and wind, as they variously wend their way into and out of work or just along terraced streets, seemingly on a road to nowhere. In his *Going to Work* (1943; Fig. 15.5) the sense of bodily motion exerted against the weather (probably a chilly wind) and 'the beast work' is notably

Figure 15.5 *Going to Work* (1943) by L.S. Lowry (1887–1976). Oil on Canvas. 45.7x60.9 cm. ©Imperial War Museum (Art.IWM ART LD 3074)

emphasized by the crowds of arriving workers, some from trams (two are drawn up, the tramline winding away from the scene), to begin their day shift at the factory gates of the long-established Mather and Platt engineering plant in Manchester. Even without the date beside Lowry's signature, the two barrage balloons in the silvery-white winter sky signify wartime conditions and seem to add some urgency, perhaps even danger, to the workforce's arrival for the day shift. The artist has mixed up his human beings in terms of age, sex and dress, yet they all walk purposively over the packed snow onwards towards the gates, their long definite strides seen sideways and then from behind as the crowd funnels in past the porter's lodge. Such repeated depictions of workers in Lowry's works encourage us to view them all as faceless proletarians rather than the real flesh-and-blood workers depicted in Wigan by Eyre Crowe. That industrial society could produce such legions of apparent automatons may also be seen as an indictment of the whole factory system. The overall effect is sombre and depressing, projecting the hard pessimism inherent in industry, especially during those terrible times of winter war.

The Mining Roots of D.H. Lawrence

Lawrence was the first to blaze the trail that moved fiction and life closer together and forward to our modern realism of novel, cinema, theatre and television. He did it as the son of a working Nottinghamshire miner, his family living in a colliery terrace row of a pit village (Eastwood) during childhood and young adulthood. The family circumstances of this background and upbringing significantly informed his early novelistic career, most notably in *Sons and Lovers* (1913) where the Morel family live out their lives in the late Victorian/early Edwardian working class culture of the rural/industrial English Midlands. It is easy to underestimate the raw force of Lawrence's prose in this early novel (he was only twenty-eight in the year of publication) and the shock that it brought to contemporary literary culture as this *enfant terrible* let his vulnerable and variously troubled and doubting working-class characters loose on the reading public one year before the outbreak of world war.

The early chapters of the novel involve a number of scenes in which the history and development of coal mining in the Nottingham coalfield come to the fore. There are also day-to-day details of the father's employment underground, which are critical in understanding the personality of this feckless and sometimes angry man and his tricky relationship with his 'genteel' wife and sensitive children. The novel begins with a concise account of what had been the local mining scene before the mid-nineteenth century:

> Hell Row was a block of thatched, bulging cottages that stood by the brookside on Greenhill Lane. There lived the colliers who worked in the little gin-pits two fields away. The brook ran under the alder-trees, scarcely soiled by these small mines, whose coal was drawn to the surface by donkeys that plodded wearily in a circle round a gin [winding mechanism]. And all over the countryside were these same pits, some of which had been worked in the time of Charles II, the few colliers and the donkeys burrowing down like ants into the earth, making queer mounds and little black places among the corn-fields and the meadows. And the cottages of these coalminers, in blocks and pairs here and there, together with odd farms and homes of the stockingers, straying over the parish, formed the village of Bestwood. [15]

This pre-industrial scene of scattered bell pits was 'elbowed aside by the large mines of the financiers' as the deep and concealed Nottinghamshire/Derbyshire coalfield began to be exploited by shaft mining in six deep pits, memorably described as: 'six mines like black studs on the countryside, linked by a loop of fine chain, the railway.'

The Morels came to live in the double rows of 'substantial and very decent' newly erected company-built terraced dwellings whose parlour views looked outwards from the hillside, yet whose chief living quarters were the kitchen living rooms that looked across the intervening ground between rows. Here were small back gardens bordering long lines of ash-pits for cinder disposal from the ubiquitous domestic coal grates and through which the back alley ran: 'where the children played and the women gossiped and the men smoked.'

Walter Morel was a miner at Minton pit, but when in his cups in the local pub he becomes disrespectful – a 'blab-mouth, a tongue-wagger'– of the authority vested in the pit-manager, whom he had grown up with. Versions of Morel's mocking tirades reached the manager's ears and, Lawrence writes:

> Alfred Charlesworth did not forgive the butty [Morel and workmate under a middleman] these public-house sayings. Consequently, although Morel was a good miner, sometimes earning as much as five pounds a week when he married, he came gradually to have worse and worse stalls [as in 'pillar-and-stall' mining], where the coal was thin, and hard to get, and unprofitable.

> Also in summer, the pits are slack. Often, on bright sunny mornings, the men are seen trooping home again at ten, eleven or twelve o'clock. No empty trucks stand at the pit-mouth. The women on the hillside look across as they shake the hearthrug against the fence, and count the wagons the engine is taking along the line [railway] up the valley. And the children, as they come from school at dinner-time, looking down the fields and seeing the wheels on the headstocks standing, say: 'Minton's knocked off. My dad'll be at home.'

> And there is a sort of shadow over all, women and children and men, because money will be short at the end of the week.

Lawrence gradually constructs a detailed picture of a coal-mining family's daily life: Morel making up his morning 'snap'; his walk to work where he eventually digs out a poor stall a mile and a half from the shaft bottom; his badly injured leg from a roof fall; his fond account given to his young son of the pit ponies that come by his stall.

The pit was thus at the centre of all his children's young lives, though they variously escaped employment in it. Yet two decades later Morel's son, Paul, now an aspiring artist of twenty-five, was walking with Clara, his new love, from the railway station (he has just met her off the train from Nottingham) through the pit village of Nuttall when:

> They came near to the colliery. It stood quite still and black among the cornfields, its immense heap of slag seen rising from the oats.

> 'What a pity there is a coal-pit here where it is so pretty!' said Clara.

> 'Do you think so?' he answered. 'You see, I am so used to it I should miss it. No; and I like the pits here and there. I like the rows of trucks, and the headstocks, and the steam in the daytime, and the lights at night. When I was a boy, I always thought a pillar of cloud by day and a pillar of fire by night was a pit, with its steam, and its lights, and the burning bank – and I thought the Lord was always at the pit-top.'

This certainty in Paul's mind and his love for his dying mother combine in the end when his mother dies and his relationship with Clara finishes. Together the events define a firm and defiant resolution to face faraway urban life full in the face and on his own terms.

Mine Disasters: Readings of Wilfred Owen and Phillip Larkin

Almost sixty years separate these two poems, each representative of the very best work of both 'modern' poets.

In January 1918, Wilfred Owen of the Amalgamated 5th (Reserve) Battalion of the Manchester Regiment had been on training leave since late November 1917 at Scarborough, North Yorkshire. He was billeted with fellow officers in the Clarence Gardens Hotel. On 13 January he read news of a colliery disaster on the previous day due to an explosion of methane gas that killed 156 men and boys at the Minnie Pit near Newcastle-under-Lyme in the North Staffordshire coalfield. A day or two after reading the news of the disaster Owen sat before the smoky fire in his comfortable five-windowed billet room at the very top of the hotel (with splendid sea views) and began to write out *Miners* over the next hour. Here are the first four verses:

There was a whispering in my hearth
A sigh of the coal,
Grown wistful of a former earth
It might recall.

I listened for a tale of leaves
And smothered ferns,
Frond-forests, and the low sly lives
Before the fauns.

My fire might show steam-phantoms simmer
From Time's old cauldron,
Before the birds made nests in summer,
Or men had children.

But the coals were murmuring of their mine,
And moans down there
Of boys that slept wry sleep, and men
Writhing for air.

His poem has lyrical and emotional depth reminiscent of his much-loved Keats – one senses a direct reference in the opening verse with the first lines to Keats' sonnet *To My Brothers* (1816):

Small, busy flames play through the fresh laid coals,
And their faint cracklings o'er our silence creep
Like whispers of the household gods that keep
A gentle empire o'er fraternal souls.

The opening imagery of Owen's own poem owes much to his rich understanding of Carboniferous coal-forming worlds, with their Lepidodendrales 'frond-forests', and of the remoteness of geological time. As a young man Owen was fascinated by geology and archaeology, and it seems likely that during his leave in Scarborough he would have visited the celebrated Rotunda Museum above South Bay, where he would have seen abundant Carboniferous plant fossils along its gallery displays. Yet, with 'the coals were murmuring of their mine...' a link is established between that recent loss of life and previous deaths in coal mining. This triggers memories of the deaths of some of his Lancashire soldiers (several had been miners) in and under the trenches, some by gas attack 'writhing for air' and how those deaths also might or might not be remembered.

Owen wrote of his efforts in a letter to his mother shortly after the poem was published in the national daily *The Nation* on 26 January 1918; his first published poem:

> Wrote a poem on the Colliery Disaster; but I get mixed up with the War at the end. It is short, but oh! Sour!'[16]

In Phillip Larkin's *The Explosion* we follow a chattering group of miners, some related, others friends, along a country lane early on a summer morning, the sun behind them as they walk westwards together behind their tall shadows towards their local pit to begin their shift:

> On the day of the explosion
> Shadows pointed towards the pithead:
> In the sun the slagheap slept.
>
> Down the lane came men in pitboots
> Coughing oath-edged talk and pipe-smoke,
> Shouldering off the freshened silence.

It is easy to envisage them walking to one of the new rural pits sunk into the concealed coalfield of a district like South Yorkshire or Nottinghamshire – the bearded men smoke pipes and wear moleskin trousers, their guise and dress perhaps placing them in the last quarter of the nineteenth century, possibly contemporaries of D.H. Lawrence's father. Their sounds disturb the early morning silence but are superseded by the antics of one who chases rabbits and finds a lark's nest with eggs in the grass – we are surprised and approving as he carefully places them back and the men carry on past the spoil tip and through the pit gates to begin their underground shift.

The terrible event of the explosion is invisible above ground, a small shake just-noticed by grazing cattle that soon continue chewing. The sun is imagined dimmed momentarily, and like a liturgical inscription the poet in the last lines makes the dead miners visible, but only to their womenfolk. They walk towards them gilded from the sun and larger than life – the inverse of that morning as they walked away from the sun and behind their long shadows to work. The poem ends with an astonishingly moving revelation – the miner, uniquely visible to his wife, who had that morning located and replaced the lark's nest intact was: 'showing the eggs unbroken.'

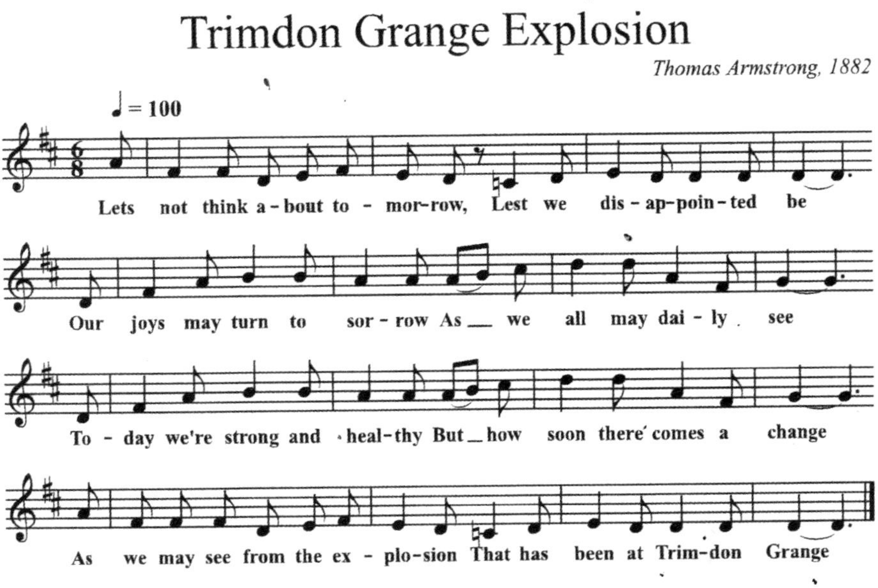

Trimdon Grange Explosion

Thomas Armstrong, 1882

♩ = 100

Lets not think a-bout to-mor-row, Lest we dis-ap-poin-ted be

Our joys may turn to sor-row As __ we all may dai-ly see

To-day we're strong and heal-thy But __ how soon there comes a change

As we may see from the ex-plo-sion That has been at Trim-don Grange

Let's not think of tomorrow,	Men and boys left home that morning
Lest we disappointed be;	For to earn their daily bread,
Our joys may turn to sorrow,	Little thought before the evening
As we all may daily see.	They'd be numbered with the dead;
Today we're strong and healthy,	Let us think of Mrs Burnett,
But how soon there comes a change.	Once had sons and now has none -
As we may see from the explosion	With the Trimdon Grange explosion,
That has been at Trimdon Grange.	Joseph, George and James are gone.

Figure 15.6 Lyrics of the first two verses and music to Tommy Armstrong's lament *Trimdon Grange Explosion* (1882). Lou Killen can be heard singing it on YouTube with music by the High Level Ranters: https://www.youtube.com/watch?v=v_rQRlf5btI

That last line ends the last poem in *High Windows*, Larkin's last collection. He leaves us in shock somehow, having joined the experience of sunlit countryside with the grieving families of the made-extraordinary men who died in the explosion. Readers familiar with his previous collection, *The Whitsun Weddings*, are not surprised by the delicate way that he treats his subject, conjuring up devastatingly powerful statements from single key images. So it was that in *An Arundel Tomb*, the last poem in the collection, the sight of the ungloved and entwined hand of a medieval knight with his lady's bare hand leads to that most memorable line, the emphatic and life-affirming: 'What will survive of us is love.'

In *The Explosion*, the miners' terrible deaths underground are set against the inevitability and ordinariness of a sunlit summer landscape where all other life carries on – cows chewing, larks singing, rabbits running. The whole effect is as powerful as that engendered by W.H. Auden's *Musée des Beaux Arts* (1939). Here, Bruegel's ploughman and shepherd are unaffected and unnoticing in the foreground, paying due attention to a plough's mould board making precise curved furrows, or tending a flock. Out to sea the flailing boy Icarus falls from the sun – white legs akimbo, splashing to his death beside a merchantman that sails blithely on. The sun is mentioned three times in Larkin's poem as the key to the exterior surface above the working coalfield deep in the dark and black-dusty underground. That world and the epicentral explosion are of no consequence to this natural sunny world where things just carry on – perhaps a metaphor for society's ignorance about miners and the mining industry.

Coda: Folk Music

We return to Kathryn Tickell's opening quote to this 'cultural' chapter by emphasizing the relevance to social history of ordinary people's music, that featuring industrial themes, times and incidents; their own self-generated popular music (known as 'folk' since the 1950s). It seems fitting to contrast the efforts of major poets like Owen and Larkin with a composition, *Trimdon Grange Explosion*, written and arranged by Thomas Armstrong, marking the deaths of sixty-eight miners in a disaster at the Trimdon Grange Colliery in County Durham on 16 February 1882. Armstrong set his lyrics to the existing tune 'Go and leave me if you wish it.' My own favourite version is a haunting, almost chanted effort by Martin Carthy, with strong and resounding guitar chords that thunder out a raw accompaniment. Figure 15.6 has the original tune and the first two verses.

PART 4

Landscapes of the Industrial Revolution

The coming of industry altered every part of our pre-industrial ancestors' relationship with their more closed physical worlds. Mobility of labour increased as horizons shrank. Cottage industries – farm dairy producers, smithies, boot and shoemakers, spinners, weavers, brick- and tileworkers, potters, metal- and woodworkers, retail businesses in small towns and villages were decimated. They were replaced by raw materials, manufactured goods, produce and products from large-scale industrial producers. These came first via a near-nationwide network of coastal shipping, the boats moored at innumerable small anchorages and quays for further local distribution by pack animals, carts and wagons. Then came canals, railways and, finally, metalled roads.

All this was superimposed on a medieval and earlier rural palimpsest that had itself grown slowly over the ages, despite terrifyingly sudden hiatuses engendered by plague, disease, war, drought and storm. Industry forced change at an unrelenting pace so that almost its total legacy was well known to my generation via the childhood memories of parents (horse-drawn ploughs and reapers replaced by tractors), grandparents (arrival of motorized transport) and great-grandparents (appearance of steam ploughs and railway locomotives). Earlier still from the oral and written history handed down from as far back as the early nineteenth century were the social consequences of horse-powered agricultural machinery.

Today we turn to innumerable national, regional and local museums, public record offices, libraries, trusts, industrial monuments, memorials, public sculptures, artwork, architecture, literature and music that enable us to see, hear, understand and appreciate our common industrial heritage. There is a strong and undeniable human trait that comes from an informed understanding of 'why' certain things came to pass at certain places at certain times – the 'Brown Cow' scenarios mentioned in the preface to this book. Addressing these human concerns is the purpose of Part Four – to provide generalized templates to help the reader place their own heritage within the events in question. We examine nine major former coal-based industrial regions, scattered places in each – Blaenavon, Kingswood, Flint, Dudley, Todmorden, Wigan, Millom, Consett, Prestonpans – to select but a random few. All have their own particular history, landscape and legacy. Hopefully any inhabitant or native of such places and of the regions at large will recognize something of their own in all this. More to the point, they may realize a communality of experience over the past three centuries, and, perhaps, of lessons learnt that might be passed down to their descendants.

16

South Wales

Maybe today they'll change the tapes!
It's the same old stories – first there was steam,
steel. Then depression, then developed bay –
Stories so fixed I can never say
more than they let me. At night I dream

these galleries shift. We open screens,
show new exhibits. The best one's my heart
in a glass case and it switches on
and off like a light bulb. This intimate room
is floodlit, is a work of art. Stop, start. Stop start. Stop, start.

The museum's night gallery curator, weary with her messages of the past, is anxious for
a spiritual shift into the living, regular heartbeat of the future. Gwyneth Lewis *The Mind
Museum, III Night Galleries* (2005)

'In These Stones Horizons Sing'

Like midland Scotland, here is a national core of industrial landscape tightly defined
and developed from its geology – much of it in upland Glamorgan and historic
Monmouthshire. The southern rural periphery includes the Vale of Glamorgan and
the westward-fringing historic counties of Carmarthenshire and Pembrokeshire. Most
invaders of southern and central Wales ingressed via these southern Marches. Here were
performed desperate martial efforts against invading Roman legions after gold, silver
and lead under Governor Julius Frontinus. These came from the aboriginal Iron Age
Silures tribe so admired by Tacitus ('validamque et pugnacem Silurum…' [1]), aided by
Welsh-speaking Caratacus and his young warriors from Essex. They were eventually
hemmed into their upland valleys and farms by the fortress and legionary garrison town
of Caerleon and its fringing valley-blocking fortlets at Cardiff, Neath and Carmarthen.

During the course of *Pax Romana* the inhabitants of at least three localities, the villas at Llantwit Major and Ely Racecourse and at Gelligaer fort, burnt coal to heat their baths and homes, also perhaps to fire tile kilns and smithies.

The Romanized lowlands were later defended by Welsh kings and princes against Saxons and Normans. This emphasis on lowland acquisition disappeared forever from the mid-1750s in the steep declivities where 'The Valleys' industrial landscape developed. Today, to stand outside the frontage of the Millennium Centre building, Cardiff Bay, with its eclectic cladding – a mix of native rocks (many-coloured slates and sandstones) and worked metal (Welsh stainless steel coated with copper oxide) – is to grasp the essence of a whole nation in the historic echo of its curved template of text. The Welsh and English words are by Laureate Gwyneth Lewis: 'Creu Gwir fel gwydr o ffwrnais awen' ('Creating Truth like glass from inspiration's furnace') and: 'In These Stones Horizons Sing'. Her words speak to a future that looks back to a pervasive industrial heritage founded on geological raw materials.

A Perfect Syncline

[2]The westward-decreasing oval of the region's centripetal-dipping Carboniferous outcrops defines one of the most obvious and definite downfolded (synclinal) patterns visible on the geological map of Britain. The Early Carboniferous limestones of the coalfield periphery corral the industrial belt – a blue, jagged-oval necklace on the geological map (Fig. 16.1). It marks the outcrops of several hundred metres' thickness of pale- to dark-grey weathering limestones – often as craggy outcrops and splendid karst country (as at the Dan-yr-Ogof cave complex in the central north outcrop) with sweet-grassed cover in Gower. It is crossed by vigorous mountain-sourced streams pouring over waterfalls whose scarps border the distant high wastes of the Black Mountains and Brecon Beacons. The core uplands are defined by massive scarps of the Pennant Sandstone Group carved out preferentially by ice and water bordering an ancient plateau surface at around 450 m elevation. Embayments and indentations mark the eroded and concealed courses of thick Late Carboniferous coal- and iron ore-bearing strata along steep-sided valleys whose broad interfluves separate adjacent rivers swift-flowing down to the Vale of Glamorgan's cliffed coastal fringes along the Bristol Channel. These southern lowlands bordering the coalfield are where softer Mesozoic rocks dotted with Carboniferous Limestone inliers and glacial deposits made rich soils and low ground. Here the Neolithic rural and agricultural hegemony of the area is implicit in the abundance of coastal and panoramic hillside monumental sites stretching from Cardiff beyond the coalfield westwards to the sacred territory of Mynydd Preseli – linked by an unknown ancestral thread to the provenance of Stonehenge's 'bluestones'.

In all some 750 m of Coal Measures strata occur in the western part of the main coalfield (Swansea–Llanelli–Ammanford area), the thickness decreasing gradually eastwards to around 450 m in the Ebbw valleys and westwards to about the same in Pembrokeshire. The conditions in which these rocks originally formed began with the

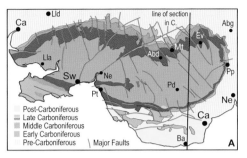

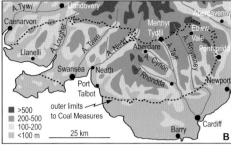

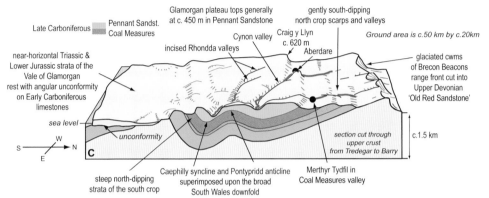

Figure 16.1 **A–C** South Wales landscapes: **A** solid geology, **B** Topography, **C** Three-dimensional panoramic section. The drainage pattern notably 'cuts across' the geological structure and the various rock divisions present in the asymmetrical Carboniferous downfold (syncline). This, together with the general concordance of most hilltop plateau elevations around 450 metres, suggests that much of the present landscape is second-hand, exhumed by Cenozoic erosion from a cover of up to a kilometre or so of Mesozoic strata now mostly eroded off, the outcrops of older units now lapping around the southern basin margin. **D** North East Glamorgan and West Monmouthshire ironworks and transport routes down to the coast. The twenty-nine

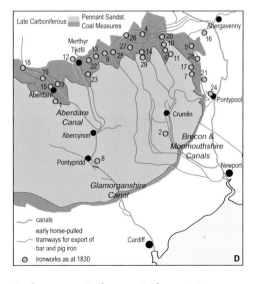

numbered historical ironworks active in 1830 are: **1** Aberamman **2** Abercarn **3** Abernant **4** Beaufort **5** Blaenavon **6** Blaina **7** British **8** Brown Lennox **9** Bute **10** Coalbrookvale **11** Cwmcelyn **12** Cyfarthfa **13** Dowlais **14** Ebbw Vale **15** Gladlys **16** Garnddyrys **17** Golynos **18** Hirwaun **19** Llwyd Coed **20** Nantyglo **21** Pentwyn **22** Penydarren **23** Plymouth **24** Pontypool **25** Rhymney **26** Sirhowy **27** Tredegar **28** Varteg **29** Victoria. Sources: material redrawn, recrafted and recoloured from: A British Geological Survey (2007a); B *Times Atlas of the World* (1987); C Trueman (1949); D Atkinson and Baber (1987).

expansion of freshwater swamp and deltaic environments between times of high sea levels, the softer mudrock strata defining what is today lower ground around the coalfield outcrop. Above there is evidence for repeated long episodes (eight or so) of freshwater forest swamp conditions, interrupted by lagoon, bay and floodplain sedimentation. The forest swamps contained peat accumulations that compacted to form tremendous coal seams up to nearly three metres in thickness, coals such as the Five Feet-Gellideg, Seven Feet, Lower and Upper Nine Feet, Six Feet and Four Feet of the Ebbw-Pontypridd area and their equivalents. In muddy iron-rich lake sediments, legions of iron-reducing microbes began to create nodules and strata of iron carbonate: the future clay ironstones known to the Welsh ironmasters as 'Welsh' or 'mine' ironstone.

It was shortly after deposition of the last-known marine ingression (the Cwmgorse marine band) in the western coalfield that the first influx of Pennant Group coarse-grained sands arrived in wide and shallow river and delta channels. Flow structures indicate that the sandy detritus came from a southern hinterland – detrital rock fragments of various kinds in addition to the usual resistant quartz. These river channels spread northwards and eastwards over the next few million years as part of a large network of riverine fans. The hinterlands were the rising front of the Variscan mountains (*see* Fig. 8.5B) whose weight pressed down the crust into broad flexures that could catch and hold the sediment. Several thick coals are present at intervals between individual sandstone members, around a dozen or so, all thinning and disappearing eastwards. These indicate that the major river channels bearing their tumultuous discharges and loads of detrital sediment were periodically diverted in that direction, so allowing the western forest swamplands to be isolated from sediment supply for longish periods.

Industrial History: Geology as a First-Order Control

Three geological features have controlled the course of economic development. First the marked asymmetry of the basin structure when traced eastwards – well seen in the three-dimensional view of Figure 16.1C. The southern portion in the east has markedly steeper dips compared to the gentler dips in the north. Early mining of both coal and ironstone in the north was made possible by driving simple, near-horizontal adit levels into valley sides. Such procedures were initially impossible in the south, the steep dips preventing economic mining until the technology of stope mining and of deep drainage by steam pump were developed. Second are superimposed smaller folds roughly parallel to the trend of the larger structure. In the upfolds (anticlines) the older and most prolific coal seams were brought closer to the surface, enabling shallow shaft-mining of their more gently dipping seams. Third, a unique range of coals were available to be mined – this thanks to the steady increase in coal rank from the south-east to the north-west – household and coking in the south east; steam in central areas, notably in the Rhonddas; anthracites in the north-west and historic Pembrokeshire. Mining of the latter was made difficult by intense stratal folding and faulting close to the margins of the Variscan Front. It was the steam coal destined to feed the railway engines and ironclad ships of the world that led to the greatest expansion of mining in the Rhonddas and massive coal exports from Cardiff and Barry docks by the mid-nineteenth century. The Admiralty tested the coal and proclaimed it superior to even the Northumberland/Durham steam coals.

Early mining communities grew and prospered on outcrops of Coal Measures at inland and coastal sites. Medieval to Early Modern times saw the growth of busy coal ports and smelters – Milford exported Pembrokeshire anthracite; bituminous coals came from Llanelli and Swansea, with Kilvey colliery large by contemporary standards as early as 1400. Neath was where the earliest copper smelters were located using imported ores from south-west England. During the mid- to late eighteenth century coal and iron ore extraction along the northern crops led to growth of towns like Pontypool, Ebbw Vale, Merthyr Tydfil, Blaenavon, Aberdare and Ammanford; in the south old Caerphilly, Pontypridd, Llantrissant, Maesteg, Port Talbot and Swansea. The easterly towns of the former group had the thickest and richest iron deposits.

Exploitation of coal and iron ore was possible due to internal and external immigration, more than half of it English. Pre-existing Welsh-speaking market towns, coastal ports and anchorages, upland villages and hamlets often grew more than fifty-fold as the tonnages of mined coal grew accordingly. The nature of the bulk of the population of the Welsh nation – its language, economy, population distribution, politics and culture – changed forever. Yet, away from all this urbanized industrial bustle the Vale of Glamorgan continued with its rural, Welsh-speaking ways, witness the account of Louis Simond on 14 July 1810 on his way west by coach with his wife:

> We are at Cowbridge, Glamorganshire. Forty miles today through Newport, Cardiff and Landaff – the country just uneven enough to afford extensive views over an immense extent of cultivation, lost in the blue distance; nothing wild, or, properly speaking, picturesque, but all highly beautiful, and every appearance of prosperity. Wales seems more inhabited, at least more strewed over with habitations of all sorts, scattered or in villages, than any part of England we have seen…The women we see are certainly better looking than nearer London. The language of the inhabitants is quite unintelligible to us; at the inns, however, all is transacted in English.[3]

The South Wales economy that grew up was based almost entirely on what used to be called 'primary production' – the *in situ* extraction and reduction of geological raw materials, with relatively little 'making' involved, unlike in several other British districts of the early Industrial Revolution, e.g. the West Country, the Black Country, south Cumberland. It involved various metamorphoses in furnace and forge (echoed in Gwynneth Lewis' words) of metal ores – iron, copper, nickel, lead, zinc and tin. First came the inland ironmasters, then coastal metal-smelters reducing copper sulphide ores from south-west England, Anglesey and elsewhere. Later came coastal furnaces and rolling mills of the Port Talbot and Llanelli steelmakers and galvanisers. That nothing much was actually made in South Wales itself (excepting massive iron chainwork, military ordnance, steam engine parts and thousands of miles of locomotive rails) didn't seem important at the time of peak production. But when the gales of de-industrialization and monetarism blew through the landscape, when chimney stack demolition became a recognized profession from the 1930s onwards, there was nothing much left to save by the early twenty-first century. The last deep coal mine closed in 2008, though the Port Talbot steelworks still survives, for years now based entirely on ship-imported high-grade iron ore and metallurgical coke – the now-famous and preserved Banksy mural (Fig. 13.2) saying nothing new as far as the local population are concerned.

Iron and Coal: Mix and Match

With the huge increase in demand for cast-iron products, adventurers raised capital from London and the chief provincial financial markets to apply the 'Coalbrookdale touch' to South Wales. Many of them turned to the Heads of the Valleys in eastern Glamorgan and western Monmouthshire (Fig. 16.1D) to try their luck where, as we have seen, plentiful and easily accessible coal and ironstone were available.[4] They found contiguous deposits of fine coking coal and several mudrock strata rich in clay ironstone exposed at outcrop in the Early Coal Measures, with limestones for flux readily available from Early Carboniferous strata just a few kilometres to the north. As important was outwards transport downvalley to the Bristol Channel and outer Severn Estuary just tens of kilometres away. But unlike the Severn downstream from Coalbrookdale, Valley rivers are steep, wild and wilful. At first packhorses transported the smelted iron out, then it was taken by horse-drawn wagons on tram rails and eventually, by 1800, multi-locked canals joined the iron furnaces to the coast, with railways following by the 1850s, first and notably that of the Neath Valley.

During the prime of the South Wales iron industry in the 1820s their products (by now mostly forged bar iron from reverberatory furnaces rather than pig-iron) dominated UK markets with forty per cent of production.[4] Some thirty or so ironworks had sprung up, from Aberdare in the west nearly to Abergavenny in the east. The first (and furthest west) blast furnace was at Hirwaun, upriver from Aberdare in 1757, but by 1790 the most productive and largest works were around Merthyr and at Blaenavon, Nantyglo, Tredegar, Ebbw Vale and Beaufort. Despite iron production steadily increasing throughout the nineteenth century to the 1870s, its share of total UK output fell inexorably. During this time the integrated water and rail transport system of South Wales started to transport rich imported Cumberland and Basque hematite iron ore northwards in ever-increasing quantities as the *in situ* supply from the valley heads dried up – the adit mining there was no longer profitable as ore grades declined and access and drainage problems increased.

From the 1860s hematite ores dominated the early steel industry, most economically set on coastal sites or at the end of railway termini for ease of import. This was followed by the 'basic' (alkaline) lining of dolomite provided by the Gilchrist-Thomas process (invented partly at Blaenaven, partly at Middlesbrough) and by similarly lined open-hearth furnaces that could tap both the immense reserves of cheap shallow ores discovered in Cleveland, North Yorkshire and, later, Northamptonshire. From then on the indigenous inland South Wales ironworks were doomed. Far from these new sources of ore they had neither the capital nor the infrastructure to adapt to mass steel production, unlike, say, inland Sheffield, whose steel 'nous', specialized forges and a long history of making saw it through the iron-to-steel transition.

Blaenavon: From Rural Hamlet to Iron Town

[5]What of the people who owned the works, planned the mining and manufacturing strategies, dug the coals, dressed the ores at pithead, stoked the boilers and fed the furnaces? Where did they come from and how did they live their lives? The owners, managers, planners and skilled operatives of the early ironworks in the Heads of the Valleys district were almost all English – hard-headed and ambitious 'techy' men from the Midlands, Bristol and London. Land was sought out: open uplands that were agriculturally worthless to aristocratic landowners. Observations made on exploratory visits identified specific sites that contained beneath them workable coals and iron ore in large quantities. These entrepreneurs had access to capital and knowledge of the latest technology – stratal geology, mining, smelting, forging, the latest steam engines and the likely courses of transport routes out and in.

Blaenavon was once a tiny hamlet at the head of the Llwyd valley when Midlands ironmasters Thomas Hill (also a banker), Thomas Hopkins and Benjamin Pratt took out leases for the largest plot of mineral rights in South Wales (some 12,000 acres) from Lord Abergavenny in 1788. Within a single year these well-prepared men had sunk a coal mine and built three furnaces, their blast bellows powered by the latest Boulton and Watt engines. The works spread in front of the furnaces, standing against a steep stone-walled bank cut into the valley side (Fig. 16.2). It was fed by the complete extractive industrial infrastructure of adit mines, roasting kilns, tramways, casting sheds, forges, brickworks, quarries for building stone, firestone (ganister for the hearth linings) and more distant limestone. The Blaenavon Iron Company soon became the second largest in South Wales (after Cyfarthfa) and by 1812 five furnaces were producing enough pig-iron to forge over 14,000 tonnes of bar iron annually, most of it shipped out on the Brecon and Abergavenny Canal.

Word spread far and wide of the new enterprise and Blaenavon grew from almost nothing to a populace of 7000 within thirty years, the immigrants coming from rural Wales, the English Midlands, Cornwall (ex-tin and copper miners), Ireland and Scotland. To house the initial skilled members of the workforce, commodious houses were built scattered around the original ironworks buildings – double-fronted four-room survivors are preserved in the town adjacent to the old ironworks in Stack Square and Engine Row.[5] The majority of the labouring workforce were given lesser housing, including early single room back-to-backs, but between 1817 and 1832 many (160) single-fronted, three-room, two-storey terraced cottages with gardens and outdoor conveniences were financed, many rows of which survive. The Company generally sought to encourage

Figure 16.2 **A–B** Part of the rich industrial legacy preserved at Blaenavon World Heritage Site (51.462130, 3.042192). **A** Two surviving sandstone-built blast furnaces from the eighteenth century lean against an excavated cut into local Coal Measures bedrock. Their gaunt, squat remains still give off an aura of massive attitude – 'Look, this is what we did here!' Over the centuries weathering and gravity have made a curiously revealing onion-like cut-away through the left-hand furnace, from the remaining massive ashlar-faced furnace walls into the rubble-work of the charging platform to the ashlar chimney stack work. Together with the mausoleum-like open maws of the furnace tapping area, it presents a timeless icon of a long-gone culture, as moving and as redolent of place and time as the valley-side tombs of ancient Thebes along the Nile Valley. Source: Wikimedia Commons; *File:Blaenafon Ironworks-two furnaces-24May2008.jpg.* Author: *Alan Stanton* **B** Head frame of Big Pit, the National Coal Mining Museum of Wales, with some of the local rail locomotives there. Source: Wikimedia Commons; *File:Big Pit, Blaenavon.jpg.* Author: *Nessy-Pic.*

Figure 16.3 *Cyfarthfa Ironworks Interior at Night* by Penry Williams (1825). Reproduced by permission of Cyfarthfa Castle Museum

house building in parishes adjacent to where the works were situated, supposedly in order to avoid paying excess poor rates at times of unemployment. Many such terrace rows were built in the 1840s–1860s by independent developers. These formed the nucleus of the eventual township and those of adjacent parishes around old chapels and other long-vanished settlement buildings. Remnants of the original ironmasters' buildings include a mansion (home of one of the founders, Thomas Hopkins, nowadays a nursing home), a delightful church (with visible structural Company ironwork and an iron font!) and the earliest known ironworks company school for boys and girls opened in 1815, the year of Waterloo. The many chapels of the town built through the nineteenth century catered for the spiritual, social and educational needs of numerous non-conformist sects. As in many industrial towns across the land, educational and social space for the workers was eventually provided by a large Workmen's Hall and Institute with reference and lending libraries, reading room, games rooms, lecture halls and smaller class and committee rooms.

Eye Witnesses at Cyfarthfa and Dowlais

For a visual record of those far-off, pre-photographic days of the early industrial Valleys we are utterly dependent upon artworks such as that illustrated in Figure 16.3, a brilliant study by a young artist, Penry Williams, who by 1825 had begun to paint industrial scenes as well as numerous inland landscapes (waterfalls along the north outcrop) of rural Glamorgan. He depicts how the pioneers of the South Wales iron industry had developed their own version of the modern linear steelworks. The cavernous open space of the works also foreshadows the cathedral-like spaces of Victorian railway sheds a couple of generations hence. It is dominated by cantilevered iron girders supporting the roof along the central hall and adjoining girdered aisle piers. The interplay of V-form roof girders and roundels gives a sense of decorative strength. These lead back with a strong sense of narrowing perspective to a distant Romanesque portal that leads into the just-visible casting floor of an adjoining blast furnace. It was from here that pig-iron emerged to source the varied activities seen in the foreground.

To the right are the basal parts of three huge square brick-built chimney stacks that serve to ventilate and take high up the smokey effluent from each adjoining reverberatory furnace. The closest has its door held open on a levered pole by one of two operatives sitting on a handcart while the skilled furnaceman is unmistakeably stirring and puddling molten iron inside the furnace into a ball-like mass – the furnace fire can be seen to the right. This is the beginning of the process of finery, the conversion of coke-smelted pig-iron to wrought iron. Beside the central furnace a steaming mass of iron has just been taken out, the hearth door half-visible. On extraction the puddled iron will be carted off to the rolling mills across and down the way to the centre and in the left aisle, where men can be seen passing iron bars through the rolling wheels to their opposite numbers. In the centre foreground cannon barrel boring is taking place, with a finished article to the right. Naval cannon had been a staple product of the ironworks during the Napoleonic wars, their accuracy, range and durability a tribute to skilled manufacturers and operatives such as these. Some sort of hammering process involving a long bar of red-hot iron is going on to the far left. All these activities would have been accompanied by colossal noise.

Finally there is something wonderful – a view through the furthest open aisle of a distant castellated building, windows lit up, set far above the works and bathed in ethereal blue light. This is what became known as Cyfarthfa Castle, the baronial-style home of the Crawshay family who owned the works (and who may well have sponsored the artist) and in whose splendid parkland it still sits as a fine museum (where Williams' painting now resides). It was newly built in 1824, so it was only a year later that the artist seized his opportunity to create a memorably subtle image of wealth and detached power located far above the night-time ironworks and its numerous toiling workers. What a view the family had of its works down below in the valley!

We may contrast this scene with that depicted twenty years later by George Childs of neighbouring Dowlais, a few miles to the west (Fig. 16.4). There is no baronial castle here! He gives us an exterior view of what had by then outgrown Cyfarthfa as the world's largest ironworks. From an elevated foreground just above a line of blast furnaces are labourers, including a female with a light pickaxe: two males hammer and chisel open what might possibly be raw coke for the furnaces below. A stout-looking wheeled tub on rails awaits the fruits of their labours. In the middle distance up to a dozen blast furnaces are pumping out smoke and flame from the works into a brisk (?westerly) breeze, joined by darker smoke from the tall chimneys of three reverberatory furnaces.

Our chief written eye-witness account of the South Wales iron industry comes courtesy of George Borrow, as recalled in his 1862 classic *Wild Wales*. The compulsive-walking, Welsh- and Romany-speaking Norfolkman reached the outskirts of Merthyr by foot from Neath in 1855. To a woman standing outside her cottage he enquired:

> 'How far to Merthyr?' Said I in Welsh. '*Tair milltir* – three miles, sir'. Turning round a corner at the top of a hill I saw blazes here and there, and what appeared to be a glowing mountain in the south-east. I went towards it…and I could see that it was an immense quantity of heated matter like lava…descending here and there almost to the bottom in a zigzag and tortuous manner…After a time I came to a house, against the door of which a man was leaning. 'What is all that burning stuff above, my friend?' 'Dross from the iron forges, sir!'[6]

George had arrived in Merthyr! The next morning was very fine:

> After breakfast I went to see the Cyfartha Fawr iron works, generally considered to be the great wonder of the place…I had best say but very little. I saw enormous furnaces. I saw streams of molten metal. I saw a long ductile piece of red-hot iron being operated upon. I saw millions of sparks flying about. I saw an immense wheel impelled round with frightful velocity by a steam-engine of two hundred and forty horse power. I heard all kinds

Figure 16.4
*Dowlais
Ironworks*
by George
Childs (1840).
Courtesy of
Amgueddfa
Cenedlaethol
Cymru —
National
Museum of
Wales

of dreadful sounds. The general effect was stunning [he means literally]…
After seeing the Cyfartha I roamed about, making general observations. The
mountain of dross which had startled me on the preceding night with its
terrific glare, and which stands to the north-west of the town, looked now
nothing more than an immense dark heap of cinders. It is only when the
shades of night have settled down that the fire within manifests itself, making
the hill appear an immense glowing mass. All the hills around the town,
some of which are very high have a scorched and blackened look.[7]

Reading these words in modern times and being of a certain age brings a dreadful
déjà vu, for many of the narrow Glamorgan industrial valleys had little space to sequester
their voluminous waste products from mining and iron production, so that the kind of
storage witnessed by Borrow on valley sides and tops was the only option open in those
unrestricted times. That this storage continued, and that it was disgracefully neglected
from the point of view of public safety by a nationalized coal industry after 1947 is well
documented. One hundred and eleven years after Borrow's encounter with his fake
volcano, a few miles down the Taff valley came the nightmare waiting to happen – a
mining waste tip collapsed along a spring line, and the stupendous crashing impact of
its huge rolling debris flow slammed into Aberfan village primary school at 9.15 am on
21 October 1966, killing 109 children and 5 teachers. The recorded after-scenes of the
disaster were cathartic: to the village; to the region and the nation of Wales as a whole.
Though innocents had been victims in underground mining disasters before, there had
been nothing remotely like this.[8] Thirty years later, had they lived, those children would
have been in the prime of their lives but would have found virtually no deep mining left
across the whole of South Wales.

Coastal Copper Smelting at Neath and Swansea

The rise of coastal copper and other metal smelting in South Wales in the later Industrial Revolution had its roots long before. Elizabeth I and her advisers turned to European expertise in order to expand commercial ventures originally set up by her father. Given the existential threat posed by Imperial Spain at the time, they were particularly anxious to secure adequate native supplies of copper to make brass cannon for the navy. A dozen or so years later we hear of a little-known Anglo-Welsh–German initiative that evidently involved intimate knowledge of, and financing for, a scheme to ship Cornish copper to South Wales for smelting – yet it was not Swansea that was the chosen place, but Neath. The metal was smelted at several localities around there by different companies almost two hundred years before the arrival of Swansea onto the smelting scene. Both towns are situated on the periphery of Swansea Bay, where thick coal seams crop out and are close to navigable estuaries facing southwards with the shortest sea passages across to south-west England.

Entrepreneurs and copper-mine owners like Thomas Williams, who owned much of the Anglesey Parys Mountain production, knew that the roasting and smelting of mineral sulphides gave off acid and acrid fumes of sulphur dioxide (*see* James Hutton's letter quoted in Chapter 4). Thus the need for very high chimneys to take away the excoriating fumes high into the atmosphere for turbulent dispersal by the wind. These were to proliferate from late-Georgian times onwards around Swansea's developing industrial districts, but despite their height the deadly fumes created a devastated hinterland leewards. Williams' copper empire included ore transporters, and with the gradual construction of an extensive canal system from Swansea he and his partners and their descendants eventually made the town the world-centre of copper smelting until the mid-nineteenth century. Louis Simond visited on 16 July 1810:

> Near Swansea we visited the copper and iron works. They were just opening a smelting furnace; the fused copper, in a little stream of liquid fire, flowed along a channel towards a cistern full of water; we saw it approach with terror, expecting an explosion; instead of which the two liquids met very amicably, the water only simmering a little. The workmen looked very sickly: we found, on inquiry, their salary was but little higher than that of common labourers. It is remarkable, that, much as men are attached to life, there is no consideration less attended to in the choice of a profession than salubrity.[9]

More extraordinary is the testimony of Charles Darwin, in the third year of his geological and biological explorations on HMS Beagle. In western Chile, having taken the Aconcagua valley to San Felipe, on the evening of 18 August 1834 he locates one copper source in the Andes and identifies the extent of the nexus of world copper trading that reached out from Swansea:

we reached the mines of Jajuel...I staid here five days. My host, the superintendent of the mine, was a shrewd but rather ignorant Cornish miner. He had married a Spanish woman, and did not mean to return home; but his admiration for the mines of Cornwall remained unbounded...These mines are of copper, and the ore is all shipped to Swansea, to be smelted. Hence the mines have an aspect singularly quiet, as compared to those in England: here no smoke, furnaces, or great steam engines, disturb the solitude of the surrounding mountains... My host says the two principal improvements introduced by foreigners have been, first, reducing by previous roasting the copper pyrites – which, being the common ore in Cornwall, the English miners were astounded on their arrival to find thrown away as useless: secondly, stamping and washing the scoriae from the furnaces – by which process particles of metal are found in abundance. I have actually seen mules carrying to the coast, for transportation to England, a cargo of such cinders. But the first case is much the most curious. The Chilean miners were so convinced that copper pyrites contained not a particle of copper, that they laughed at the Englishmen for their ignorance, who laughed in turn, and bought their richest veins for a few dollars.[10]

By the 1870s, the copper mines of Anglesey and south-west England were approaching exhaustion, their profits declining. Around this time additional super-rich copper discoveries began to be made in the North American Rockies. The huge Swansea smelters gradually starved to death as the low-grade 'porphyry' ores of those places were mined opencast by steam-shovel and smelted *in situ*. By the early 1900s the Swansea near-monopoly was obviously declining, and by the 1920s and the continued rise of smelted production from North America and central African colonies (former Northern Rhodesia and the Belgian Congo) the whole enterprise was comprehensively over. Here, as with the ironworks, we see that although entrepreneurship was all very well, it was still 'primary production' and had no lasting worth in the wider scheme of things – no copper-dependent manufacturing concerns could carry it on.

Yet today much of the nineteenth-century terraced housing built in the lower Swansea Valley remains: estates and streets still bear their paymasters' names in Grenfelltown and Trevivian adjacent to the long-disappeared smelting complexes of Hafod, Morfa, Upper Bank, Middle Bank and White Rock.[11] Surviving boundary walls in these estates feature unrendered quoins of characteristic dark, vesicular copper slag – pointing proudly to their provenance. Sight of them recalls Glamorgan poet Idris Davies' numerous elegies for South Wales extractive and processing industries, in particular the ironic Wordsworthian take in:

> The daffodils dance in the gardens
> Behind the grim brown rows.
> Built among the slag heaps,
> In a hurry long ago.[12]

17

England's West Country: Somerset and Gloucestershire

> Much of the Somerset coalfield is a low plateau deeply cut by valleys,
> a pleasant rural area with occasional black tips along the hillsides but
> with none of the grimness of most coalfields.

A.E. Trueman, *The Scenery of England and Wales.* Trueman does not inform his readers
that all of the British coalfields away from the bigger industrial towns and cities were,
in fact, 'pleasant rural areas' and if there was an admixture of industry it was often
welcomed from the point of view of people's employment, despite its 'grimness'. (1938)

Of Place and Time

The geographic proximity of north Somerset/south Gloucestershire and the Vale of
Glamorgan is notable, the outfall of the River Avon only thirty or so kilometres across
the Outer Severn estuary from that of the Taff below Cardiff. Inland it is just double that
distance from the Forest of Dean in west Gloucestershire to the South Wales coalfield
margin at Pontypool. As we shall see, the geological similarities are also strong. Yet
the historic and economic legacy could scarcely be different. On the one hand there is
Cardiff's relatively late (mid-nineteenth century) emergence as both the administrative
and cultural capital of Wales and as a world-class exporter (with Barry) of Rhonddha
steam coal. On the other, Bristol is a Norman-founded city bordered by a coalfield within
an ancient Royal Chase whose enormous pastoral hinterlands fed woolsacks and cloths
to Europe through its port, the second in England. It also dealt with the processing and
export of lead and zinc smelted from ores mined from the Mendip orefield and, like
South Wales, copper from south-west England. Most importantly, after colonization of
the Americas its vigorous mercantile marine quickly became the dominant focus of cross-
Atlantic trade in produce (tobacco, sugar) and, later, to its now-recognized detriment,
part of the African trade in human beings. The Forest of Dean had no such maritime

connections, but its iron industry was long established. Coal mining here had a unique medieval legal construct, that of the 'Freeminers' – their right to mine independently as determined by fellow miners – which lasted until reorganization, mechanization and the onset of deep-shaft mining in the late nineteenth century.

It is easy to see from geology and topography why the Bristol and Somerset Coal Measures were exploited before those of upland Glamorgan. Although the hinterland was much smaller it was largely lowland, drained by slow-flowing channels of the Avon catchment. In its pastoral vales the gently dipping synclinal Coal Measures were unconformably overlain at sometimes shallow depths by soft Mesozoic mudrock. Consequently there was much early shallow shaft mining to several tens of metres depth. Canals were easily cut, eventually connecting the navigable Avon above Bristol eastwards to the Thames catchment. Its succession of Measures include a much-attenuated presence of the landscape-forming Pennant Sandstone Group – there is no equivalent of the Glamorgan Plateau with its high, wet and craggy wastes.

Geological Background

Coal Measures outcrops occupy only some twenty per cent by area of a large subcrop (Fig. 17.1) in several downfolded (synclinal) structures (Radstock, Pensford, Coalpit Heath, Easton). These contain thick (up to 3000 m) successions that overall thin northwards. The downfolds are bounded to the south by the Early Carboniferous limestones of Mendip – a fold-and-thrust belt of spectacular upland-forming anticlines – asymmetric upfolds whose steeper-dipping strata often face northwards.[1] Coal Measures successions are similar in many ways to those of South Wales. There is a lower 6–700 m with marine bands and several thick coals (though with few workable ironstones compared to Glamorgan) of which the Main and Kingswood Great are the thickest, up to 2.6 m, and the discontinuous but thick (nearly 2 m) Coking Seam (formerly mined at Mackintosh Pit). Overlying is the 800 m Pennant Sandstone, as in South Wales a succession of river channel deposits with a southerly derivation. Several coals towards the base were worked in the Bristol area, notably the Parrot, Buff, Millgrit and Rag, some being notably good for coking. Finally there is a thick uppermost 1200 m succession of coal-bearing mudrock strata in the Grovesend Formation with ten or so workable but mostly thinnish (≤ 0.7 m) seams that tend to split and diminish northwards.

The Late Carboniferous Forest of Dean coal and iron ore field is a heart-shaped outcrop with apices roughly at the historic towns of Coleford by the River Wye and Cinderford and Lydney on the Severn. It forms a plateau rising to 500 m elevation between the two rivers – a synclinal inlier with encircling outcrops of Old Red Sandstone and Early Carboniferous limestones and sandstones. It was in these surrounding limestones that the iron ores resided as relatively high-grade lenses and fissure-fillings concentrated along bedding planes and enlarged joints. This was particularly so at the junction between limestone formations having different structures and permeability, as within the Crease Limestone at its upper junction with the Whitehead Limestone.

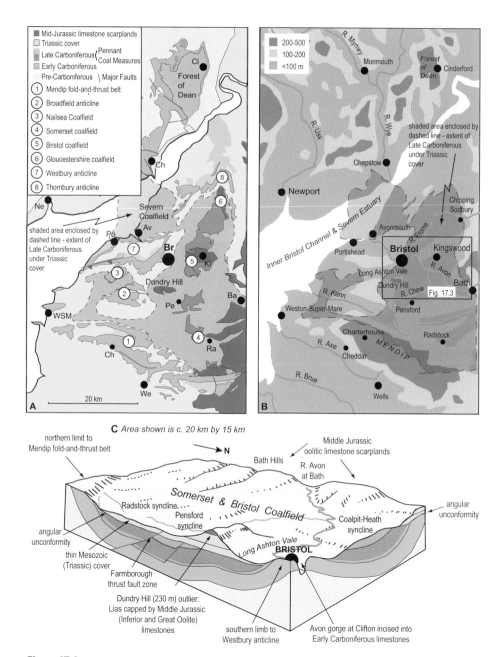

Figure 17.1 **A–C** West Country landscapes. **A** Solid geology, **B** Topography, **C** Three-dimensional panoramic view. The rectangle in B is the approximate area of Figure 17.3. As in South Wales the drainage pattern 'cuts across' the geological structure and the various rock divisions present in the Carboniferous fold arrays that characterize the region – most notably and spectacularly as the River Avon cuts a gorge through the Carboniferous limestones at Clifton. Again, this suggests that much of the present landscape is second-hand, exhumed by Cenozoic erosion of Mesozoic strata, the older outcropping units of which now lap around the folded Carboniferous structural margins with unconformity. In addition the highest points of the Mendip plateau contain fissures in the Early Carboniferous limestones filled with a variety of Mesozoic deposits and fossils, from Triassic to Upper Liassic in age, proving conclusively they were once overlain by Mesozoic strata. Sources: material redrawn, recrafted and recoloured from: A the British Geological Survey (2007a), B *Times Atlas of the World* (1987), C Trueman (1949).

The ore was largely composed of the mineral goethite, believed to have formed by downwards percolation of iron-bearing solutions of acidic groundwaters that dissolved out the more permeable limestone, replacing it with ferric iron precipitate.[2] This mode of mineralization, *per descensum* (from above), is borne out by the general decrease of ore with depth in the limestone strata. It is possible that the leaching process involved a cover of iron-rich Triassic strata, and that it took place during a period of hot, wet tropical weathering in the Early Cenozoic era. At that time much of western Britain was uplifted and its thick Mesozoic cover stripped away, as also happened in South Wales.

Like South Wales, the coalfield structure itself is asymmetric, with a steeply inclined eastern limb (along Staple Edge) opposite a gently inclined western limb whose structure enabled easier access via surface adits, and where working was long concentrated. The four hundred or so metres of Coal Measures strata are the equivalent to the Pennant Sandstone of Glamorgan. They rest unconformably upon the Early Carboniferous limestones. There are no equivalents here to the Middle Carboniferous and South Wales Coal Measures. A lengthy time gap of perhaps fifteen million years at the unconformity suggests that a pervasive uplift and erosional event occurred before Pennant times. The lowest 100 m or so of the coarse-grained and pebbly basal sandstones include two widely worked and thick seams, the Trenchard and Coleford High Delf (up to 1.6 m), both high-volatile, gas-prone and general purpose/domestic coals. Initially mined from numerous peripheral adits around the western outcrop, the seams were the target of deep shaft mines sunk much later in the late nineteenth and early twentieth centuries in the centre of the coalfield. The Pennant Sandstone here, as in South Wales and Bristol, was the sedimentary product of southerly-derived river channels. It is overlain by a group of coals within a mudrock-dominated succession. These form lower ground around the inner periphery of the central basin. The eight coals, their thickness varying spatially and often at the limit of workability (around 0.6 m), have evocative names – Churchway, Breadless, Rockey, Starkey, Parkend, Twenty Inch and Crow. One or more of these must have been a decent coking coal, because several Darby-style ironmasters, including Robert Forester Mushett, who made valuable contributions to the Bessemer process (*see* Chapter 20), set up furnaces and forges in the coalfield in the nineteenth century. Above these youngest coals the outward ramparts to the field are high scarps of massive, reddened sandstones.

Bristol Coal Mining

The 'top ten' cities of England by population in the 1660s are based upon extrapolations from Hearth Tax records. They were, in descending order, London, Norwich, York, Bristol, Newcastle-on-Tyne, Exeter, Ipswich, Great Yarmouth, Oxford and Cambridge. Of these only Bristol and Newcastle owed their economic growth and domestic comfort to local availability of cheap coal. In the Bristol area coal and coke was to pioneer and power the growth of metal refining and making (copper, zinc, tin, brass), glass and pottery manufacture, and brewing and sugar refining over the next century and more.

The early history of Bristol coal mining centred on the Kingswood anticline dividing the Gloucester and Pensford coal basins.[3] Here, thick seams in Lower and Middle Coal Measures outcrops extended in shallow subcrop under the city suburbs. The outcrop included the 3400 acres of the Royal Forest of Kingswood Chase whose records for 1223 reveal charges made for persons there digging 'sea cole'. A further record in 1371 mentions it dug by permit from Edward III. These references imply that in addition to domestic consumption and a local ship-borne south-west England trade, there may have been also a trade with London. Coal exports figure prominently in Bristol port accounts from the sixteenth century onwards, but generally for relatively small cargoes of up to a few score tonnes. J.U. Nef (1932) has explained these by stressing the varied nature of the Bristol exports and that European shipping bringing in imports (Scandinavian bar iron for re-export, wine, spices, sugar and tobacco) would leave with West Country cloths and manufactured articles, the coal serving as saleable ballast. A 1615 survey of the Royal Forest provides one of the first written threats of deforestation due to coal mining in England, reporting that:

> the cole mines also devour the principal hollies in all parts of the Forest for the supportation of these pittes...

After the Civil War Kingswood Chase was privatized into 'liberties' under a succession of rival wealthy local landowners who leased mining rights to master colliers in return for fifteen per cent or so of the mined coal's value. By 1670 there were seventy-two pits, increasing to over a hundred by 1790. In the early 1700s there were over a thousand miners (perhaps including cognate trades like the many wagoners observed by Celia Fiennes) living in the area. They were independent and determined men who flexed their industrial muscle during disagreements on coal pricing with the leaseholders and against the establishment of toll gates along the roads where their coal was destined to pass in wagons to the Bristol depots and wharfes[3]. One can only assume their forcible (and successful) endeavours were encouraged by both mine captains and by the wagoners themselves. Most of the later larger pits nearer Bristol were sunk around the turn of the nineteenth century, with the last two big collieries at Deep Pit and Speedwell in 1850–60, these closing 60–70 years later. The earliest pits in Somerset were in the southern outcrops south of Midsummer Norton with concealed measures to the north developed from 1730 – the deep Old Pit in the centre of the Radstock syncline was sunk in 1763.

Bristol Metal Smelting: Lead, Copper, Zinc and Brass

[4]Lead ores were mined from pre-Roman times in the Mendip orefield just twenty or so kilometres to the south of Bristol. Charterhouse (Fig. 17.2) was the oldest established site, dating to very early Roman times. Recently, traces of its *in situ* smelted lead (the galena ore was roasted to oxidation in open wood-fired hearths) have been detected in nearby cave stalagmite deposits.[5] Local coal was first used in an innovative manner

Figure 17.2 Worked-out lead/zinc veins at Charterhouse-in-Mendip (51.180001, 2.432543) show up clearly in the grassy limestone landscape as hummocky ('gruffy' in local dialect) ground in linear surface depressions ('rakes' as they would be called in Derbyshire), excavated along the surface outcrop of lead veins by miners from Roman times onwards. Charterhouse was a major centre for the mining and smelting of silver and lead from the earliest years of conquest. Industrial archaeological surveying reveals that the area features black glassy slag heaps, round stone-lined 'buddle' pits used for washing the ore, smelting plants, old flues and a complex network of dams, leats (artificial water channels) and old mineshafts. Mining was resumed in 1283 by a cell of Carthusian monks from Witham, near Frome, who obtained a royal grant for the purpose. The origins of their order at Chartreuse is the source of the name of the modern hamlet, where the remains of mining activity are now proudly preserved as industrial heritage. Many similar Roman and medieval mining localities about the Mendip were revisited from the seventeenth century onwards for their discarded zinc carbonate minerals, whose smelted metal formed the essential component of brass alloy and which led to the centring of the brass industry in Bristol. Source: Wikimedia Commons©NotFromUtrecht.

in the Bristol area in the 1680s for smelting lead at the Stockley Vale lead works in an area now known as Nightingale Valley near Clifton Suspension Bridge. Here there was a reverberatory furnace (locally called a 'cupilo') fuelled by a coal-fired heat source, a chimney providing the intense up-draught necessary for the passage of the hot air over a bed of pulverized galena ore. John Coster II was one of the lead smelters who went on to develop and patent the rather involved coal-based reverberatory technique to smelt copper-iron (chalcopyrite) sulphides in 1687–88. This was later developed in the 1690s in the Forest of Dean (where Coster's father had been an iron smelter) at the Redbrook Works west of Coleford (towards the Wye) using local coal and Cornish copper ore. By the late 1690s the operation there had become the largest copper smelting operation in the country with around 1000 tonnes of ore being processed each year. The new smelting process was subsequently taken up back in Bristol by Gabriel Wayne on a site by the Avon at Conham, a couple of miles upstream from the city (Location 2, Fig. 17.3). John Coster's son Thomas became MP for Bristol and invested in copper smelting interests at Neath and built the White Rock Works in the Lower Swansea Valley, signalling the beginning of the aforementioned South Wales hegemony of copper smelting.

By the early 1700s coal had enabled a large range of industries to be set up in the Bristol area. The rather specialized requirement for coke in the Bristol brewing industry – providing a cheap, steady, low-level and non-sulphurous source of heat for malting barley – led indirectly to Abraham Darby I's decision to try the same method (though at much higher temperatures) for iron smelting. Meanwhile the Bristol Brass Company started by Darby and his associates at the Baptist Mills on the River Frome (Location 3 on Fig. 17.3) in the early 1700s flourished. The brass furnaces provided the raw material that water-powered trip hammers milled and noisily pounded out (thus the name 'battery' mills) into the increasingly popular alloy used for the local manufacture of

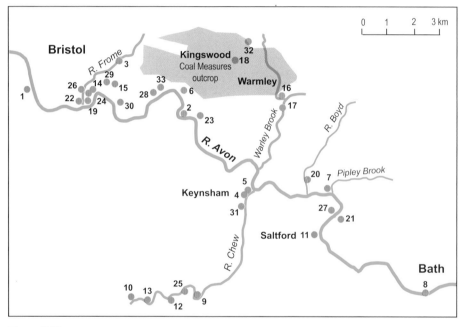

Figure 17.3 Sites associated with brass making, manufacture and marketing in the Bristol–Bath area along the River Avon and tributaries, numbered in chronological order from the 1680s onwards. The location of the map is shown in Figure 17.1B. **1** Stockley Vale/Rownham Copper Works **2** Conham Copper Works **3** Baptist Mills **4** Chew Mill, Keynsham **5** Avon Mill, Keynsham **6** Crew's Hole Copper Works **7** Swinford Rolling Mill **8** Weston Brass Mill, Bath **9** Woodborough Mill **10** Bye Mills & Belton Mills **11** Saltford Brass Mill **12** Publow Mill **13** Pensford Mill House **14** The Brass Warehouse, Queen Street **15** Babers Tower, Bristol **16** Warmley Works **17** Hole Lane Mill **18** The Cupola, Kingswood **19** Small Street Counting House **20** Bitton Mill **21** Kelston Mill **22** St Augustine's Back Warehouse and Foundry **23** Hanham Smelter Works **24** Corn Street Counting House **25** Woolard Mill **26** Lewin's mead Wireworks **27** The Old leather Mill, Saltford **28** Netham Brass Works **29** Redcross Street Warehouse **30** Cheese Lane Works, Bristol **31** South Mill, Keynsham **32** Soundwell Brass and Zinc Works **33** Blackswarth Lane Works. Source: redrawn and coloured after original map by Day (1988) with the addition of the extent of the Coal Measures outcrop around Kingswood.

domestic, decorative, musical and practical (clocks, compasses, surveying levels) items. A total of over thirty historic sites associated with copper smelting and brass making, manufacturing and marketing along the Bristol–Bath axis are spread out along and adjacent to the River Avon and its several tributaries. The industry grew up here because Quaker entrepreneurs like Darby and his successor Nehemiah Champion had high levels of practical and technical 'nous'. They took advantage of the combination of water power and the availability of local Mendip zinc ore, together with copper smelted in novel and innovative ways (see below) from rich sulphide ores transported by ship from South Devon and Cornwall.

The particular zinc ore required for brass production was sourced from the long-worked Mendip orefield where large reserves of zinc carbonate (calamine, now known as smithsonite) occurred. Mendip was almost unique in this respect in Britain, for in other metal mining fields the zinc occurred predominantly as the mineral sulphide, 'zinc blende', (sphalerite), usually with only minor amounts of the oxidized carbonate. Sphalerite could not be used in the brass-making furnaces at the time. They used a historic method dating back to the Romans and rediscovered in medieval Germany (pictured in Agricola's sixteenth-century opus *De Re Metallica*). This was the almost metamorphic reaction of vaporized zinc released slowly from powdered and calcined calamine (as zinc oxide) interlayered with fragmented red-hot sheet copper. Such copper 'brassification' was known as cementation, carried out over periods of up to nine hours in seven or eight crucibles arranged above a hot coal fire in special beehive-shaped furnaces a couple or so metres high and about half as wide. Remains of such furnaces were discovered in the 1980s by industrial archaeologists amongst the ruins of the Warmley Works of the Champion family (*see* below). Also preserved today to some extent, chiefly at Saltford (Location 11, Fig. 17.3), are large coal-fired annealing furnaces of the numerous battery mills along the Avon. These were for the heating-up and softening of partially finished, hammered hollow-wares produced by the trip-hammers, which hardened the cooling brass by their frequent and violent percussions.

In 1738 after a long tour of European brassworks, Nehemiah Champion's youngest son William patented a new process of industrial-scale zinc smelting based upon a furnace whose reduction of zinc oxide and charcoal at around 1000 degrees centigrade in sealed pots was accompanied by the transfer of zinc vapour downward via iron tubes into water within a condensation chamber. This distillation process avoided contact of the zinc vapours with atmospheric oxygen, which would revert them to the oxide. It produced around 400 kilograms of zinc per 70 hour charge from six loaded crucibles in the furnace, using an astonishing amount of coal in the process. Champion's initial works were on the Old Market in Bristol City, but in the face of understandable public complaints about the fumes emitted, in 1742 he set up a huge integrated copper and brass works at Warmley on family land (Location 16 on Fig. 17.3). Eventually this had over thirty furnaces and a large reservoir for water-powered machinery replenished by a Newcomen engine. The works included his own house in landscaped gardens and houses for his several hundred workers, the whole site being one of the largest industrial complexes in Europe at the time. It had the further and most decidedly overwhelming economic advantage of being sited immediately adjacent to the Kingswood Coal Measures inlier and its concentration of collieries, thus saving five or so miles of expensive wagon transport for the large quantities of coal required to power the engines and furnaces in the works.

Around 1758 Champion also set up furnaces, smelters and wire mills in Flintshire. When his zinc patent ran out, other brassworks used his basic furnace design, particularly around Liége, where abundant calamine ores were found. The undercapitalized Champion could not compete and became bankrupt in 1769. Mendip calamine may also have been worked out by this time, necessitating expensive imports from the Belgian orefield. William's elder brother John had previously adapted their father's process for smelting copper sulphide ores to those of zinc sulphide, whose reserves across English and Welsh mining districts were very large and almost undeveloped. Like copper-iron sulphides, the zinc sulphide ores were calcined to oxides by driving off the sulphurous constituent in a reverberatory furnace. By 1781 the Champion brass process became redundant with the invention and patenting of the direct alloying of metallic zinc with copper by James Emerson.

Graffin Prankard: A Quaker Iron Merchant in the Slave Trade

[6]Bristol's varied trade in the early- to mid-eighteenth century included the vigorous import of Swedish bar iron via the Baltic port of Danzig which, together with Russian and Spanish iron, formed an important part of the international iron trade of the time. The primacy in the eighteenth century of Swedish bar iron (*see* Fig. 20.4C) smelted from low-phosphorous magnetite ores has already been noted and was the *sine qua non* of serious steelmakers in Stourbridge, Birmingham, Sheffield and Newcastle at that time, as well as the steel furnaces at Keynsham near Bristol itself. By 1700 Britain was Sweden's chief customer for its iron, taking nearly half of its total annual output of around 15,000 tonnes. By the 1730s, iron imports were equal to the entire home production, a new locus being the Urals ironworks of Tsarist Russia. From the 1760s to the outbreak of the Napoleonic Wars these various sources dominated British imports. Disruption during the wars, the imposition of import tariffs, and home-produced bar iron from the new Cort-process fineries (puddling and rolling mills) eventually reduced post-war imports. An important exception was the traditional preference of Sheffield master cutlers for Swedish *Öregrund* to make their special steels.

Although most Swedish bar iron came into port at London, Bristol in the 1730s began to take iron directly from Stockholm and Gothenburg together with Russian iron. The details of the trade were revealed by the survival of accounts and letterbooks of Graffin Prankard, described as the leading iron merchant in western Britain, whose large bar-iron imports into Bristol in the 1730s were disseminated via his myriad of contacts throughout the Severn basin and further afield. The records:

allow the links in the commodity chains that stretched from *Bergslagen* (the mining district of central Sweden) to the steel furnaces of Birmingham, the boatyards of west Walian river estuaries, or the slave markets of West Africa to be put together.[6]

It is the latter link that surprises the modern reader, for we also hear that Prankard was from an old Quaker family who had married into another such. Further we discover that he was a close associate of Abraham Darby's circle of Quaker merchants in Bristol, indeed to the extent that he was a founding partner in Darby's Coalbrookdale Iron Company. The extent of his involvement in the slave trade (the 'African Trade') involved control of Swedish *Öregrund* cast into precise masses (25 pounds) of a certain bar size. This was the currency of choice in the bartering process that involved the interaction between indigenous African slave procurers, iron workers and Anglo-American slave ship agents. Such iron currency was ordered yearly by Prankard via his Stockholm agent, one Francis Jennings, from the forges of Uppland and brought over to Bristol, where it was sold on and then re-exported by the numerous slaver partnerships there at the height of their infamy. Prankard monopolized the supply of this fine-iron 'currency', widely and euphemistically known as 'voyage iron'. All this was a generation or two before Quakers shifted their moral beliefs and responsibilities to become the most prominent of abolitionists.

Coda: Bristol Coal Measures and the Development of Stratigraphy

The long tradition of coal mining in the Bristol Coalfield led to key contributions to geology in general, and to the Carboniferous in particular, by John Strachey of Sutton Court in south Somerset (1671–1743).[7] He provided a detailed depiction of the likely nature and extent of faulted underground rock in a paper published by the Royal Society in 1717 (Fig. 17.4).[8] His magnificent engraved section (he does not mention Sinclair's work) is fine by any standards, but for the time it is exceptional. It shows that his own land was underlain by the same coal-bearing strata that were being mined on his neighbours' estates. He shows sets of strata passing upwards from depth and cropping out (basseting) at the surface. One group of strata tilted downwards to the south-east and contained successive coal seams (his *veyns*). Another group of Triassic (Strachey's 'Red Earth') and Liassic ('Lyas') strata drawn above the coal-bearing strata are only gently tilted and are discordant upon them: an unconformable relationship documented and explained by James Hutton seventy years later. The stippled zone in the centre of Strachey's section is a faulted zone (one of George Sinclair's 'Troubles') that displaces the coal-bearing strata. Strachey describes this as:

> A ridge which breaks off the veyns and makes them trap down or trap up from their regular course.

The famous pioneer of British stratigraphy and geological mapping, William Smith, when a young man, learnt much from (but acknowledged nothing of) Strachey's work on the Somerset Coal Measures as he undertook his first commissioned fieldwork in the

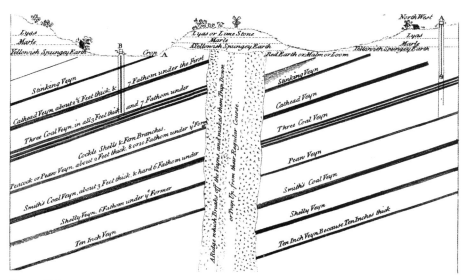

Figure 17.4 James Strachey's original geological section of 1717, nominated as the first ever 'modern' such construction. He shows seven named coal seams (his 'veyns') dipping away south-eastwards from their outcrops or from under younger overlying Triassic ('Red Earth') and Jurassic ('Yellowish Spungey Earth, Marls, Lyas') strata. The vertical extent of coal-bearing strata recorded is somewhat greater than thirty-four fathoms (*c*.64 m) of which the two mine shafts shown only penetrate into the top a little greater than fourteen fathoms (26 m). The deeper parts of the strata must have been penetrated elsewhere. The fault (his 'Ridge') throws the north-west strata downwards by about fourteen fathoms (26 m) with respect to the south-eastern strata. The section is an obvious improvement upon George Sinclair's more abstracted geometries of fifty years earlier, but Strachey seems unaware of Sinclair's work. The lively coal seam names do not survive to the present day, but were used into the 1950s when certain seams were still known locally as 'veins', with the thick and hard Smith's Coal Veyn (as the Smith's (or New Smiths) Seam) still proudly bearing its ancient moniker. Source: Strachey (1717).

area in 1791–92. We now know that he possessed a copy of Strachey's section, torn out of a copy of the Royal Society's Transactions and annotated in his own handwriting. It was found at the Sedgwick Museum, Cambridge, in the 1930s, loose in a trunk of papers belonging to his nephew and biographer, John Phillips (Fig. 17.5) who was Professor there in his latter years. Smith had evidently seen Strachey's section and the information that the coal-bearing strata contained diagnostic fossil forms, for above the mined coal seam known as the 'Peacock or Pean veyn'. Strachey had written:

> The Cliff [shale] also over this vein is variegated with Cockle-shells and Fern-Branches, and this is always an indication of this Vein.

This was not only the first published observation concerning the existence of fossil plants associated with coal-bearing strata, but also the first use of fossils in identifying

Figure 17.5 William Smith's uncle and biographer, John Phillips, as portrayed in a rather frightening image on the Yorkshire Geological Society's Phillips Medal. Source: Author.

particular strata – Strachey's 'Cockle-shells and Fern-Branches'. Smith has always been attributed as the originator of this approach, but clearly wasn't. Further, in a later paper of 1725 Strachey identified twelve distinct groups of strata that he had traced across England from Wessex to Yorkshire and on to the Northumberland coast, including, from young to old, with the bracketed words as the modern nomenclature of periods or epochs: Chalk (Late Cretaceous), Freestone (Late and Middle Jurassic), Lias and Marl (Early Jurassic), Red Earth (Triassic), 'Clives' and Coal (Late Carboniferous) and Limestone with Lead (Mendip, Derbyshire etc); (Early Carboniferous). Again, Strachey was the first to notice this swing of distinctive Mesozoic strata across England, and his observations imply a great deal of travelling and field observations to achieve such a synthesis. His hypothesis to explain the 'swing' involved an imaginative, and to us today, unlikely, Newtonian mechanism involving forces acting on the strata arising from the Earth's rotation soon after the Creation.

In his biography of his Uncle William[9], John Phillips elides all mention of Strachey, his pioneering papers and the fundamental stratigraphic advances showing dipping strata, outcrops and faults made seventy years before his uncle's first foray into mineral surveying. The nephew-biographer (perhaps too family-close to be an objective one) quotes Smith remembering those revolutionary years of 1791–92, taking credit himself for Strachey's section and conclusions:

> Coal was worked at High Littleton beneath the 'red earth' [Triassic marls], and I was desired to investigate the collieries and state the particulars to my employer. My subterraneous survey of these coal veins, with sections which I drew of the strata sunk through in the pits, confirmed my notions of some regularity in their formation.

Today we would call this sort of thing plagiarism – a regrettable conclusion considering Smith's own later addition to the type of geological mapping presaged and influenced by Strachey's sectional approach. Yet science must recognize and acknowledge priority of discovery as an absolute, regardless of time and reputation.

18

North Wales

The marsh along the Dee estuary is an ambiguous place: half solid, half liquid: part Welsh, part English…After a day of full exposure the sun is setting over the hills of Flintshire, the tumbled blocks that were once Denhall Quay provide a warm viewing spot. In the 18th century a busy colliery sat here alongside the River Dee, but gradually the coal ran out, the estuary silted up and grasses began to take hold. Now the quayside borders Parkgate, a vast saltmarsh RSPB reserve…

From Ella Davies' atmospheric *Guardian Country Diary* for 5 October 2019. Joseph Turner provides the ultimate sunrise for Ella over the silted-up Deeside moorings (*see* Fig. 18.6).

Looking Up from Deeside

Looking up to the west and north-west from Deeside one sees the limestone backslopes of the Clwydian Range and Halkyn Mountain (Fig. 18.1). At their eastern limits the strata of the North Wales coalfield and its overlying Permo-Triassic cover crop out. These descend, sometimes troublesomely faulted, to the estuary under a cover of Pleistocene glacial deposits. The North Wales peoples own a strong sense of their pre-industrial, Welsh-speaking history. Europe's largest and most complete Bronze Age mine workings for copper are at Ormes Head and the unique and marvellous gold Mold Cape, now in the British Museum, was buried hereabouts. The area also has numerous Iron Age hillforts. Deeside had Flint as its original seaborne outlet after the Romans had established their legionary stronghold at Deva (Chester) in the AD 70s. The latter's position and strength controlled the Dee waterways and the fertile lowlands to the east, separating and containing the stroppy upland hill tribes of North Wales (Ordovices) and the Pennines (Brigantes) to the east. Conquering the homelands of both had the promise of rich bounty from mineral veins of gold, copper, zinc and lead – these became rapidly

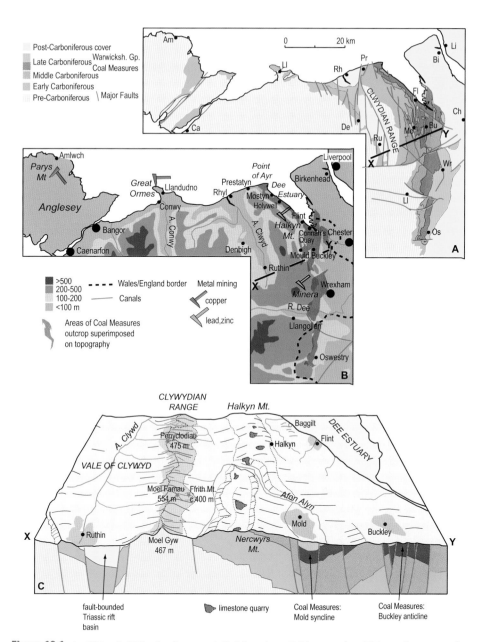

Figure 18.1 **A**–**C** North Wales landscapes. **A** Solid geology, **B** Topography, **C** Three-dimensional panoramic view taking in a geological section from Ruthin to Buckley. Source material redrawn, recrafted and recoloured from: A British Geological Survey (2007a) B *Times Atlas of the World* (1987). C is a three-dimensional panoramic view by the author with a geological section adapted from material in Davies *et al.* (1999).

exploited by legionary military engineers, forerunners of modern engineering geologists. It was lead ore from Halkyn that first provided payback to the Roman incomers, with early smelting facilities excavated near Flint and ingots surviving from Chester with the inscription DECEANGL. This names the Iron Age Deceangli tribe that called north-east Wales its home, and who doubtless provided the slaves needed to work the mines and to endure the privations involved in charging the open-hearth lead smelters, which emitted terrible acidic and sulphurous fumes. As we shall see, the Halkyn orefield became the deepest, largest and best-run in Britain by the mid- to late-nineteenth century – courtesy of Quaker and Unitarian mining entrepreneurs, engineers and social pioneers.

Late-medieval developments in trade, mining and industry[1] followed Edward I's invasion of North Wales, his policy of stern containment witnessed by the ring of huge castles at Beaumaris, Conway, Caernavon and at Flint itself. Local Carboniferous sandstone was used here, and many thousands of tonnes of local coal were transported to the other sites to fire the mortar used in their construction. Two building accounts remain – for half of both 1283 and 1295 – a total recorded carriage of 3000 tonnes from Deeside ports. The founding of Flint led to the spread of English settlements and included skilled lead miners allowed land allocations and independence to mine as they pleased – free independent men from Derbyshire whose Saxon surnames (Nuttall, Redfern, Bagshaw, etc.) still feature large in records after the Civil Wars and Restoration. By then Flintshire coal supplied local needs as well as sourcing a long-founded export trade with Dublin – not always reliable given weather conditions in the Irish Sea. It also satisfied the domestic and manufacturing needs of Chester, connected by the rapidly silting but still navigable Dee estuary channel and, later, the north-western termini of the Midlands canal network. A post-industrial heritage is preserved across the area in spectacular and celebrated aqueducts, heritage museums, industrial monuments, pottery, sculpture, nature reserves and accessible mine workings.

The worked coalfield in historic Flintshire stretched from Mostyn (named after coal-owning gentry whose mines dominated seventeenth-century production) southwards via the Greenfield valley at Holywell to Flint, Mold and Buckley. The Denbighshire portion lay across the Bryneglws Fault to Broughton, Wrexham, Bersham, Ponciau, Ruabon and Cefn Mawr, hence to Oswestry in northern Shropshire. The coalfield formed a two-way geological, cultural and economic borderland. It was the western limit to the lowland spread of the Triassic-floored Cheshire Basin and the eastern limit to a much-faulted Carboniferous hinterland that limits the Lower Palaeozoic North Wales massif and the Silurian bedrock of the Denbighshire Moors. This two-sided arrangement separated entire historical cultures whose often uneasy juxtaposition echoes down the ages in the Saxon/Cymru dichotomy made explicit by the preserved northern outposts of Offa's and Wat's Dykes.

Carboniferous Geology

The Carboniferous succession here is over two kilometres thick, deposited on the subsiding northern flank to the Wales–England–Brabant landmass. It was deposited in a tectonic environment distinct from that of South Wales and the West Country. Equally

Figure 18.2 Panoramic view looking north-east of the very northern part of the Clwydian Range scarp, believed to be near Trefnant, south of St Asaph (*c*.53.124116, 3.244281). It shows well-bedded crags of Early Carboniferous limestone resting unconformably upon Lower Palaeozoic (Late Silurian) slaty mudrock and siltstone strata. The Triassic bounding fault to the Clwyd rift valley lies between the vista and camera position – it throws up the Palaeozoic rocks against softer Triassic sediments of the valley floor (*see* Fig. 18.1). Source: Author.

distinctive were the effects of later rifting tectonics that ruptured midland and north-western England, the Irish Sea and south-west Scotland in the Permo-Triassic. Major north–south trending normal faults bounding new rifts like the Cheshire basin cut across older Carboniferous structures. The chief such fault in north-east Wales creates that most memorable of tectonic landscapes – the Vale of Clwyd (Fig. 18.2). This chops across the whole Carboniferous succession and throws part of it downwards under the Vale (where it is overlain by Permo-Triassic sediment) and part upwards along the northern slopes of the Clwydian Range. The latter was later eroded into a tremendous arcuate fault-line with solid, craggy, Lower Palaeozoic outcrops and to the east a subsidiary scarpland featuring the various formations of the one-kilometre-thick Early Carboniferous Clwyd Limestone Group. For thirty kilometres these pale grey-cream limestones memorably stride across the landscape as stepped crags down the dip-slope eastwards to the Cheshire plains. The limestone with its extensive lead/zinc mineralization and cave-ridden karstic underworld unconformably caps the Lower Palaeozoic core to the Clwydian Hills, including the peak of Moel Famau (554 m) north-east of Ruthin. In the south, after a right-jog over the major Bryneglws (Llanedilan) Fault just south of Caergwrle along Nant-y-Ffrith, the limestone scarplands and dip slope continue southwards to Llangollen in the spectacular Dee Valley via the mountains of Esclusham and Eglwyseg (511 m).

The coalfields themselves record a maximum thickness of some 900 m of Late Carboniferous strata. The lower two-thirds are the much-faulted Pennine Coal Measures Group; the remainder chiefly comprise red/ochre-coloured mudrocks and thin sandstones of the Warwickshire Group, generally devoid of mineable coals. The Coal Measures themselves thicken somewhat from south to north with numerous workable coals throughout: in the Wrexham area of the Denbigh Field especially the Nant, Fireclay, Main, Two Yard, Powell, Drowsell, Wynnstay Five Foot, Upper Stinking and Bersham

Yard Group seams. Many of these were fine steam-generating coals, but locally intense faulting sometimes made mining difficult and gave rise to unpredictable methane pockets, and hence the region's tendency to explosive disasters. Household coals also feature, yet the seams at Bersham must also have contained enough good coking coals to nourish the pioneering local ironworks in the 1720s. Eight marine bands occurring in the lower three-quarters of the Coal Measures gave rise to underlying sulphurous coals (eg 'Stinking Coals') with a thick sandstone member (the Cefn Rock) featuring towards the top.

Bersham was the last deep-mined pit in the Denbigh coalfield. It closed in 1986, with Point of Ayr the last mine in Flintshire a decade later. A rail link from the latter fed the coal-fired Connahs Quay power station whose exterior was all-too-familiar to anyone travelling westwards from the Lancashire and Merseyside conurbations to climb, hike or geologize in the mountains of North Wales. In an irony of economic history repeated time and again in Britain since the 1960s, offshore natural gas from the Irish Sea now flows into a terminal on the demolished mine site – the treated methane is pumped eastwards to Connahs Quay, where a gas-powered electricity generating plant stands in place of the former landmark concrete cooling towers.

Industrial History

Entrepreneurs and their accompanying capital poured into the convenient arc of outcropping and concealed Coal Measures of North Wales from the 1690s onwards. There came mineral leasers, metal masters and smelters (brass, copper, iron), then textile engineers, pottery technocrats and glass manufacturers. Later came a chemical industry based upon alkalis generated from coal-bearing acids, shipbuilding concerns, and all manner of cognate trades spawned by the Industrial Revolution. *In situ* resources like the variegated mudrocks in the Lower Coal Measures of the Buckley area east of Mold, together with clays of glacial origins, were extensively quarried and mined for flourishing potteries from the seventeenth century. These included distinctive and beautifully playful *sgraffito* decorated slipware (Fig. 18.3).

Many of the incoming industrialists had their major interests elsewhere, and

Figure 18.3 Slip-decorated earthenware dish with sgraffito design of bird and worm, made at Brookhill Pottery, Buckley, *c*.1640–1670. 'The sgraffito technique is produced by applying a slip coat of liquid clay over the whole of the body of the pot. When it is leather hard, a design is scratched through the slip to reveal the contrasting body of the pot beneath.' Source and quote: People's Collection Wales (2019) National Museum of Wales © Buckley Heritage Centre. Reproduced from Buckley Heritage Centre, Creative Archive Licence.

in moves that seem like very early rehearsals for modern-day globalization/outsourcing trends, they set up their satellite companies around Deeside. In this way better profits could be made from the close availability of cheap labour, cheap energy and a ready seaborne trade – nationally via coastal vessels and inland by the Midlands canal network. International trade was facilitated by the proximity to burgeoning Merseyside just the other side of the Wirral. The subsequent demise of the Deeside ports in the eighteenth century was caused partly by the growth of Liverpool, but mostly due to siltation of its own muddy estuary channels.

So it was that late-seventeenth and eighteenth-century Bristol and Somerset Quaker metal producers and manufacturers came here to exploit metal ores of lead and zinc, notably the London Lead Company, with the aforementioned William Champion looking to the abundant Halkyn zinc ores as his Mendip sources became exhausted. The blast furnace at Bersham[2] was the first in Wales to use coke to smelt local clay ironstone *a là* Coalbrookdale. The founder of the enterprise was Charles Lloyd, a Quaker blacksmith and friend of Abraham Darby with a shared background. The following terse diary entry (by a local man, one John Kelsall) on 3 December 1721 informs us of this historic first for Wales:

> I understood Bersham Furnace ceased this day blowing with charcoal and went on blowing with coakes for potting.[3]

Lloyd made hollow ironware (pots, pans, fireboxes, etc.) in sand moulds, Darby-fashion, from his coke-smelted iron but lacked the raw material and furnace skills to achieve a uniform quality product. After his bankruptcy Darby's son-in-law John Hawkins took over the plant, and by arrangement with Coalbrookdale concentrated on making hollow ware while the Shropshire plant specialized in castings for steam engines.

Native north Welsh copper barons like Thomas Williams of Anglesey invested heavily in Flintshire smelters to process the treated Parys ores from Anglesey and used rolling mills to make copper sheet from Swansea-refined copper. By the end of the eighteenth century, Lancashire and other non-local merchants and financiers set up the huge Cotton Twist Company at Holywell, the yarn mills using water power and the latest machinery along the steep gradients of the Greenfield valley.[4] Eventually, in the late nineteenth and early twentieth century, the Courtald family of Essex became the largest manufacturer in North Wales after they set up alkali-based chemical industries around Flint and Holywell including their own patented rayon yarn, a semi-artificial fibre manufactured using cellulose from wood pulp.

Nowadays the largest post-industrial concern involves the manufacture of Airbus aircraft wings at Broughton. Unlike their ancestral industries 200 years ago, they depend neither on local raw materials nor location, situated as they are at the forefront of technology more than a thousand kilometres from the Toulouse assembly lines to which their wings are flown in a special transporter. Their employees may wonder at

Figure 18.4 *Y Bwa* (The Arc) by David Annand (1996). Installed in central Wrexham, a miner and a steelworker in bronze are balanced on brick plinths at full leverage, each trying to bend steel pipe segments into one continuous arc. The work is full of both physical and dramatic tension. It deals with the efforts of unionized industrial labour to achieve symbiotic existence – the powerful miners' union (plus transport workers and railwaymen) formed 'The Triple Alliance' to help fight the labour conflicts of the 1920s. A poem in Welsh by Myrddin Ap Dafydd of Llanrwst is engraved along the arc segments. It explores the close bonding involved between all who have to undertake the manual work needed by society and industry. Source: Wikimedia Commons, Rept0n1x.

the extinct symbolism involved in the sculpture Y Bwa (The Arc) by David Annand in central Wrexham (Fig. 18.4)

Halkyn Mountain Orefield and its Long Adits

Halkyn Mountain and neighbouring Minera just to the south of the Bryneglws Fault (Fig. 18.5) were the pre-eminent Welsh base-metal mining fields from the early seventeenth to well into the twentieth centuries, their renown earned not just through tonnages of lead and zinc metal concentrates mined (second only in Britain to the North Pennines) but also in the manner of that mining.

This was due firstly to the London Lead Company, which from the 1690s sought out working prospects, took over existing concerns, paid close attention to ore preparation at the minehead (water-powered stamping and ore-separating machinery) and put the welfare of its workers nearer to the centre of its company philosophy than most others at this time.[5]

Secondly, in 1813, from the copper mines of Devon came the brilliant mining entrepreneur and engineer, Norfolkman John Taylor, with a huge reputation for hands-on management and the establishment of efficient mechanized mine procedures[6]. He took up as agent with full control of the rich lead mines owned by Lord Grosvenor, introducing the latest Cornish pumping and ore-dressing hydraulic machinery. His Cornish mine-

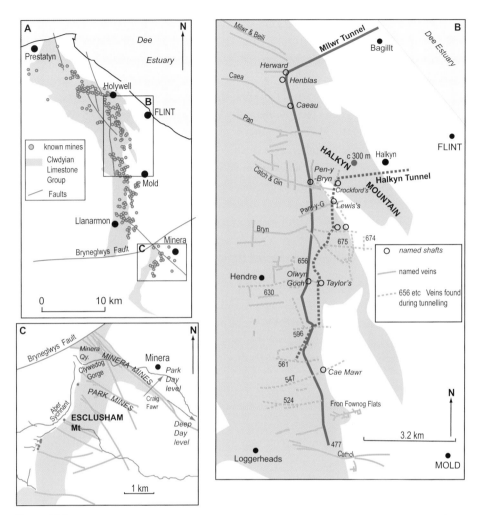

Figure 18.5 **A–C** Halkyn Mountain and Minera lead/zinc mineralization, mine localities and drainage adits. Sources: Redrawn, recrafted and coloured from material in British Geological Survey (2007a) C. Ebbs (2008; 2nd edn) and Appleton (2013).

captain, one Absalom Moore, was put in charge of day-to-day organization and of the rigorous worker supervision that characterized the professional and fair-minded south-western mining philosophy developed by Taylor. The ores at this time were shipped out of Flint for smelting, either to Bristol or to Chester. Taylor's technocratic approach included the drawing up of detailed mine plans with accurate and informative geological information as aids to mine planning and development, seen to great effect in his report concerning the Bryncelyn mine (Pen-y-Fron) at Mold.[7]

Thirdly, from High Victorian times into the modern era (to 1957), an amalgamation of mine company interests got together to form tunnelling concerns that single-mindedly forced some of the world's longest adit tunnel drainage schemes into fruition over a period of almost eighty years – the Halkyn and Milwyr Tunnels. The former was started by the Grosvenor estate in 1818, and we must assume that John Taylor would have had much to do with its engineering and geological aspects throughout the 1820s when he was successfully guiding the Grosvenor mines through a difficult period of low metal prices. Over the next fifty years the tunnel was driven through on the south-east side of the Halkyn Mountain to successfully drain the mines there for deeper working, the project taken over in 1875 by an amalgamated drainage company. Given the success of the Halkyn tunnel in accessing new and rich reserves, a second tunnel was planned to drain the deeper parts of the central mining area. It was begun in 1897 at sea level near Baggilt on the Dee estuary and driven up inland intermittently over succeeding decades at a uniform gradient of 1:1000 for a distance of sixteen kilometres, finally reaching Cadole, west of Mold in 1957. It drained more than fifty veins and intersected several rich discoveries on the way, forming a labyrinth of over ninety kilometres of interconnected passageways. A historian of the tunnel, Cris Ebbs writes:

> At a time when low ore prices were crippling the industry nationally, the working of new and existing ore deposits during the 1930s ensured that up to 650 men were employed by the mine and tunnelling records were being broken.[8]

Halkyn and Minera were both limestone-hosted orefields whose main minerals were sulphides of lead (galena) and zinc (blende) accompanied by an enveloping matrix rich in calcite and/or quartz. Ore veins were mineralized, steeply inclined, podded to tabular bodies, decimetric to metres wide and up to a few kilometres long. They occupied open fractures (many of them faults) that cut across the limestone strata of the Clwyd Limestone Group. The veins formed along open cracks by percolation from below (*per ascensum* mode) of warm-to-hot saline and mildly acidic fluids that precipitated out the ore and matrix minerals. In both orefields the veins have a persistent east–west to north-west–south-east strike, with steep dips up to thirty or so degrees either side of vertical, i.e. inclined either northwards or southwards – they were formed as the result of a regional stress field set up by Late-Variscan tectonics, *c*.300–290 Ma. Although most mineralization occurs as veins, there are also accumulations that follow stratal bedding of the host limestones and which are seen to replace it by chemical alteration. When such 'flats' occurred parallel to gently tilted bedding they were a joy to mine along road-like levels leading from a shaft. Veins were also followed along levels but were mined upwards from level floors in stopes, the ore falling down under gravity to be collected and run along the level floor to shaft bottom. Some flats in Halkyn and Minerva were fabulously rich, as were certain veins whose surviving metres-wide stopes extended tens of metres vertically and hundreds of metres along their whole strike length.[9]

Coda: The Deeside Sunrise of Joseph Turner

According to the monumental Tate Gallery catalogues of JMW Turner's works, the artist visited the area at least three times on various sketching tours. Flint Castle (Fig. 18.6) is his magnificent mature work of 1838, the ruined western side in shaded silhouette in front of the wondrous sky-filling sunrise over the dark, indigo waters of the Dee estuary. Picked out on the horizon to the left is the rising land of the Wirral's Triassic sandstone ridges. The human side to landscape is never far away in most of Turner's paintings, and so we see four beached craft on the Dee shore at low tide; two small square-riggers, a half-hidden schooner and an unrigged vessel with its cargo being unloaded onto horseback in the foreground. A disused mooring buoy of some size is cast on one side in the left foreground. The figure in the extreme right foreground is mending nets, the second figure to his left carries a cross-shaped item with the faint suggestion of netting around the cross-part, and carries a wicker creel on his back. This would be a salmon fisherman returning with his catch from a low tide excursion into the treacherous tidal channels of the estuary. The implement would be a haaf-net, a word of Norse extraction and extensively used down to the present day in north-west England and south-west Scotland.

The whole composition is a desultory reminder from Turner that the once strategic stronghold of the castle and the vital port of medieval to early industrial Deeside had fallen on hard times, though salmon still plied the shallow estuary channels. Yet the sunrises and sunsets still astonish!

Figure 18.6 *Flint Castle* by J.M.W. Turner (1838). Watercolour. Source: Private Collection, Japan. Image: https://www.wikiart.org/en/william-turner/flint-castle.

19

English South Midlands

The glowering carnival: nightly solar-flare
from the Black Country; minatory beacons
of ironstone, sulphur. Then, greying, east-northeast,
Lawrence's wasted pit-villages rising early,
spinning-wheel gear-iron girding above each
iron garth; old stanchions wet with field-dew.

Geoffrey Hill, CXXXVII, *The Triumph of Love* (1998)

Cradle of Revolution

The region became the cradle for the Industrial Revolution, its incandescent geography vividly recounted by Bromsgrove-born Geoffrey Hill in the excerpt above. Here was landlocked red-soiled lowland country founded mostly on Triassic marls and sandstones with an overlying fertile veneer of soils on glacial sands, silts and clays. Five coalfield inliers amongst this rich pastoral countryside contained the mineral riches that allowed alternative or tandem ways (when combined with agriculture) for families to make a decent living (Fig. 19.1). Here was mining of coal and iron ores, smelting, forgework and the manufacture of iron goods of every conceivable description, from weapons to wire to pins and nails – including all the widgets of their time. Two great navigable river outlets facilitated the export of both agricultural and artisanal manufactures. In the west the Severn drained down to its long estuary into the Bristol Channel, hence to the Western Approaches, Azores and Americas. To the east the Trent flowed up to the Humber where eastwards across the southern North Sea were the richest markets in Northern Europe – the old Hanseatic connections of eastern England's hinterlands of wool production and cloth making.

Although geology, geography and climate set up the basic necessities for the region there were also intertwined historical and economic factors in play. First, after the

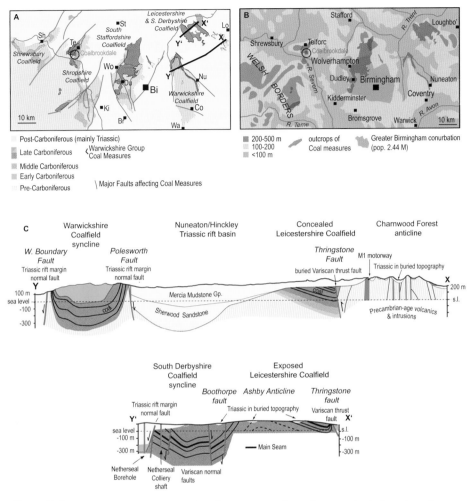

Figure 19.1 **A–C.** South Midlands solid geology, topography and two geological sections from the eastern coalfields. Sources: material redrawn, recrafted and recoloured from: A British Geological Survey (2007a) B *Times Atlas of the World* (1987) C British Geological Survey (2007a); Mitchell *et al.* (1954).

Civil War it hosted many independent-minded nonconformists whose business and technical acumen in the metallurgical trades originated from family forge and foundry enterprises (the Darbys were one such). Second was the mid-eighteenth-century alliance between, on the one hand, adventurous Whig landowners hungry for rental income from the subsurface coal seams invisible under their estates and on the other, pioneering industrial entrepreneurs like Wedgwood and Boulton. Such groupings of men forced through the legislation necessary for construction of the entire Midlands canal network as explored in Chapter 5.

Five Coalfields

Two, Shrewsbury and Shropshire, are fringing structures (like North Wales), situated almost astride the ancient rocks of the Welsh borders. Their productive Coal Measures are often concealed beneath thick and mostly unproductive Warwickshire Group strata. Shrewsbury would wait its turn for deep-mining development in the nineteenth century. The gently dipping measures at Coalbrookdale in Shropshire provided the majority of output from the sixteenth century well into the eighteenth – this from predominantly adit mines that enabled the first development of longwall extraction in Britain.[1] Through the efforts of families like the Darbys these became conjoined with the new techniques of coke-based smelting pioneered in Coalbrookdale from the earliest eighteenth century onwards. The three remaining fields, South Staffordshire, Worcestershire and South Derbyshire/Leicestershire are of a particular kind. Each is in the form of an inlier – folded and faulted Carboniferous and older strata (some Precambrian) surrounded and partially overlain by an unconformable cover of younger and softer Triassic marls. The inliers poke out like isolated islands, the remains of an eroded geography carved out in the Permian period when Britain was part of semi-arid eastern Pangea. The largest and most westerly field of South Staffordshire (Fig. 19.2) had its stupendous Thick Seam (up to ten metres) that provided an abundant source of cheap energy. This enabled a rich heritage of diverse medieval to early modern metal-working to prosper. Later on precision machine manufacturing came courtesy of families like the Boultons and Wilkinsons, the latter originally from Cumbria.

Some Geological History

During Late Carboniferous times the west-to-east spread of the five separate coalfields we see today was one continuous belt of swamp forest, mires, lakes and deltaic distributary channels. It lay twenty or so kilometres south of the main axis of crustal subsidence in south Lancashire and north Derbyshire that accommodated their sedimentary deposits. Yet the area had little history of earlier Carboniferous subsidence, for almost everywhere we see Late Carboniferous strata resting upon either very thin successions of older Carboniferous (as in Coalbrookdale) or with marked unconformity upon much older Palaeozoic and Precambrian rocks (Fig. 19.2) as in the country around Dudley, Nuneaton and Charnwood Forest. Further, the total thickness of Coal Measures strata was only a fraction, sometimes only a small fraction, of the fifteen hundred metres reached to the north in the South Pennine coalfields. Nevertheless the percentage of strata that was coal increased.

What was happening here to cause such marked geographic variation?

One idea is that the 'thermal sag' subsidence experienced by Late Carboniferous northern Britain spread southwards at reduced rates onto the flanks of what had been a persistent upland (the Wales–England–Brabant massif) of largely unrifted crust during the Early Carboniferous. In the first half of the ten million year course of the

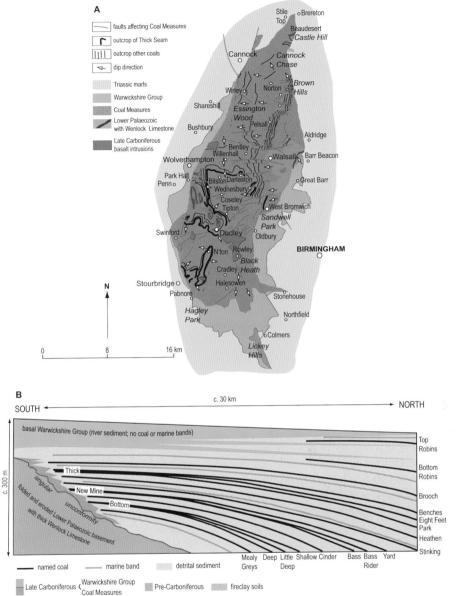

Figure 19.2 **A–B** South Staffordshire coalfield **A** Redrawn version of Jukes' (1859) pioneering hand-coloured map of the entire South Staffordshire coalfield showing individual seams, faults and dip directions. **B** Schematic south to north section showing the repeated trend of 'seam-splitting' as seen in the South Staffordshire coalfield, and as explained more generally in Figure 9.7. Source: material redrawn, recrafted and recoloured from Hains and Horton (1969).

Late Carboniferous the land surface here was subsiding at only a fraction of the rate further north. This gradient of subsidence encouraged the wandering sediment-laden delta distributaries of the Pennine River to occupy deeper bays and lagoons to the north and led to preferential accumulation on the flanking land surface of both swamp forest peats and of clay ironstone deposits in intervening mudrock sediments. Consequently the proportion of coal in the Coalbrookdale Coal Measures exceeds ten per cent (with eight seams over a metre or so thick) and that of clay ironstone around seven per cent. Both figures are extraordinarily high compared to the typical average of three or so per cent in the main subsiding Pennine basin. This concentration led to the easy exploitation of multiple seams (Fig. 19.2B), first from outcrop, then from shallow adits and finally from shafts drained by Newcomen engines by the mid-eighteenth century.

The situation just described did not last into the latter part of the Late Carboniferous because of the arrival into the Midlands area of the effects of a new tectonics induced by a mechanism that both massively increased crustal subsidence and brought new south-sourced sediment into play. Such were the effects described previously that had also led to the arrival of the vigorous Pennant rivers into South Wales and the West Country – the loading of the Welsh and English crust by Variscan mountain-building tectonics. Over the next few million years great thicknesses of the Warwickshire Group would be laid down over the entire south Midlands, much of it devoid of forest swamp environments. A notable exception was deposition of the Halesowen Formation, which contained significant peat deposits, but none on the scale of the earlier Coal Measures.

Why Coalbrookdale?

The wider Coalbrookdale district, in particular the area south of the Severn gorge surrounding Broseley, Benthall and Willey, had long contained some of the most productive and profitable coal mines in England. This graphic quote from the Victoria County History of Shropshire explains the situation pertaining in pre-industrial times:

> From Benthall the seams of coal are continuous through Broseley, where they lie so near the surface in places that the inhabitants not infrequently meet with it in their gardens and cellars, and here on the upper side of the Broseley fault their depth seldom exceeds twenty yards.

It was this accessibility, created by low stratal tilt, combined with easy access to the nearby navigable River Severn, that led to its pre-eminence – as historian John Hatcher nicely puts it: 'Here was a lesser Tyneside.'[2] Given these simple geological conditions it is no surprise that three Coalbrookdale collieries between them had a total annual output in the seventeenth century comparable with that of the North East. The parish of Broseley was the leading centre as early as 1575, with the local manorial lord, one James Clifford, as chief entrepreneur. The coal was carried downhill from the mines along wooden wagonways, Tyneside-style (the idea borrowed from there) for a kilometre or so to the many wharfs along the Severn gorge. This involved such minimal transport costs that the price downstream at Tewkesbury, fifty miles away at the confluence of the Severn and Avon, was only six shillings a tonne in 1670. This was about one-third of the London price, since it was exempt from the hefty tax on coastal transport. As such it was widely competitive right down to Bristol itself and even beyond into north Devon.

As a prominent brass manufacturer and sometime brewster in Bristol, Abraham Darby would have known of the good coking properties of certain of the Broseley coals. Yet, as we have seen (Chapter 3) Darby was not interested in setting up on the Severn's right bank simply as a coal or iron ore producer and merchant. His chief interest was iron smelting and the direct casting into sand moulds of sturdy and practically beautiful domestic ironware – the Le Creuset of its day – particularly the lovely three-legged, broad-bellied cooking pots to be seen today as exhibits in the Coalbrookdale Museum. So it was that he settled upon Coalbrookdale, a left bank tributary to the Severn, its drainage catchment comprising steep-gradient streams that had cut down through the Coal Measures to the main Severn gorge from a broad 200–250 metre elevation plateau to the north. The local geology of the catchment provided a cornucopia of raw material riches spread compactly over some twenty square kilometres (Fig. 19.3).[3] Topography and geology combined to create a natural gravity-driven 'vertical' supply chain. On the valley sides and plateau tops, coal and iron ore were mined. Most important to Darby were the coking properties of several of the coal seams – the two low-sulphur Clod Seams in particular – grist to the mill for Darby's inquisitive and experimental mind, which would lead him quickly on to achieving his ambition of smelting local ironstone with his own coke.

The ironstone ore itself was standard Coal Measures clay ironstone, here comprising two varieties, one the nodular 'Baull stone' and the other a more stratified 'Flat stone', the former mined or quarried from above the 1.2–1.7 metre thick Top Coal and the latter from above the 0.9–1.8 metre Big Flint Coal in outcrops above the Dale.[3] There is a unique artwork showing womenfolk picking ironstone nodules from shale at the ironstone mines of Madeley Wood (Fig. 19.4).[4] The ore was gathered into layers on the ground, where it was weathered by repeatedly turning. Any remaining adhering claystone that lowered the ore grade was chipped off by hammer, an outdoors, labour-intensive job for both the women and children. Finally the ore was calcined (the carbonate driven off by roasting) prior to incorporation in the furnace charge. Local Silurian-age Wenlock Limestone was used in the charge as a flux to melt and meld with remaining clay impurities in the ironstone to form a fusible calcium silicate slag (a sort of artificial metamorphic rock), of the kind witnessed by George Borrow streaming over the hillside at Merthyr.

The wagon ways that brought the coking coals and ironstone downhill to the Severn passed adjacent to Darby's smelting site (the pre-existing furnace dated from the seventeenth century) where they were joined by the limestone transported from quarries along the banks and valley sides of the main Severn gorge. Downstream the gradient lessened markedly and the valley floor widened sufficiently (Fig. 3.1) so that

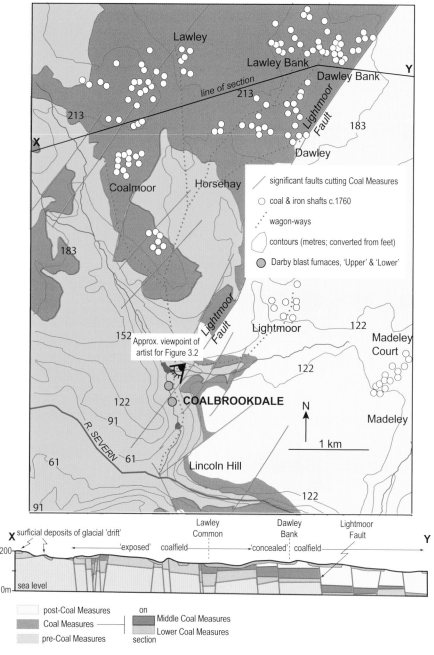

Figure 19.3 Simplified 'solid' geology (omitting superficial glacial deposits), topography, mining locations and geological section for the Coalbrookdale catchment near modern Telford, Shropshire. Redrawn, modified and crafted by the author from British Geological Survey map sources, chiefly the 1:25 000 Telford sheet, and from historical middle-eighteenth-century data from Raistrick (1989).

Figure 19.4 *Adam's Engine and Ironstone Pit, Madeley Wood, Shropshire* by Warington Smyth, 1847. Reproduced by permission from the copyright holder: *The Sir Arthur Elton Collection, The Ironbridge Gorge Museum Trust.* This engaging watercolour painted by a well-known mid-Victorian geologist has staged innocence: female manual work in a landscape with machinery – an artistic combination rarely attempted. Madeley Wood had been a chief source of clay ironstone ore for over a hundred years when the painting was composed. The 'Adam' in question was Adam Heslop, a local man who in the 1790s improved on the economy of the Newcomen (and the Boulton/Watt) steam engines by placing cylinders on each beam linkage, a cold condensing one of atmospheric type and a hot one operating under higher than atmospheric pressure as a single steam-powered cylinder. This 'Heslop Engine' was a more portable affair, built on a reduced wooden frame. No stone engine housing was needed: the boiler is the domed structure in front of its chimney and was separate. The beam linked to standard gearing drove horizontal motions that enabled its use in winding ore and workers up and down the two pit shafts, the derricks and wheels running cord through windlasses. The driving wheel and its linkage may also just be seen. More problematic is the role of the female workers, for they are depicted picking at a surface outcrop for siderite ironstone nodules when at that moment a carrier full of ore ready to dress would seem to have appeared from below the left-hand shaft head, accompanied by another female operative. Perhaps they were staged so by the artist to illustrate a former mode of working, with females not only dressing ore but collecting it. Nevertheless the scene is unique and precious in both the industrial and social history of British iron-ore mining.

furnace and forge pools could be excavated to provide the necessary reservoirs and sluiceways for hydraulic working purposes – such as the operation of the all-essential furnace bellows.

Arthur Young's visit to Coalbrookdale in 1776 provides us with a parting vignette that adds some interesting economic facts:

Crossing the ferry where Mr. Darby (the III) has undertaken to build a bridge of one arch of 120 feet, of cast iron [the Ironbridge, opened in 1781], I passed to his works up Colebrook Dale…Pass his new slitting mills, which are not finished, but the immense wheels 20 feet diameter of cast iron were there, and appear wonderful. Viewed the furnaces, forges, &c. with the vast bellows that give those roaring blasts, which make the whole edifice horridly sublime… Mr Darby in his works employs near 1000 people, including colliers. There are 5 furnaces in the Dale, and 2 of them are his: the next considerable proprietor Mr Wilkinson, whose machine for boring cannon from the solid cast is at Posenail and very curious. The colliers earn 20d. a day, those who get lime stone 1s.4d. the founderers 8s. to 10s 6d a week. Boys of 14 earn 1s. a day at drawing coal baskets in the pits. The coal mines are from 20 yards to 120 deep, and the coal in general dips to the south east.[5]

South Staffordshire: Why the 'Black Country'?

The name is supposedly a reference to either the black ground due to the sinuous outcrop of the Thick Coal (Figs 19.2) and/or to the tremendous smoke pollution induced by hundreds upon hundreds of works chimneys over the same outcrops and subcrop that sprang up as the Industrial Revolution ran its course. Here is Thomas Carlyle visiting the area in August 1824, although his frequent use of 'they' seems casual – 'they' could not possibly all have been earning 40/- per week:

A dense cloud of pestilential smoke hangs over it forever, blackening even the grain that grows upon it; and at night the whole region burns like a volcano spitting fire from a thousand tubes of brick.…here they were wheeling charred coals, breaking their ironstone, and tumbling all into their fiery pit…rolling or hammering or squeezing their glowing metal as if it had been wax or dough. They also had a thirst for ale. Yet on the whole I am told they are very happy: they make forty shillings or more per week, and few of them will work on Mondays. It is in a spot like this that one sees the sources of British power.[6]

The Thick Coal, the thickest in Britain and easily mined along its gently inclined stratal course, was accompanied by other mineable seams both above (Brooch) and below (Heathen), with accompanying clay ironstone strata of variable thicknesses. Rich developments of the latter included the Diamond, Silver Threads and Blue Flats seams below the Deep/Bottom Coal towards the base of the Coal Measures and the Pennystone and New Mine seams between the Stinking and Heathen Coals at the base of the Middle Coal Measures.

The spread of iron smelters followed the outcrop of the main clay ironstone seams through coppiced woodland that provided charcoal. Elizabethan law laid it down that such wood must be fourteen-year-old growth, never more than a foot square at the base.

It was brought convenient distances to furnace sites dotted over particular countryside estates. The rise in the number and power of professional ironmasters before and after the Civil War led to landowners eventually selling their charcoal to them directly – the large quantities required locally (the friable charcoal could not be transported far) leading to the rationalization of iron-making into agglomerated concerns. It was upon such a palimpsest of pre-coal smelting that mid-eighteenth-century industrial production became concentrated by the mining of the Thick Seam and others suitable for coking. Ironmasters could now seek out locations for their larger furnaces that were independent of landowning interests (the canal system helped here) and could sell on their pig- and bar-iron to the legions of ironmongers who now manufactured their special products in their own special ways – chain manufacture was long a speciality of the district, culminating in the manufacture in Dudley of the gigantic anchors and chainwork for the *Titanic*.

In the sixteenth century, mining was mostly from outcrop in acre-scale plots or in shallow adit workings drained where necessary by inclined tunnels (soughs). Much of the outcrop area of the Thick Coal and associated seams was within and below aristocratic estate land whose manorial tenants held copyhold tenure that in many, but by no means all, circumstances entitled them to mine for their own profit since their rents were otherwise fixed. In other cases copyhold tenants reached agreement with landlords concerning a division of mining profit. These coal- and iron-working clans had initiative and drive to obtain lucrative non-agricultural income in rents and royalties – witnessed by much internecine rivalry, legal and illegal competition for fuel, ore and finished iron (sometimes illegally bought up wholesale for monopolizing), with often lengthy spells of furious litigation and occasional riot and affray. Historic information comes from cases examined by the Court of Chancery in London, as well as a few surviving account books.[7]

The efforts of a dozen or so such ironmaster families (for example the Pagets, Willoughbys and Dudleys in late-Elizabethan times; Dudleys, Foleys, Knights, Parkes, Middletons, Goerings, Colemans, Chetwynds in the next century) eventually led to a pre-industrial iron-making tradition based on water power and charcoal. It was forged, cut and moulded into a myriad of manufactures – from humble but all-essential iron nails (from rod iron made in specialized splitting mills) to military armaments that flashed and fired across the length and breadth of the land during the Civil War. The Dudley–Wolverhampton–Walsall–Birmingham quadrilateral became the chief English locus for the manufacture of all manner of wrought iron goods.

Professional coal masters took over the originally feudal mining businesses by the seventeenth century. As the mines went deeper after the Thick Seam subcrop, the necessity of better drainage was provided by Newcomen engines (which made their world-debut at Dudley) in several pits. It was not until the canal network was developed that a wider market for Black Country coal developed and deep mining spread into the northern parts of the coalfield basin.

Coda: Five Centuries of Iron Working

The long story of Midlands iron working is encapsulated by one heroic archaeological excavation at a now-levelled site – the former Wednesbury Forge in the Tame Valley[8]. It began life as a late-Elizabethan water-powered enterprise, surviving as an edge-tool (for cutting, digging, pruning and planting) manufacturing company until the early twenty-first century – surely one of the longest-lived manufacturing centres ever recorded. Although the original position of the forge site on the outcrop of the Thick Seam witnesses the importance to forging of abundant local (and therefore cheap) coal for furnaces and ovens, the actual layout was determined by hydraulics. Excavations revealed splendid (and unique) remains of two Elizabethan oak-lined tailraces that took Tame water down from separate reservoirs into wooden undershot water-wheels that rotated in wood-lined and carefully curved base-pits. This essential layout for the forges' hydraulically driven trip-hammers was 'fossilized' by all subsequent peripheral developments (drives for grindstones, iron-boring, metal shaping and so on) over the centuries, as seen in a fine engraving from the 1860s (Fig. 19.5) showing a busy water-wheel still turning even when coal-fired power was available.

The earliest documentary reference to the forge site is to a 'riotous and unlawful assembly' in 1597, when workers armed with shovels and axes from a rival local

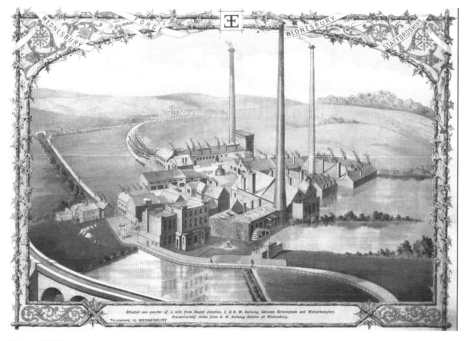

Figure 19.5 Wednesbury Forge complex in 1869 with its splendid water-wheel amongst all the chimneys. The location of Wednesbury astride the Great Seam is shown in the central part of Figure 19.2A. Source: Belford (2010).

concern, once the manorial corn mill, descended with wrecking intentions on the new enterprise. This bears out the nature of the intimidation that Dud Dudley felt he had suffered in South Staffordshire, noted in Chapter 3. Having survived this attack, in 1606 it was leased by the Lord of the Manor to one Walter Coleman. By the late 1650s it belonged, lock, stock and barrel, to the ubiquitous West Midland ironmasters that made up the Foley partnership and who later sub-let it to various tenants. In 1704 it was let to John Willetts, beginning four generations of occupation and ownership by that family, who manufactured, amongst other things, saws and guns. The slump in demand for firearms in general, and for flintlocks in particular at the end of the Napoleonic wars saw cessation of small-arms manufacturing, while by mid-century saw-making suffered when Sheffield crucible steel captured the market. In 1817 the forge was let to Edward Elwell, an edge-tool maker, who bought the site outright in 1831. Even to the present day the site is locally known as 'Elwell's Forge'. Over the centuries water power had remained and was enhanced by rebuilding and enlarging the races and pits in brick. An auxiliary windmill was even built in the mid-eighteenth century and eventually water turbine pits were constructed in steel. Elwells made over 1200 items (sickles, scythes, billhooks, axes, spades, shovels, forks, hoes, shears, etc.) – many were remarkably specialized products (such as their renowned cocoa pruner!) produced for over 100 years. In 1970 the forge was taken over by Spear and Jackson, who continued to manufacture edge tools there (including the cocoa pruner) until closure in 2005.

Archaeologist Paul Belford of Ironbridge Gorge Museum Trust, who managed the excavations at Wednesbury, writes elegiacally about the industrial tradition set up and maintained there over five centuries:

> This small but powerful enterprise was built with trees that were growing when men were fighting at Agincourt. By the beginning of the 17th century it was already a substantial enterprise, and from the early 1700s developed as an integrated factory making a diverse range of products that were sold around the world. Wednesbury-made saws, guns and edge tools were literally at the cutting edge of imperial expansion in the 18th and 19th centuries… The story of Wednesbury Forge is a truly global one. Yet the site is also firmly rooted in its locality. Generations of Wednesbury families were associated with the forge during five centuries of iron making, and it is their hard work, enterprise, courage and skill that are reflected in the archaeology… Wednesbury forged the modern world.[8]

20

East of the South Pennines
(Yorkshire/Nottinghamshire/Derbyshire)

Dissecting corpses with Keats at Guy's,
Leeds-born Thackrah shared the poet's TB.
Cadavers that made Keats poeticize
Made Thackrah scorn the call of poetry.
…
But there are pentameters in Thackrah's tract,
The found iambics no prose can destroy,
Which want to stop the heart with simple fact:
We do not find old men in this employ.

The first and third (last) verses of Tony Harrison's *The Ode Not Taken*. Keats and Thackrah (*see* also Chapter 13) had studied at Guy's Hospital, London from 1815 to 1816 under Sir Astley Cooper. From: *Under the Clock* (2005)

General

This is the country where a West Cork acquaintance engaged in the fish trade of his youth did the M62 run from Manchester to the eastern English coastal fishing ports. He always wondered on his journeys why such dense populations had originally settled in West Yorkshire's high, chilly, wet moorland country so far from the sea. After the long climb up from the Manchester and Rochdale basins through cuttings of Millstone Grit, his route through West Yorkshire took him past the hilly outskirts of Huddersfield, Halifax, Elland, Bradford and Leeds. So the answer to Peter's puzzlement is that these were the northern outposts of Britain's largest and structurally simplest exposed coalfield (Fig. 20.1).

Descending towards the A1 our acquaintance crossed the eastern limits to the Coal Measures outcrop marked by rolling hills with only occasional sandstone 'edges'

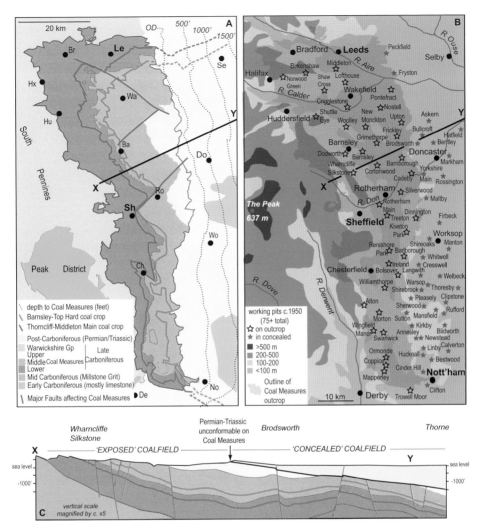

Figure 20.1 **A–C**. East of the South Pennines: **A** Solid geology. **B** Topography and location of historic collieries. **C** A geological section along the line XY marked on the maps. Sources: material redrawn, recrafted and recoloured from: A. British Geological Survey (2007a), Edwards, W. (1954); B. *Times Atlas of the World* (1987) and Edwards (1954); C. Edwards, W. (1954).

standing proud. It ended at the longitude of Castleford with a perceptible scarp formed by the unconformably overlying Permian Magnesian Limestone (Fig. 20.1), seen briefly in roadside cuttings as fresh pale-grey to yellowy dolomitic limestone, weathering cavernous and browner with age. These signal entrance into the Mesozoic lowlands of the Vale of York, across to Hull and Humberside. For fifteen or so miles, as signs were passed for places like Selby to the north and Doncaster to the south, he drove above the subcrop of the Pennine Coal Measures Group. Often deeper than 1500 feet below, legions of men and women worked out their daily shifts as cutters, mechanics, drivers, drillers, engineers, surveyors and geologists – getting coal out of some of Europe's most advanced deep mines.

However, before the deep mining industry, folk came first to textile factories, the looms worked by water power along the deep, steep-gradient valleys and numerous tributaries of the Aire, Calder and Colne. Later, steam power was provided by locally mined coal. The millhands and miners lived in terrace rows along the valley sides spilling upwards onto the 'tops'. Changing economic circumstances had led them away from their homes and cottage industries based in seventeenth- and eighteenth-century hamlets and villages around the margins and on the summits of numerous grit-fringed plateaux. Such were townland settlements like Sowerby, Heptonstall, Luddenden and Warley, high above the Calder and its many tributaries. The Early Modern master-clothier, rural-based woollen trade of the West Riding was itself the 'outland' successor to the guild-dominated York master-clothiers with all their restrictions on output, style and marketing. This was originally based on wool from vast monastic sheep flocks across the breadth of England east of the Pennines. Freed from such hindrances the West Riding trade, from raw wool to hand-loom woven 'pieces', was carried out along countless miles of packhorse routes, the cloth sold to factors in places like Halifax's famous Piece Hall. The trade was early-mechanized in the pioneering woollen mills of Calderdale and Airedale. By the late eighteenth to early nineteenth centuries the burgeoning woollen and linen industries and, further south and sixty years later, the Sheffield steel industry, grew to world domination.

The concealed coalfield stretched eighty kilometres further south as the crows fly to the Trent valley east of Nottingham and to the Brinsley pit where A.J. Lawrence (father of D.H.) dug coal before the Boer War. All along its length were the great pits, many being 'million-tonne-a-year' concerns with their huge black tips of mudrock spoil left surplus after the longwall coal-cutters had passed noisily below along their black-seamed walls – invisible harvesters cutting their paths to and fro in a dusty labyrinth of well-organized mechanized labour. Although the industry is now entirely gone, tens of thousands of retired miners and their countless descendants remain scattered over the half-wedge of coal-bearing outcrop and its concealed extension. For many the trauma of mining, the manner of its going, and its aftermath for children and grandchildren will never go away.

A Gentle Structure

[1]The coalfield lies on the eastern flank of the Pennine axis defined by a regional anticline, the chief landscape-defining structure of northern England's core (Fig. 20.1). The rugged Pennine interior of Millstone Grit scarps and basinal mudrocks envelop the sweet limestone pasturelands of the Peak District in its southernmost part. The structure had its origins in the end-Carboniferous inversion tectonics that transformed this part of the northern English rift province into a fold-and-thrust belt. During the Early Carboniferous the area was a complex series of inter-connecting rift basins whose shallow platforms

and ramped margins sustained prolonged calcium carbonate deposits that compacted and cemented into the limestone beds we see walking through the Castleton area and the Manifold valley. Thick successions of Bowland Shales were subsequently deposited from suspension and by bottom currents as organic-rich muds during periodic highstands of sea level. Eventually the deepish waters of this Central Pennine Basin were filled by copious deposition from sediment pouring out of the river-mouth conduits of the Pennine River. Its many subaerial channels and delta-front submarine channels deposited thick pebbly sand deposits that form the gritstone scenery epitomized by the ridge-forming Kinderscout Grit. The final encore for this vigorous drainage was deposition of that most widespread of all the Pennine gritstones, the sheet-like Rough Rock (rough because it is often a pebbly coarse sandstone, much-loved by boulderers and climbers for its good holds). This regional landscape-former, from Wharfedale to Derbyshire, features formidable scarps and slab-like dip slopes that dominate the scenery bordering the Coal Measures country of the fringing Pennines. The Pennine River's great extent and activity in depositing the Rough Rock was compared by C.S. Bristow to the Brahmaputra River of Bangladesh where he had done comparative fieldwork.[2]

Figure 20.2 Surviving pillar-and-stall workings of a small, shallow depth mining operation that probably dates to the ?early nineteenth century, and which won coal from the prodigious Barnsley-Top Hard seam somewhere in the general Barnsley area close to its outcrop. The well-stratified sandstone forming the roof over the two-metre-thick seam is clearly seen, with some roof-fall evident over succeeding years. In such stalls (though not with seams as thick as this) D.H. Lawrence's character Walter Morel of *Sons and Lovers* worked in the 1880s. Source: lost website of an unknown South Yorkshire underground exploration group, the location kept anonymous.

During the Late Carboniferous, sediment deposition and peat accumulation reached a thickness of around 1300 metres at the latitude of Sheffield, thinning eastwards and southwards towards Mansfield. Though there are numerous coals present, the bituminous majority preserved in the Lower and Middle Coal Measures have an aggregate thickness of only around three per cent of the total accumulation, the majority of seams being unworkably thin (usually less than 40 cm). The names of the dozen or so thicker examples form a litany of coal-mining history. The most famous and widespread was the 3 m-thick Barnsley/Top Hard seam, which had the admirable accompaniment of a flat-bedded sandstone more or less immediately above it providing ideal roof conditions for mining by both pre-mechanized pillar-and-stall (Fig. 20.2) and, later, longwall methods. But coal thickness itself was by no means the paramount consideration in the mining field – the gentle and predictably uniform easterly dip of just a few degrees, combined with a general lack of unpredictable large faults, sharp folds and igneous intrusions, made it an ideal location for mechanized longwall mining by the beginning of the inter-war years of the twentieth century. Luckily one whole mine complex, complete with headframe and winding gear – Caphouse Colliery east of Huddersfield – has been lovingly preserved as the National (English) Coal Mining Museum.

Historic Iron Working

The simple geological structure of the exposed coalfield combined with abundant timber resources in the wooded hill country of its western rim gave ample opportunities for ironstone mining and for charcoal-based iron smelting at multiple localities. Roman workings in this Brigantes tribal heartland are documented around modern-day Bradford and Sheffield. As at many localities across the Romanized island, the essential metal was in common usage and it was widely evident that:

> iron-working was carried on locally in connection with forts, camps and villas… the Roman *faber ferrarius* smelted the iron from the ore and worked it into useful shapes as successive stages in one operation.[3]

Such was the case in the headwaters of the many river tributaries issuing from the steep-sided and high-gradient gritstone valleys of the South Pennines. Here a particular combination of factors led to the construction, beginning in the 1130s, of numerous monastic Cistercian bloomeries and forges for the centres and ancillary houses of Rievaulx, Fountains and Kirkstall, all with their associated paraphernalia of holding ponds, water-wheels and trip hammers.[4] The monks and lay brethren of this industrious order built close to the outcrop of the thicker clay ironstone horizons such as the Tankersley bed at Flockton, by Wentworth Castle, probably at Rockley, and in the general Barnsley area. Here are the evocative words of one 1940s geologist on the latter:

> In general the ironstone has been worked by bell-pits; here and there old slag heaps indicate the sites of bloomeries; little vegetation grows on them except

Dog's Mercury; consequently a characteristic crunching sound is produced when one walks over them.[5]

Other worked ironstones were associated with the Hard and Soft Bed coals around Bingley and elsewhere. All the aforementioned localities were well away from the hunting forests of lowland Yorkshire, Derbyshire and Nottinghamshire, where strict manorial and royal conditions for charcoal-burners were introduced in later medieval times in an effort to conserve the great trees of the hunting grounds – culminating in the radical (and to us today, environmentally sound) strictures on coppicing in force by the first Elizabeth's reign. The documents written by the accountant for the Rothwell forges near Leeds during later Edwardian times harked far back to give an idyllic sense of memory:

> a glimpse of the inevitable changes in the wooden banks of Rothwell when the blaze of the furnaces and the ring of hammers invaded the seclusion [and]…the memory of the parker went back to the days of his youth when the herons had nested on Rothwell Shaw, and the sparrow-hawks had been reckoned of value. Now it was hardly worth while to look for honey or wax, and the herons nesting there had been frightened away by the forges.[6]

The early ironstone mines were concentrated at outcrop edges where, after open quarrying of the ore, adits would be driven into the hillsides to pursue it further. Thicker and richer nodular ore seams up to fifteen or so metres below surface were also mined systematically in bell-pits. Here shafts were sunk to ore level from the surface and then extended radially, the ore being dragged back to the shaft bottom in baskets for ascent to the surface via a windlass ('as in a well', as Celia Fiennes wrote). Excavation of ore ceased when the bell-shaped working cavity reached a certain diameter and the roof became increasingly unstable due to rockfall. The miners then moved the required distance away to begin sinking a new shaft, and so on through hundreds of repeat operations. The great extent of such former shallow workings is documented in central Leeds, where a nodular ironstone above the Black Bed Coal was selectively extracted (Fig. 20.3) over an area of some two square miles from modern Briggate to Burmantofts. Comparably dense networks of bell-pits are documented across Low Moor, Bradford where they reached ironstone seams nearly fifteen metres below surface.[7]

The widespread nature of the coal and iron working is clear from the words of the indefatigable William Cobbett, writing vividly in 1832 on the last of his great 'Tours' through Britain on the eve of the passing of the First Reform Act:

> All the way along, from Leeds to Sheffield, it is coal and iron, and iron and coal. It was dark before we reached Sheffield; so that we saw the iron furnaces in all the horrible everlasting splendour of their everlasting blaze. Nothing can be conceived more grand or more terrific than the yellow waves of fire that incessantly issue from the top of these furnaces, some of which are close

Figure 20.3 A unique image from Briggate in central Leeds taken when deep foundations for redevelopment were being dug shortly after the First World War. It shows a section through a perfect medieval bell-pit whose lower limit, around two metres below the then ground surface, was marked by the top of a coal seam, the Black Bed Coal, seen here at the base of the photo. The ironstone miners had no interest in digging out this later much-valued, 60 cm thick, low-sulphur, coking coal, probably because abundant charcoal for smelting was still available in the general area at the time. Source: annotated by the author, photo in Kendall and Wroot (1924).

by the way-side. Nature has placed the beds of iron and the beds of coal alongside of each other, and art has taught man to make one operate on the other, as to turn the iron-stone into liquid matter…from this town and its environs go nine-tenths of the knives that are used in the whole world.[8]

Although the iron smelting and working as described by Cobbett were widespread over the Coal Measures ironstone outcrops of West and South Yorkshire, the more specialized cutlery and implement trade had grown up in and around the district of Hallamshire, of which the village of Sheffield grew to become the chief centre. As historian David Hey writes:

Sheffield was a cutlery town and a smoky centre of industry long before it became Steel City. In 1608 a friend of Gilbert Talbot, seventh Earl of Shrewsbury and Lord of Hallamshire, wrote about his forthcoming visit to Sheffield, joking that he expected to be 'half choked with town smoke'. A century later, Daniel Defoe observed that 'The town of Sheffield is very populous and large, the streets narrow, and the houses dark and black, occasioned by the continued smoke of the forges, which are always at work.'[9]

The earliest sign of this activity is the mention of a named cutler who in 1297 paid taxes in Sheffield, one Robertus le Cotelere (i.e. Bob the Cutler). A century later we learn of the existence of a Sheffield-made knife, a dagger known as a 'thwitel'. This was strapped hidden under hosiery on a person's lower leg, much like the Scottish Highlander's *Sgean Dhu* but pre-dating it by several hundred years. We know of this local weapon courtesy of Geoffrey Chaucer's *Canterbury Tales*. In his 'Reeve's Tale' he equips his braggart Cambridgeshire miller with: 'A Sheffield thwitel baar he in his hose.' Chaucer was a much travelled man who had lived in Hatfield near Doncaster and so had knowledge of such things.

Hearth tax and probate returns and records[10] inform us that mid-sixteenth to mid-seventeenth century cutlers partook of a 'dual economy', combining smithy-based precision ironwork, probably mostly out of season, with outdoor work on small farms. Iron ore was had locally from outcrops such as that of the aforementioned Tankersley Seam. It was the type of economy widespread in many areas of pre-industrial Britain in which raw materials (grown, mined or quarried) were converted into finished goods in domestic locations where the whole family provided labour. In some Hallamshire villages there was a smithy to every third household. Here, wrought iron bars were either beaten and drawn into nails or beaten into edge tools like scythes, also into the finer cutlery items required by higher status households. By the 1670s the craftsmen of certain parishes would seem to have achieved a higher state of income than that pertaining generally – their probate inventories were as valuable as those of certain yeomen and husbandmen, and there were few among them who could not pay tax for at least one or two household hearths.

The Rise of Sheffield Steel – From Cementation to Crucible

Water power from five main rivers, charcoal supplies and metallurgical expertise in foundry and forge were more than adequate to nurse into being specialized steel manufacturing concerns by the early eighteenth century, in particular that of sharp-edged steel cutlery and tools. But such steels could not be made using iron smelted from the local clay ironstone. This was inferior stuff, due as we now know (from mid-nineteenth century chemical analyses) to the ubiquitous presence of small amounts of phosphorus in the ore[11] that could not be removed by the smelting and refining techniques of the time. Instead, in common with steel workers elsewhere, the Hallamshire cutlers got their iron and steel from various continental Europe steel producers as part of the aforementioned Europe-wide trading network that formed the basis of the late-medieval iron trade (*see* Chapter 17). As David Hey again writes:

> before the Cutlers' Company was formed in 1624 local craftsmen were working with foreign steel, which they obtained from the Basque Country via Bilbao, from Germany via Cologne and the Rhine, and from Sweden via Danzig and other Baltic ports…Such steel was a superior form of iron, which had carbon added to make it harder, more malleable and easy to grind to a cutting edge, and to hold that edge once it was made.[12]

It was therefore a tribute to the skills of the early Sheffield cutlers and of their mercantile steel importers (mainly through the port of Hull and down the navigable Trent) that the use of expensive imported low-phosphorus steel ingots still enabled their products to compete in the market places of Britain and Europe. Thereafter they embraced the cementation process of steel making themselves (originally developed at Liége). This involved the intense heating of low-phosphorus bar iron with charcoal for many hours, a process that further purified the iron, changing it to 'blister steel'. The process was analogous to the early zinc smelting 'cementation' process outlined previously in Chapter 17. The first *in situ* South Yorkshire steel making by this process probably began in the mid-seventeenth century. The oldest surviving cementation furnace is in County Durham (*see* Fig. 23.3) and dates from early in the eighteenth century.

Figure 20.4 **A**-**C** The priceless relics of crucible steel smelting preserved at Abbeydale Industrial Hamlet, Sheffield (53.200244, 1.304344). **A** The first floor of the eighteenth-century furnace. The five individual crucible holders are bedded into a stonework base and back and were originally closed by pivoting iron lids. The coal-fired furnace was situated on the ground floor below, vented by bellows driven by one of the hamlet's two water-wheels. A typical clay crucible (made on-site: a YouTube video depicts the making, in an adjacent pot room) stands to the left. The crucibles were filled with an exact mix of blister steel fragments and coke in the adjacent Charge Room before being let down into their nests for firing until the molten mix had equilibrated ready for moulding and/or forging. **B** Part of a group of 1900s furnace men from an unknown site; the aproned two on the left sport thick multi-layer overshoes and trousers that enabled them to stand close enough on the furnace top to lift out the +1000°C crucibles on long tongs and to pour the molten steel into moulds nearby. **C** A fragment of blister steel made by the cementation process at Abbeydale and featured in the Abbeydale exhibition space. This is low-phosphorus bar iron, the ingot stamp indicating its provenance from Swedish öregrund works: a priceless and tangible link with the industrial past. Sources: A, C Author, B from a photo exhibit at Abbeydale.

It was a process invented in Sheffield itself that was to be the town's first world-breaking incursion into steel smelting. This was achieved in the 1740s by one Benjamin Huntsman, a Doncaster clock maker of Quaker parents who moved to the Handsworth part of the wider Sheffield area. He saw the need in his own and other professions for very high quality steel, especially for use as springs in clock mechanisms. He developed a way of making such steel from the refined ends of cementation steel bars melted in clay crucibles with powdered coke (Fig. 20.4). This superior 'crucible steel' established Hallamshire as both the source of the world's finest mass-produced steel and the leading centre of British cutlery and edge tool production. Not bad for a small Pennine town – soon to be 'Steel City' of the world.

Henry Bessemer's Converter

By the mid-nineteenth century there were numerous British steel-makers using low-phosphorus iron derived from the indigenous high-quality hematite iron ores of West Cumbria – the Glamorgan and Monmouth iron-makers had largely switched to this source by this time. Such iron had previously been supplied by the aforementioned Swedish imports, but these had become overly expensive following tariffs put up during and after the Napoleonic Wars. At the same time the production of steel was limited by the relatively small scale of the crucible techniques now employed in all steelmaking centres. Yet there was a growing and insatiable demand for large quantities of cheap general-purpose mild steel that could be used for the manufacture of load-bearing structures (bridges, girder frames for buildings, rails, chains), ships and, especially, large-scale armaments (siege and naval guns). That such cheap, abundant and strong steels could eventually be made from indigenous Carboniferous clay ironstone ores was due to a combination of inventions over a hectic period of twenty or so years that saw the world's emergence into the 'Steel Age' with Sheffield as its capital.

The first essential breakthrough came courtesy of Henry Bessemer.[13] This inventive and driven man had the idea of passing a hot air blast through molten pig-iron that would utilize the air's oxygen to oxidize impurities like excess carbon, silicon, manganese and sulphur, so converting the pig-iron into cheap mild steel. He knew that the mineral reactions of the oxidation process would give out enough heat to keep the pig-iron thoroughly molten during the process – in modern scientific parlance they were exothermic ('heat-giving') reactions. By 1855 Bessemer had designed, patented and built a converter for this purpose in Sheffield (Fig. 20.5). Its swivelling, squat carronade-like steel barrel was lined with silica-brick to protect the carapace. The converter proved capable of quickly achieving steel production using iron forged from Cumbrian hematite ore with the cognate production of a calcium silicate slag. But the steel had a tendency towards brittle properties since the speed of the conversion process (only a dozen or so minutes for several tonnes) made it difficult to retain the exact amount of carbon needed to make a reliable product.

After receiving a sample of such brittle converter steel from a friend, Robert Forester Mushet (whom we left smelting iron in the Forest of Dean in Chapter 17) solved the

Figure 20.5 The magnificent Bessemer Converter preserved at Kelham Island Museum in Sheffield (53.232246, 1.282096) – apparently one of only three surviving worldwide. It was used by the British Steel Corporation in Workington, West Cumbria until 1975, producing the last Bessemer steel made in Britain. The tilt mechanism about the central coaxial casing is clearly seen, as is the open eccentric spout into which molten pig-iron was poured, together with an aliquot of manganese and carbon before a blast of air was blown through from the base. The spectacular discharge of flames and fountains shooting out of the top of the converter were the 'minatory beacons' of poet Geoffrey Hill (quotation to Chapter 19). After a period of a dozen or so minutes the air blast was stopped and the converter tilted again, forwards this time, so that the seven or so tonnes of newly-made steel could be run out. Source: Wikimedia Commons, File Bessemer 5180.jpg, author *Chemical Engineer*.

carbon problem by proposing that the converting process be prolonged by several minutes until all carbon had been oxidized away by the hot air blast.[14] Then the exact required quantity of carbon (1–2%) could be added to the molten iron charge to turn it to mild steel. From his many experimental results Mushet also proposed that the addition of particulate manganese would aid the various oxidation reactions and improve the malleability of the steel. This solution to Bessemer's problems was successfully carried out in 1856 by the use of a ferromanganese alloy, spiegeleisen, which Mushet had come across from north Germany, smelted originally from a mineral carbonate of iron and manganese. Spiegeleisen was subsequently produced on an industrial scale by adding manganese mineral oxides to previously smelted iron in a furnace. Although Mushet had applied for patents covering his metallurgical technique, he was unable to keep up payments for them and they lapsed, bringing him nothing in return for his brilliant and original work. A belated (and perhaps grudgingly given) annual pension had to be begged in person from Bessemer himself by Mushet's daughter, Mary.

Despite carrying all before it as the first industrial-scale source of cheap but reliably strong mild steel during the 1860s, the Bessemer Converter with Mushet's modifications could still only deal with low-phosphorus iron – this despite Bessemer undertaking many unsuccessful experiments to solve the phosphorus problem with leading metallurgists and chemists of the day. A solution eventually came from two young Welshmen, Sidney Gilchrist Thomas and his cousin Percy Carlyle Gilchrist.[15] Thomas was working as a clerk in London and, as a keen amateur chemist and mineralogist, had taken up evening classes in natural sciences at Birkbeck College in 1870. Three years later he had obtained two first-class degrees (in mineralogy and inorganic chemistry) from the Royal School of Mines (later to become Imperial College). Apparently first attracted to the phosphorus problem in steel-making by the comments of one of his lecturers, he devoted himself to its solution. He realized that he had to find a way of taking out the element while it was in the iron melt by combining it with a suitable substance to form a neutral compound. Calcium was the obvious answer, readily available in nature from limestone or dolomite – it could combine with the phosphorus to precipitate calcium phosphate as part of the converter's silicate slag. After discussions with his metallurgist cousin, Gilchrist, the two joined forces in a series of experiments and converter trials over several years (some at Blaenavon, the final successful trial in 1879 at Middlesbrough) that ended up with them lining a Bessemer converter with bricks composed of a patented mix of fragmented dolomite and sodium silicate, amongst other ingredients. By also adding a certain amount of calcium oxide to the molten iron, they brought about a spectacular climax to their long labours – complete phosphorus removal by an afterblow, producing the required pure steel and its phosphatic slag. All iron ores were now amenable for steel-making in both the Bessemer Converter and the newly emerging Open Hearth furnaces by what we now know as the Thomas-Gilchrist Basic Process. The slag was a real bonus – a ready source of cheap industrial agro-phosphate that enhanced world agriculture by replacing expensive Chilean guano (bird dung) and home-grown Cambridgeshire phosphatic coprolites (fossil dung).

Coda: Lead Mining in 1720s Derbyshire

We journey once again with Daniel Defoe, this time amongst Carboniferous Limestone and Millstone Grit landscapes of the Peak District. He gives an unforgettable account[16] of the lot of a Derbyshire lead mining family in the 1720s when he and his group arrived 'a little on the other side of Wirksworth…' at the beginning of Brassington Moor searching for yet another Peak District 'wonder', the Giant's Tomb. They are distracted by discovering a cave-dwelling family whose circumstances and demeanour determine that Defoe will tell something of their story rather than concentrate on yet another dubious and tiresome 'wonder'. He intends to set a moral example (although accepting the *status quo*):

to show the discontented part of the rich world how to value their own happiness, by looking below them, and seeing how others live, who yet are capable of being easy and content…

The woman of the family had a lead-miner husband (himself the son of a miner) and five young children whose home in the large limestone cave was divided into partitions by hung curtains. The place was orderly, clean and tidy with shelved pots, earthenware and pewter utensils and flitches of bacon hanging in a chimney constructed as an excavated shaft above a fireplace. The dwelling was fronted by a small field of barley and just by the 'door' a small cow grazed, with a sow and piglets also running around. Defoe was curious as to the nature of the income for the family's obviously adequate subsistence, and asked the woman how much her husband could earn a day in the lead mines:

she said, if he had good luck he could earn about five pence a day, but that he worked by the dish (…in proportion to the ore, which they measure in a wooden bowl…). Then I asked, what she did, she said, when she was able to work she washed the ore. But, looking down on her children, and shaking her head, she intimated, that they found her so much business she could do but little…But what can you get at washing the ore, said I, when you can work? She said, if she worked hard she could gain three-pence a day. So that, in short, here was but eight-pence a day when they both worked hard, and that not always, and perhaps not often, and all this to maintain a man, his wife, and five small children.

Leaving the woman tearfully happy with a half-crown whip-round, the party made off in the direction of the lead mines on the sides of an adjacent hill. Here:

there were several grooves ['rakes'; worked-out surface traces of steeply-dipping mineral veins], so they call the mouth of the shaft or pit by which they go down into a lead mine; and as we were standing still to look at one of them…a hand, and then an arm, and quickly after a head, thrust up out of the very groove we were looking at…the man was a most uncouth spectacle; he was clothed all in leather, had a cap of the same without brims, some tools in a little basket which he drew up with him…he [also] brought up with him about three quarters of a hundred weight of ore…

At this point Defoe gives a sudden aside, jolting readers of his narrative with a sharp rebuke:

If any reader thinks this, and the past relation of the woman and the cave, too low and trifling for this work [ie his Tour], they must be told, that I think quite otherwise; and especially considering what a noise is made of wonders in this country, which, I must needs say, have nothing in them curious, but much talked about, more trifling a great deal.

So there! Defoe reveals his common touch and humanity. He goes on:

> We asked him, how deep the mine lay which he came out of. He answered…
> he was at work 60 fathoms [*c*.112 m] deep, but that there were five men of
> his party, who were, two of them, eleven fathoms, and the other three, fifteen
> fathoms deeper. He seemed to regret that he was not at work with those three;
> for that they had a deeper vein of ore than that which he worked in, and had
> a way out at the side of the hill [i.e. via an adit], where they passed without
> coming up so high as he was obliged to do [via a ladder into the deep shaft].

Defoe's account is a unique document of early eighteenth-century mining life,
enriched by his irrepressible practical curiosity. For the record, the eight pence a day
combined wage for the lead-mining husband and his ore-dressing wife was about one-
third of that for a skilled tradesman at the time.

21

West of the South Pennines (Lancashire/North Staffordshire)

I have never seen a class so deeply demoralized, so incurably debased by selfishness, so corroded within, so incapable of progress, as the English bourgeoisie;…I once went into Manchester with such a bourgeois, and spoke to him of the bad, unwholesome method of building, the frightful condition of the working people's quarters, and asserted that I had never seen so ill-built a city. The man listened quietly to the end, and said at the corner where we parted: 'And yet there is a great deal of money made here; good morning, sir.'

Frederick Engels: *The Condition of the Working-Class in England*. (1845)

Perspectives

West of the Pennine divide, Lancashire developed a unique industrial scenario in the eighteenth century. Centred around Manchester, water-powered mills in scores of west Pennine valleys spun and wove raw cotton, a procedure utterly unknown in the rest of northern Europe. As noted previously (Chapter 13) this had originally arrived in Liverpool docks from Bengal, later from the slave plantations of what was to become Confederate America by the mid-nineteenth century. It harnessed the full potential of revolutionary mechanical inventions in textile-spinning and weaving put successively into place by Hargreaves, Arkwright and Crompton.

Yet a glimpse at the geological map (Fig. 21.1) shows that the majority of these new cotton towns north of Manchester – Burnley, Accrington, Blackburn, Bolton, Bury, Rochdale, Oldham, Wigan – were located within or close to the Lancashire coalfield. In contrast to the higher peat-ridden gritstone moors the gentler slopes of valleys through the Coal Measures provided tractable water power. Also, valley-side terraces provided decent agricultural grazing potential. Cheap local coal was also of benefit for domestic

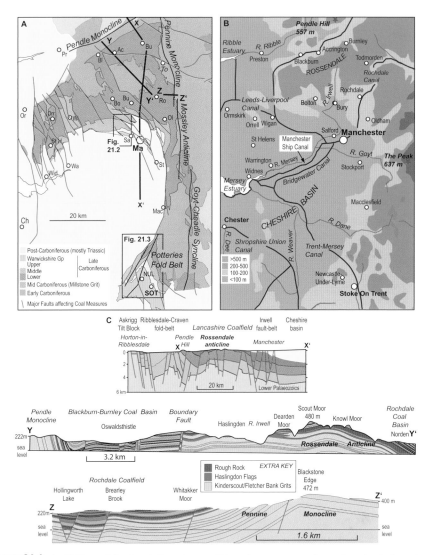

Figure 21.1 **A–C** West of the South Pennines: solid geology, topography and geological sections along the lines X–X', Y–Y' and Z–Z' marked on A. Sources: material redrawn, recrafted and recoloured from: A British Geological Survey (2007a) B *Times Atlas of the World* (1987) C Aitkenhead *et al.* (2002); Wright *et al.* (1927).

use by the growing urban workforce, for industrial heating and to power the steam engines that replenished mill-race reservoirs.

Manchester, soon to be labelled 'Cottonopolis', was the nexus and central depot of the trade, located as it was on the boundary between the upland Coal Measures and the fertile breadth of the Cheshire Triassic basin. This geographical locus was in exactly the right (flattish) place for both enlargement of navigable rivers and the construction of canals. Both aided low-cost imports of raw cotton and exports of finished cloth via the Mersey estuary and Liverpool's burgeoning docksides and warehouses. The first industrial city's population was warmed by cheap coal from the Duke of Bridgewater's estates at Worsley via his eponymous surface canal. Similarly the bottle kilns of the North Staffordshire potteries were nurtured by their own rich *in situ* reserves of coal and the import and export of their goods and raw materials by construction of the Trent–Mersey canal, aided by the efforts of chief potter, Josiah Wedgwood.

Later, as mines deepened, pit coal would power the pumping engines that kept the mines workable and would also furnish the heat and power needed to establish the iron and steel manufacturing trades of South Lancashire. These provided the precision engineering products needed for the construction and maintenance of the mechanized mills. By the mid-nineteenth century coal would also power the steam-driven mills that grew up across the whole region, their locations now independent of hydraulic considerations as second generation technology led to huge increases in cotton cloth output in response to world demand. The demise of water power took a long time coming but in the end, as Andreas Malm has emphasized[1], the immense portability of coal power (courtesy of James Watt's rotary engines) into suburbs and lowlands decided its fate.

Inverted Coal Basins

[2]One might initially suppose that the Lancashire, Cheshire and North Staffordshire half of the Pennine Axis might be a mirror image to that revealed in Yorkshire and Nottinghamshire. If so, the Coal Measures should decline gently westwards towards the Irish Sea basin under the softer Permo-Triassic rocks of the Cheshire, Merseyside and Fylde coastal lowlands – just as they did eastwards under the Vale of York on their relaxed subsurface journey towards the North Sea basin. Yet a close look at the accompanying section of Figure 21.1C reveals a different picture. The gentle dip of the eastern Coal Measures and Millstone Grit continues far to the west of the axis of their symmetrical outcrop. There is then a sudden increase in dip accompanied by sometimes vigorous faulting and tight folding, followed by a diminution further westwards, best seen in section Z–Z' across the Pennine Monocline. In addition there is an abundance of rather meaty-looking faults, most obvious to the north-west and south-east of the Manchester–Salford metropolis. These slice up and displace the Coal Measures outcrop from the longitude of Bury in the east through Wigan to St Helens in the west. Significantly, most are normal faults that trend in the north-east by north-west quadrants and which cut clean across Triassic strata as well as disrupting the Carboniferous.

How are we to make sense of these observations? First, that the Pennine Axis formed by late-Carboniferous inversion tectonics is an asymmetric structure – the structural axis of the Pennines is shifted way to the west of its geographic axis. This is because it is a monocline (a step fold). After its rather gentle entrance from the east, the strata suddenly flip over westwards at dips of up to fifty degrees, often with the accompaniment of intense faulting. We also notice the southward tapering and disappearance of the Coal

Measures outcrop on the eastern flank of the field from Macclesfield southwards until its reappearance in North Staffordshire. Here a tell-tale chevron of outcrops defines a pairing of acute folds that define the Potteries coalfield. Such structural features mean that the mining of coal and clay ironstone in East Lancashire and North Staffordshire was usually more complicated and expensive than that in the great coalfields to the east. Folded outcrops were often so steeply dipping that stoping methods identical to those practised in metal vein mining had to be employed to get the coal and ironstone out.

Second is the previous observation from North Wales, that the Cheshire and East Lancashire Triassic basins that bound the Lancashire/North Staffordshire Coal Measures to the south and west were rifts in their own right. The rifting explains why the Pemberton faults that bound the Orrell and Wigan coalfield segments in South Lancashire have throws of 4–5000 metres westwards. Those in the east create a faulted contact between very thick Triassic rocks downthrown against the Coal Measures, as in the Irwell valley. As we shall see below, serious problems of mine drainage arose because of these major faults.

These are the chief reasons why the West Pennine coalfields are so very different from those to the east. The next chapter sees that such comparisons persist northwards, as between west Cumbria and Northumberland/Durham to the east.

The Worsley Navigable Levels: a Unique Endeavour of Resource Exploitation

The Worsley Navigable Levels are entirely hidden from view below the Third Duke of Bridgewater's former estates at Worsley Old Hall, a few miles west of Salford. They deserve to be better known as one of the most spectacular practical solutions devised to mine geological raw materials during the course of the Industrial Revolution. In the words of historian Hugh Malet:

> The output of coal in the Worsley area about 1757 was considerable, but production and profits at the Duke's two mines had declined as the seams delved deeper and began to fill with water. As a result, he commissioned one of his junior agents, John Gilbert, to carry out a survey. To Gilbert is due the credit for solving three major engineering problems with one brilliant idea. If a tunnel, called a sough or adit, large enough to carry canal barges, could be drilled into the Worsley mines, boats would be able to transport coal direct from mine to market, the springs inside the hill would provide a head of water for a canal, and the flooded coalmines could be drained.[3]

The logic to John Gilbert's plan becomes clearer as one looks at the geological map and section (Fig. 21.2). Upper Coal Measures form the bedrock to the Bridgewater estate – the Duke was sitting on a huge reserve of some of the finest coal in Britain. The ground surface rises gently but steadily in altitude northwards from about 25 metres OD to the former village of Farnworth at around 100 metres OD. At Worsley Delph the moderate southward

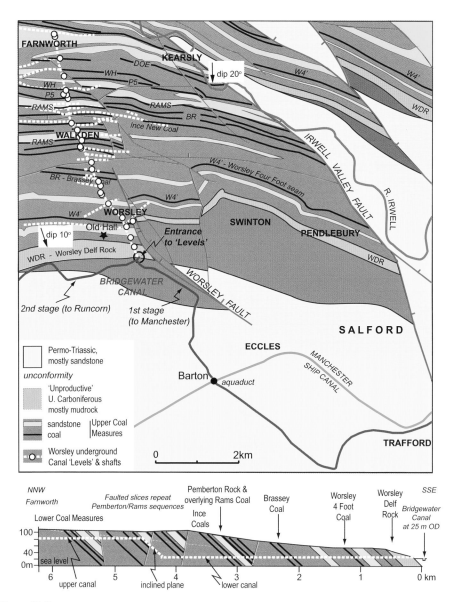

Figure 21.2 Simplified 'solid' geology (ignoring Ice Age superficial deposits) of part of Greater Manchester in south Lancashire to show location for the Bridgewater Canal and the 'Worsley Levels' underground canals with their twenty or so associated mining shafts. The geological section is taken along the rough line of the Worsley–Farnworth trace of the canal levels on Bridgewater lands. Sources: redrawn, modified and crafted by the author from British Geological Survey map sources, chiefly the 1:50 000 Manchester sheet 85 and from historical middle-eighteenth- to nineteenth-century data available online at various sites. The geological section drawn by the author also gives a schematic sectional view of the course of the underground level from Worsley to Farnworth with the lower canal leading straight up along the incline to an upper canal. In fact the lower level probably also carried on underneath but has been omitted for clarity.

dip, ten degrees or so, of the Coal Measures on the southern flanks to the Rossendale Anticline brings the Worsley Delf Rock to its prominent outcrop. Below this thick sandstone the chief workable seam was initially the Worsley Four Foot. More important to the longer-term development of mining on the Duke's estates was the presence of five more underlying workable coals, one of which was two metres thick. These lie below a further thick sequence of sandstone, the Nob End Rock, making a total of six thick workable seams in the twelve or so square kilometres of the Bridgewater estate.

The thick sandstones in the lower Worsley stratal succession were to blame for the original mine-water problems, for they had decent permeability and flow. But the geological map reveals an additional reason – a major fault line, parallel to the aforementioned Triassic rift fault in the Irwell valley, cuts through the Worsley succession. This Worsley fault lies immediately adjacent to Worsley Delph and throws down 'tighter' impermeable Warwickshire Group mudrock strata against the Coal Measures. The greatest egress of water into the Duke's adjacent mine workings would have taken place by transmission from the faulted Worsely Delf Rock aquifer.

It seems obvious that Gilbert possessed the necessary geological understanding to appreciate the three-dimensional form of the Worsley Coal Measures strata. As Hugh Malet indicates, his original concept began with a tunnel level cut back into the Delf Rock sandstone at Worsley. By driving horizontally northwards into the underlying Coal Measures at an elevation of around 25 metres the shallow level, 2.4 metres from roof to bottom, eventually intersected the Worsley Four Foot seam in 1761 at a horizontal distance of about a kilometre. Over the next hundred years its continuation intersected all the underlying workable seams in the Upper Coal Measures succession.

Once any seam was intersected at 25 metres OD it was worked laterally by tunnelling new level branches along the strike, roughly normal to the trend of the main canal. Successive shallow shafts for ventilation and access were sunk along the future trace of the main level and the channel extended gradually northwards. In this way coal could be brought out from the successive shaft bottoms into suitable narrow-waisted and shallow-keeled barges, called 'Starvationers' (Fig. 21.3). The slow but inexorable outflowing mine-water kept the main canal filled along its whole length into Manchester, the entire underground network of levels a *de facto* drainage adit. Gradually over the next century the levels spread their watery tentacles in a rectilinear grid, the main trunk orientated only slightly oblique to the stratal dip with periodic offshoots far along stratal strike (around 2.8 kilometres for the Worsley Four Foot seam).

A higher underground level was eventually joined to the main lower level by a steep counter-balancing rope-connected incline. This enabled coal to be mined at a shallow depth from the surface, then taken down to the main canal in a barge, at the same time as propelling an empty barge upwards. Here was the perfect '*incliné souterraine*', as the Duke's eccentric heir termed it in a pamphlet describing the system (written in French) published in 1812[4]. By 1887 the levels had reached a staggering seventy-four kilometres in length. Together the underground level system and the open canal (subsequently

Figure 21.3 Historical engraving of Worsley Delph (53.300334, 2.225242); date unknown but probably late eighteenth to early nineteenth century. It shows the business-end to the Navigable Levels with their entrance to the right and exit to the left (cf. Fig. 5.2). The moorings are busy with long, narrow-waisted ('Starvationer') barges used for transporting and offloading Worsley coal to larger barges for transport out on the main Bridgewater canal system westwards to Runcorn and eastwards to Trafford and Manchester. The central wharfside area features piled ridges of coal presumably recently offloaded from exiting Starvationers (perhaps by the lifting contraption seen at the extreme left) ready to be transferred to larger barges. An empty Starvationer awaits entrance to the right. Source: Malet (1990).

known as the Bridgewater Canal) eventually enabled carriage of coals to Salford and Manchester in quantities of over 100 thousand tonnes a year. In 1761, the very first year of production, it cut the cost of coal there by fifty per cent.

Tunnelling in North Staffordshire: the 'Harecastle Nightmare'

The James Brindley we met in Chapter 5 was a fearless individual in the face of seemingly insoluble logistical obstacles. He had dewatered mines by inverted siphons, constructed canals on aqueducts and developed puddling-clay liners. Yet the Harecastle tunnel leg of the Trent–Mersey Canal under Goldenhill almost defeated even his brave endeavours. Though only about one per cent in length of the whole course, it took eleven years to excavate; indeed his early death in 1772 meant that he never saw it completed. Josiah Wedgwood had captured the spirit of the man, also his vulnerability, when he had written five years previously, somewhat despairingly (and sanctimoniously, in view of his own disposition to overwork) about his friend to Thomas Bentley in March 1767:

> I am afraid he will do too much, & leave us before his vast designs are executed; he is so incessantly harassed on every side, that hath no rest, either for his mind, or Body, & will not be prevailed upon to have proper care for his health…[5]

The 2800-metre-long tunnel is situated below Goldenhill west of Tunstall, at an elevation of around 125 metres (Fig. 21.4). It connects northwestwards through an exit in Kidsgrove and a flurry of locks to the main floor of the Cheshire basin. Its lengthy construction represented a long and costly delay. For example, until finished in 1777, Wedgwood's entire pottery output destined for the Mersey terminus was offloaded at Tunstall (together with the horses, since the tunnel had no towpath) and carted over Harecastle Hill via the 'Boathorse Road' to the Kidsgrove exit below to the north of Clough Hall estate.

What was it about the excavation of the Harecastle tunnel that proved so difficult, and, in a little-asked question, why did it have to be constructed at all?

Geological factors severely affected Brindley's decision to proceed by the Harecastle route. Engineering geology problems beset his tunnellers as they drove its planned course obliquely through the Potteries fold belt. Here in the local Coal Measures and Warwickshire Group there were steep stratal dips and rapidly varying alternations of permeable sandstone and impermeable mudrock with coals. This caused the lack of a coherent and continuous roof rock and the necessity of time-consuming brick-lining. In addition, the abundance of open joint fractures associated with the folding ensured that there was plentiful egress of groundwater from the several sandstone bodies where tunnelling was in any case very slow. Evidence that Brindley was aware of at least some of these problems is provided by the plan of construction eventually put into effect. This involved the demarcation of a line of survey across Golden Hill's Latebrook–Ravenscliffe

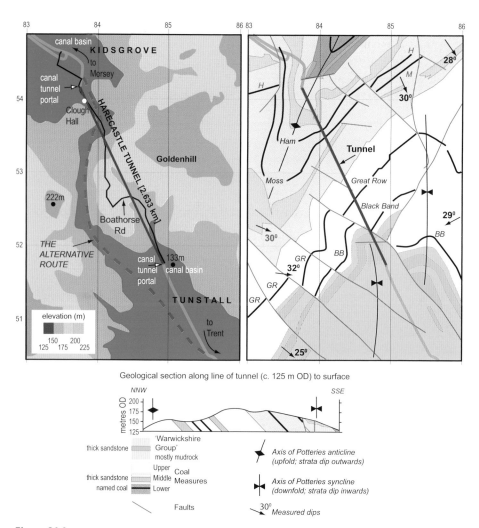

Figure 21.4 Geological, topographic maps and a geological section of the Harecastle Tunnel area and the tunnel course (53.050435, 2.143972). Source: Map material redrawn, recrafted and recoloured from British Geological Survey using the 1:50 000 Stoke-on-Trent sheet 123 and from Ordnance Survey of Great Britain 1: 25 000 Stoke-on-Trent sheet 258. The section is along the line of the tunnel (*c.*125 m OD) to surface. Source: Author.

ridge 55 metres higher than the proposed canal level. Along this line were cut no fewer than fifteen shafts every 180 metres or so. From these, separate headers along the line of transect allowed tunnelling to proceed in both directions, the tunnel 3.5 metres high at its tallest point and 2.75 metres wide at its widest.

Despite this careful plan, the first sign of trouble was the need to place pumps and Newcomen pumping engines together with ventilation measures at certain shafts to deal with drainage and roof problems.[6] The average rate of tunnelling was only around 20 metres per month. Brindley's initial optimism may have come from accounts of the history of progress of the main Worsley navigable level – this had taken a year to progress around one kilometre, and at that rate the Harecastle tunnel would have taken only three years to cut. But the geology at Worsley was very much simpler (compare Figs 21.2 and 21.4), with uniform and shallow-dipping strata providing predictable roof conditions.

Given the immense labours and costs involved in completion of the tunnel (the second longest ever cut in Britain at the time) the question arises as to why Brindley might have chosen the route he did. Inspection of the topographic map (Fig. 21.4) shows that the 55 metres of relief on the Latebrook–Ravenscliffe ridge could have been avoided if a sinuous contour course deviating west of Tunstall from Chatterley to Clough Hall (along the course of a later and still extant rail route) had been chosen. The topographic rise of a few metres could easily have been overcome by construction of just a single lock. Yet it is difficult to believe that a man of Brindley's insight might not have considered such an option. One then thinks that perhaps a problem with an unsympathetic landowner might have caused him to choose the tunnel option *faut de mieux*. But the history of the Clough Hall estate establishes that at the time of canal construction its resident owner was a certain Thomas Kinnersley who was:

> for many years an eminent banker in Newcastle-under-Lyme…a leading local corporator and businessman, closely connected with the Trentham interest of Lord Stafford [aka the Earl Gower]. [7]

Kinnersley would have known of the possibility of coal under his own lands from Brindley himself or from Stafford's agent Thomas Gilbert. More importantly, he would have hardly refused permission to cut the canal when his interests were so closely involved with Stafford.

It remains to consider an unpalatable truth – that Brindley's self-interest persuaded him that whatever impediment there may have existed to prevent him taking the far easier 'contour option' it would prove more financially beneficial to himself and his friends, with the successful Worsley endeavour in his mind, to simply go for the tunnel option. We can take this line of analysis because we know that he was familiar with the locality, since he owned land and property where he and his young wife lived above the tunnel's course. We also read in Cyril Boucher's biography[6] that he and Thomas Gilbert both had 'interests' in coal mines on Goldenhill. As Boucher suggests, it may be that he saw a Worsley-type operation bolting on to the main business of the Trent–Mersey canal, with navigable levels leading off from the canal tunnel to mine shafts in the area.

An Exceptional Account of the Orrell Coalfield

The history of the Orrell segment of the South Lancashire coalfield west of Wigan has been lovingly served by the most complete and impeccably interdisciplinary account of any British coalfield known to the author. Published in 1975 it was written by Donald Anderson, a local man, son of colliery managerial staff and a chartered surveyor and engineer in his own right with huge mining and surveying experience both in Britain and abroad in gold and coal mines[8]. Uniquely qualified, Anderson takes us on a historical, geographical, sociological, geological and technological tour of this coalfield over the hundred years or so from 1740 to 1850, with glances further backwards and forwards as he goes. He knows what he is talking about in all of these fields, not least from a broad interest in the history of his native soil, but also with a deep and penetrating personal knowledge of the developing engineering and geological sciences of the times.

We learn of the chief Georgian landowners and the exact locations of their plots of land; the sites and targets of the many pit shafts, drainage soughs and railway branches; the transport of mined coal along the Douglas Navigation from the Ribble estuary to Wigan (from 1742), the role of the southern part of the Leeds–Liverpool canal from Liverpool to Wigan (from 1774) that cut through the northern part of the coalfield and the coming of the Victorian railway built by the Lancashire and Yorkshire Railway Company. In between we are informed of technical mining, drainage and geological developments, the economics of coal marketing, social conditions and the economic and financial history of the principal colliery concerns. As would be expected from a chartered surveyor there are many original maps, drawings and technical accounts. Reflecting on the few surface remains of mining activity and its long-ago reversion to agriculture, Anderson concludes his history in a moving way:

> Probably the only local people to benefit financially [from the mining] were the landowners, a few engineers and surveyors whose skill and knowledge made the operations successful and the shopkeepers and other tradespeople… The greater part of the wealth produced went to build up the fortunes of the merchants from Liverpool and Bradford who owned the largest concerns… Reflecting upon the laborious conditions in which the coal of Orrell was produced the words written by Thomas Carlyle spring to mind:

> 'Venerable to me is the hard hand, venerable too is the rugged face…For us was thy back so bent, for us were thy straight limbs and figure so deformed.'

Coda: The Goniatites of Rossendale

W.S. Bisat's use of fossil goniatites found in marine bands to precisely subdivide and correlate Middle and Upper Carboniferous strata was highlighted in Chapter 9. Their first recorded practical use was to help in the mapping of an exceptionally confused jumble of faulted sandstones in the upland fells of the Rossendale Anticline in north-east Lancashire. Three years after the end of the First World War, four Geological Survey officers had been tasked by their District Geologist, W.B. Wright, to undertake the necessary fieldwork. The nature of the task before them was starkly outlined by Wright in his introduction to the resulting memoir and map of 1927. He writes that, despite the simple overall structure present in the area (*see* Fig. 21.1C, section Y–Y'):

> in following a series of escarpments along a hillside, the eye is soon baffled by their complete disappearance and replacement by another set. The faults…vary enormously in their influence on the scenery, being often marked by splendid features, but as often devoid of any effect on the surface of the ground.[9]

The only answer to mapping in such poorly exposed and peaty fell terrains was to locate the black shales of the marine bands and identify their goniatites. In that way a sensible local stratal succession could be established and then traced on the ground into adjacent areas and beyond.

In his introduction to the fossils of the Rossendale area, Wright graciously acknowledges W.S. Bisat's role in those successful Geological Survey mapping campaigns of the early 1920s:

> The majority of the geologists on the Lancashire unit of the Survey have been in repeated contact with Mr Bisat during the progress of the work and have obtained valuable guidance and instruction from him. They have made themselves familiar with the zonal forms [of goniatites], and have thus been able not only to test the accuracy of the zonal system, but also to use the fossils as a direct guide to mapping where independent evidence was weak.[10]

In this way the publication of the Rossendale memoir was epoch-making and a world first. The Middle and Upper Carboniferous stratal successions therein delineated by goniatites could now be used by coal geologists and mine-captains and visited by academic specialists from all over the world, and who could be sure that a secure succession of strata was being examined. For example, previous palaeobotanical investigators in north Lancashire had to painstakingly judge from their own observations of local stratigraphy to determine whether coal ball samples collected from different localities were coeval with one another or distinct, either younger or older. Marie Stopes and David Watson, featured in Chapter 10, in fact got it right. For later generations, from the 1960s onwards, this made possible detailed sedimentological explanations for the extraordinary stratal successions of the South Pennines.[11]

22

West and South Cumberland

A hundred years of the Bessemer process—
The proud battery of chimneys, the hell-mouth roar of the furnace,
The midnight sunsets ladled across a cloudy sky—
Are archaeological data, and the great-great-great-grandchildren
Of my grandfather's one-time workmates now scrounge this iron track
For tors and allies of ore bunkered in the cinders and the hogweed.

From: *On the Dismantling of Millom Ironworks* by Norman Nicholson in *Sea to the West* (1981). The whole poem is a sustained elegy, perhaps his finest work. *See* also Chapter 15 on Wordsworth.

Contrasting Views

On the face of it, the coastal lowlands of West and South Cumberland (I use the historic county name throughout for sentimental reasons) flanking the Lake District mountains seem unlikely sites for the growth of an industrial enclave that continues to this day to provide heavy engineering work on a large scale – the Barrow-in-Furness shipyards and wind turbine plants and the more techy Windscale nuclear repository being examples. The arcuate onshore outcrops and extensive offshore subcrops of the Coal Measures between Whitehaven and Maryport are one reason (Fig. 22.1), while to the south the now worked-out hematite iron ore mines around Egremont, Cleator and Furness provide the other. Here came Britain's entire output of high-grade, low-phosphorus iron ore from the nineteenth century well into the twentieth. It fed both local iron and steel plants and, via the ports of Whitehaven, Workington and Maryport, specialized steel furnaces and forges across the whole country.

Today, from elevated ground in the lower fells of the northern Lake District we can look over the former coalfield across the Solway Firth to rural Galloway, past the squat mass of Criffell mountain towards the decommissioned nuclear plant at Chapelcross near

Annan. From here one gets a feeling of unstoppable environmental and technological advance: views of the future as the slowly revolving blades of scores of wind turbines in Scotland's biggest offshore windfarm on the sandbank of Robin Rigg glisten in the early spring sunshine, generating renewable electricity for the national grid. Similar sights from the southern foothills feature the Barrow Offshore Wind Farm complex, including the world's largest offshore field off Walney Island. Yet it is hard not to avoid recalling Furness poet Norman Nicholson's reflections on the extinction of his native town's industrial birthright. His own grandfather was 'foreman of the back furnace', sluicing out molten steel smelted from Millom hematite in a Bessemer converter.

Geological Settings

Geological events generated the valuable raw materials that made the region what it was and is: more than just an extractive economy – a 'making' one. The events in question came in two distinct deliveries, widely separated in geological time. Long after the deposition and tectonic preservation of Coal Measures strata came the chance juxtaposition of iron mineralization that gave rise to much higher grades of ore than are found in clay ironstone. The hematite ore is aptly named Red-Ore (from the colour of its powdered state – its 'streak') in the accounts books of the Glamorganshire iron smelters who imported it in the early nineteenth century.[1] Although hosted in Early Carboniferous limestone strata, it was, as we shall see below, entirely unrelated to any Carboniferous geological process.

Preservation of the coal resources here was due both to the wider pattern of regional subsidence that came as the finale to Early Carboniferous rifting along the Northumberland–Solway basin and to subsequent inversion tectonics. Here we are well to the north of the axis of maximum Upper Carboniferous subsidence and so, in common with the far larger Northumberland–Durham coalfield directly to the east, the total thickness of coal-bearing strata is only a fraction of that recorded in South Lancashire, around 500 metres in all. Despite this there are several thick seams of general-purpose bituminous coal, some with reasonable but not outstanding caking behaviour that nourished early coke-based iron smelting. This gradually took over in the later eighteenth century from charcoal-fired furnaces as the Lakes valleys were almost denuded of their sylvan glory (the role of nasty coke in preserving what was left of Britain's upland forest cover is often ignored). A century later the local coking coals could not compete in quality and quantity with those from south Durham, which could be transported by rail already coked into the main steel-making centres.

Ten or so major seams were worked. The oldest was the Harrington Four Foot – thick in the south but often sulphurous due to its close juxtaposition with an overlying marine band. The Upper and Lower Three Quarters were worked in the north, the latter above a thick and valuable fireclay. The Lickbank Six Quarters had its upper part valued in the Whitehaven area. The Little Main was consistently good and extensive but with high phosphorus making it a poor source for metallurgical coke. Finally, the youngest,

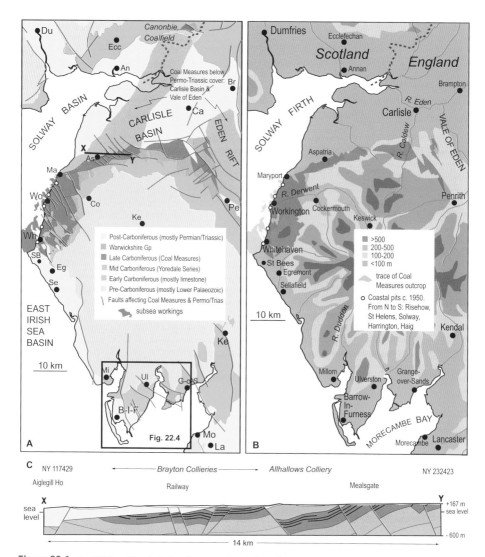

Figure 22.1 A–C West Cumbria landscapes: **A** solid geology, **B** topography, **C** geological section along the line X–Y, marked on A. Sources: material redrawn, recrafted and recoloured from: A British Geological Survey (2007a) B *Times Atlas of the World* (1987) C British Geological Survey 1:50 000 Cockermouth sheet 23.

the Main Band of the Middle Coal Measures, was the thickest and most extensive of all, especially sought after and valuable in subsea workings up to 5.5 metres thick, thinning and splitting northwards. Above a remaining half dozen or so worked seams there are up to 300 metres of sandstone-dominated 'Red Measures' of the Whitehaven Sandstone Series (equivalent to the Warwickshire Group of Midland England, and similarly southerly-derived) with just a couple of economic seams.

Tectonics has not been kind to the West Cumberland Coal Measures. Faulting during tectonic inversion and deep erosion of the Lake District dome left the remaining measures forming a flanking north- to west-dipping synclinal structure. Subsequently the aforementioned Permo-Triassic rift of western Britain cut up from the south (as seen in the North Wales and Lancashire coalfields) which further dismembered the successions along north–south trending faults. The most prominent of these defines the markedly linear NNE–SSW trend to the coastline south of Whitehaven down to Barrow and marks the *de facto* eastern margin to the northern segment of the East Irish Sea rift basin (Fig. 22.1). As an aside, all this faulting and the later flow of mineralizing fluids (*see* below) has led some geologists to criticize plans for enlargement and deepening of the existing Sellafield nuclear repository.

To the east the remarkably parallel linear trend to the margin of the Vale of Eden rift is also caused by a major normal rift-bounding fault, the magnificent west-facing scarp of Cross Fell rising high above the rift basin lowlands. Similar trends mark the courses of many of the smaller faults that cut the coalfield Coal Measures outcrops, picked up again to the north-west across the Solway Firth in Galloway by the linear western faulted margins to the Dumfries and Lockerbie Triassic rift basins.

Early Days of Coal Exploitation

Large-scale exploitation of the Coal Measures came after the Civil War.[2] Carlisle excepted, local domestic needs from the sparse rural population were minor. Even in the early 'coke era' there was enough timber available from charcoal to satisfy the ironmasters who had leased mining rights (some being incomers from as far away as the English Midlands and South Wales) for hematite ore. So it was that canny aristocratic landowners with an entrepreneurial bent saw clearly that their west-facing aspect and geographic position conspired to make burgeoning Dublin city the most obvious large market for their coal. Chief among the pioneers of this trade was Sir Christopher Lowther of Whitehaven, who set up Dublin trading links using ample-sized colliers and improved docking facilities at Whitehaven harbour – transporting several thousand tonnes a year across the Irish Sea.

After Lowther's early death in the 1640s his son John set to the family business in earnest, vastly improving mine drainage in its properties around St Bees, buying up freehold rights to extend mine properties and further improving the harbourage on this often treacherous windward shore. By the dawn of the Irish Wars in the late 1680s the Lowther mines, most notably at Howgill, were supplying nearly 40 000 tonnes a year to Ireland in sixty collier vessels from Whitehaven and Workington. Cumberland now dominated the Irish trade over lesser outputs from the North Wales, Lancashire and Ayrshire coalfields. After a lull in the early eighteenth century the burgeoning demand for coke from the ironmasters led to the construction of the first coke-smelted ore at Little Clifton in 1723 using coal from the Frizington area.[3,4] The owners followed the Coalbrookdale 'integrated practice' of mining their own coal locally and converting it to coke in the open, covered with a thick layer of soil. By 1752 a furnace at Maryport roasted its coal to coke in brick-built 'beehive' ovens with a further iron-making enterprise using coking ovens established at Barepot near Workington in 1763.

The route to exploitation of new coal reserves followed the sinking of the first coastal colliery in 1731, the Saltom Pit at Whitehaven. In the subsequent decades and centuries subsea mining dominated coal production, with the Main Band the chief target. By the 1950s five coastal pits (Haig, Harrington, Solway, St Helens and Risehow) had extended their underground workings up to eight kilometres offshore – amongst the furthest limits reached in any British coastal coalfield (Fig. 22.1B).

Iron and Steel from Hematite Ore

Charcoal smelting of hematite ore was favoured by William Rawlinson in the Quaker family business of the Backbarrow Company whose first blast furnace was completed in around 1711[5] and which is still, miraculously, extant (Fig. 22.2), though since much-modified.

Figure 22.2 William Rawlinson's Backbarrow Company furnace in South Cumbria (*c*.54.152069, 2.591719) was originally built in 1711, just three years after Abraham Darby I (a Quaker friend of Rawlinson, *see* text) had rebuilt his own Coalbrookdale furnace. It was rebuilt in 1770 and in 1856 was recorded as one of only three charcoal-fired furnaces left active in Cumbria – it had been given a steam-driven hot blower (still visible inside) in 1824. Backbarrow Company went bankrupt in 1903 and reopened as the Charcoal Iron Co in 1918. They made Valley brand cold blast pig-iron smelted from local hematite to order. The furnace was closed in 1967 and has been sadly abandoned to the weather ever since (*see* https://www.28dayslater.co.uk/threads/backbarrow-ironworks-cumbria-aug-2013.83650/). Source: Wikimedia Commons: Backbarrow1.jpg by Peterrivington.

It was he whom the ultra-secretive Abraham Darby I indirectly approached by personal letter trying (unsuccessfully) to interest him in his 'new method' (i.e. coke smelting). Darby tempted Rawlinson by mentioning he could save £700 a year by the switch from charcoal, and that in exchange for giving Rawlinson the details he would ask for a cut of one-eighth of the savings made. Rawlinson's reply to this letter is not preserved, though doubtless it was negative since he stuck with charcoal. Eventually this traditionalist went so far as to build a new furnace in the timber-rich Scottish Highlands at Invergarry, to which he shipped his Cumberland hematite for charcoal smelting – so valued was that ore.

From the beginning of the nineteenth century Cumberland low-phosphorus hematite (0.01–0.03% phosphorus by weight) had become the prime iron ore of Britain. Until invention of the Gilchrist-Thomas process in the late 1870s, such ore was required for reliable steel production in the Bessemer Converter, and from the later Siemens-Martin open hearth process. It was shipped out to blast furnaces throughout the land, with South Wales and Staffordshire (via Chester and Runcorn into the Midlands canal system) being the chief purchasers, the former predominant. Local steelworks constructed at Harrington and Workington were undoubtedly among the first to use Bessemer converters in 1856/7.

There was a massive growth in ore production from the 1850s with the discovery of major new ore bodies, like the Burlington deposit near Askham in 1851 and the giant Hodbarrow discovery of 1855, which spawned Millom town and its ironworks. The former established Barrow-in-Furness as a major iron and steel producer under the aegis of Henry Schneider who had arrived in the area in 1839, setting up as an entrepreneurial private explorationist for iron ore and as an iron producer.[6,7] He initially took over the Whiteriggs mine, while also developing other deposits. His company, Schneider, Hannay and Company, was founded in 1859, becoming the Barrow Hematite Steel Company Ltd in 1865. By 1866 the Hindpool works had expanded to include all of ten blast furnaces and eighteen 5-tonne Bessemer converters. Further expansion of its output for shipbuilding steel in the ever-expanding Barrow Vickers yards and the worldwide demand for tough steel rails made it the biggest steel mill in the world by end-century. The town had grown rapidly in forty years (from a population of 3000 in 1841 to 47,000 in 1881, mostly by immigration from Lancashire) and continues today as a major site of precision heavy engineering, a rarity in modern Britain, despite having to import its entire steel needs, since local production stopped in the 1980s.

With the growth of steel output in the late nineteenth century it was natural that Cumberland ore exports dropped drastically in volume. Welsh and other customers switched to foreign imported low-phosphorus hematite from the Basque Country. Hematite continued to be mined, though in ever-decreasing quantities as reserves declined, only steadying or slightly increasing by superhuman effort during two world wars. It was finally used only by specialist alloy steel makers from Sheffield and for the manufacture of varied ochreous paint pigments, dyes and cosmetics – the last ore for the latter purposes dug from the Florence mine at Egremont as recently as 2007.

The Origin of Cumberland Hematite

[8]The ore was invariably hosted in Early Carboniferous limestones along up-faulted margins to the East Irish Sea sedimentary basin (Fig. 22.3). The ferric iron of the hematite was sometimes seen to have replaced the calcium carbonate making up the limestones – witness once-calcareous-shelled fossils changed into perfectly formed hematite casts. The mineral ores occurred in three structural and geometric forms.

First were elongate veins that grew along pre-existing normal fault fractures, widening along fissure openings as steeply dipping faults brought together thick limestone beds along their depth and length. They were mined by conventional stoping methods from different levels running parallel to the veins – the mined-out veins creating linear depressions across the modern landscape.

Second were the much sought after bed-like 'flats' running parallel to the bedding courses of the host limestones – mining was by the pillar-and-stall method – easier and more efficient than that of stoping. It created extensive areas of subsidence, as around Hodbarrow.

Thirdly were occurrences of cavernous 'sops', a miner's term (as in 'milk sop') denoting their pancake-like form – thick and flat in the middle and tapering rapidly outwards. Mining of sops by opencast quarrying created circular areas of subsidence, now lake-infilled, as around Park Farm close to the Park fault. Their cavernous three-dimensional form suggests that they originated as limestone solution (aka karstic) features before the onset of mineralization.

The favoured mechanism for iron mineralization involves heated subsurface formation water rising from the depths of the East Irish Sea basin within porous and partly iron-cemented Triassic strata (such as seen splendidly at outcrop at St Bees Head). The acidic waters stripped out part of the ferric iron coatings formed originally by precipitation around the silica sand grains below the contemporary Triassic desert surface. The upward-percolating formation waters were slow-moving and, as fluid inclusions (*see* Chapter 23) indicate, low-temperature by the time they precipitated out their hematite close to the contemporary land surface. The timing of the hydrothermal mineralization was most probably during the early Cenozoic era (around 60 Ma) during the general uplift of north-west Britain.

Coda: 'Our Mining Geologist'

As we have seen, the heyday of hematite mining was set in motion by the new Bessemer converters. It also coincided with broad and far-reaching plans by the then director of the Geological Survey, Sir Henry de la Beche, to set up an applied geological teaching and research institute that might match long-lived continental examples like the Freiburg Mining Academy and the École des Mines in Paris. Such a project greatly appealed to the inclusive and interdisciplinary mind of Prince Albert, whose ear De la Beche had cultivated. Here is a later Director of the Survey, Sir Edward Bailey, writing of this time:

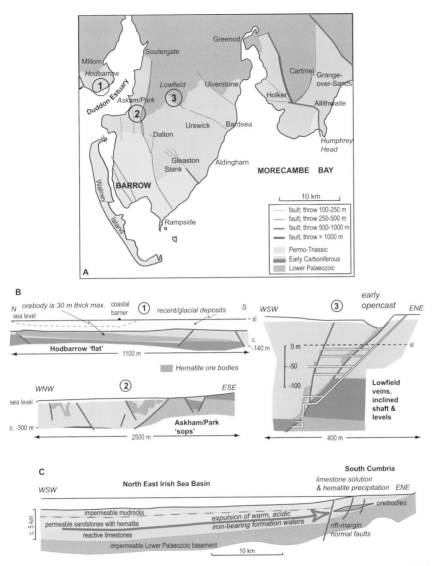

Figure 22.3 A–C The South Cumbria (Furness district) iron ore field. **A** General simplified geology. **B** Representative sections to show the three chief modes of occurrence of the iron ore deposits. Locations 1–3 in A. **C** A regional schematic section to show the inferred mode of formation of Furness hematite. Sources: material redrawn, recrafted and recoloured after originals by Rose and Dunham (1977).

The year 1851 brought to both these great men the consummation of their dearest hopes: to De la Beche, the opening of his new Museum of Practical Geology at 28, Jermyn Street, off Piccadilly, with accommodation not only for the Geological Survey and Mining Record Office, but also for a Government

Figure 22.4 Unique engraved image of workings at one of the large hematite 'flats' discovered at Todholes, Cleator, a few miles south-east of Whitehaven in the 1850s, as illustrated and described by Warington Smythe (1856). The resident iron-ore master at the time was one John Stirling, there being nine others in the general Whitehaven district. The main orebody (labelled '3' on the image) is usually 5–10 m thick – solid hematite overlying red/white mottle mudrocks ('almost of the nature of a fireclay') of the Early Carboniferous close to its unconformity with underlying Lower Palaeozoic rocks. A large adit opening into the 'flat' is seen, together with cognate opencast workings below and above with a tub and tram line at the higher level. A distant steam engine implies there was also a deeper shaft mine into the orebody where it would have been worked 'pillar-and-stall' style.

School of Mines and of Science applied to the Arts; and to the Prince, the triumphantly successful Great Exhibition of the Industry of all Nations.[9]

One of the first six professors and lecturers at the School of Mines was one Warington Smythe, whose brief was 'Mining and Mineralogy' and it is from his pen that we have the first coherent account of the Cumberland hematite mines in 1856 at the very beginning of their rise to prominence. In his introduction to the Geological Survey memoir, *The Iron Ores of Britain*, De le Beche's successor, R.I. Murchison, proudly writes of its great national importance in contributing to the knowledge of the chemical composition of British iron ores and that:

The clear and accurate description, by our mining geologist, Mr Warington Smythe, of the various rock formations in which the ores occur, forms a valuable portion of the Memoir.[10]

Concerning Cumberland iron ore, Warington Smythe wrote of the geometry of the various orebodies, contemporary mining methods (quarrying, pillar-and-stall), mine hazards (water influx from unconsolidated sands overlying sop orebodies), production details (over half a million tonnes produced in 1854), transport costs (the great expense of rail; the use of carting in Cleator to the north), the low market price for these premium ores, the high royalties given to landowners (*c*.10%) and the names and affiliations of the chief mining consortia and their managers. It is thanks to Smythe's unique engraved field sketch that we can admire the scale and working of a typical opencast 'flat' (Fig. 22.4) and also learn of the unique survival of charcoal smelting in an age of the almost universal use of metallurgical coke:

> A small amount of the Ulverstone hematite is still smelted with charcoal at the furnaces of Newland, Backbarrow (*see* Fig. 22.2), and Duddon, one of which only is in blast at a time [probably because of scarcity of sufficient charcoal]: and this forms the only relic left in England of the old mode of production, so completely has the introduction of coal swamped the use of vegetable fuel.[10]

23

Northumberland and Durham

But who knows East Durham? The answer is – nobody but the people who have to live and work there, and a few others who go there on business. It is, you see, a coal-mining district. Unless we happen to be connected in some way with a colliery, we do not know these districts. They are usually unpleasant and rather remote and so we leave them alone. Of the millions in London, how many have ever spent half an hour in a mining village?

From J.B. Priestley *English Journey*, p. 302 *et seq.* Just two years after Priestley's bitterly ironic comments the Jarrow marchers visited London – the North East was saying 'Here we are: help us!' (1934)

The North Pennine Template

From Cross Fell in the west to the Cheviot in the north, the North Pennines of Northumberland and Durham define a declining plateau cut by glaciated dales – tributaries of the Aln, Coquet, Blyth, Wansbeck, Tyne, Derwent, Wear and Tees – eight swift-flowing rivers (Fig. 23.1). Their steep-gradient courses cut deeply into mostly Carboniferous bedrock – generally easterly stratal dips eventually bringing Coal Measures to outcrop. Here is an exact, coincident and sharp increase in settlement, population and road density in a zone running approximately from Bishop Auckland in the south via Consett to the Tyne west of Newcastle. To travel today in these parts along the ridges and valleys breasted by the A68 with its many adjacent wind turbines is to traverse a gradient of cultures spanning many centuries. Spreading out to the west are stone-walled pasturelands whose sweet grass is nourished by glacial deposits resting on limestone bedrock – rich Yoredale strata. To the east is the old exposed coalfield whose lanes and byways feature long-abandoned pit sites, many of them quite local affairs with just a few score miners and their families at each. All traces of their surface structures are now gone, but there remain associated beehive coking ovens

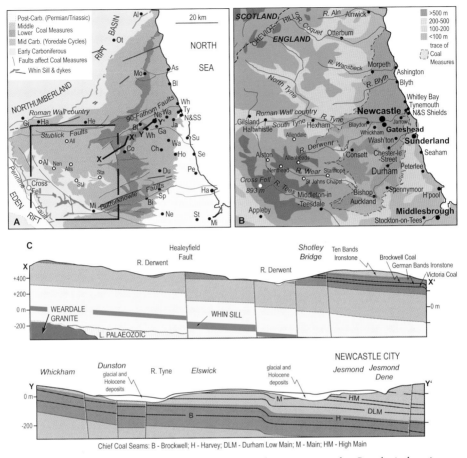

Figure 23.1 A–C Northumberland/Durham: **A** solid geology, **B** topography, **C** geological sections along the lines X–X' and Y–Y' marked on A. Sources: material redrawn, recrafted and recoloured from: A British Geological Survey (2007a). Rectangle shows approximate location of Figure 23.4A. B *Times Atlas of the World* (1987), C British Geological Survey 1:50 000 Newcastle-on-Tyne Sheet 20.

mostly hidden by vegetation on the valley sides. The settlements themselves survive – usually a linear row or two of snug terraced housing, frequently retaining the suffix 'Colliery', as at Hedleyside and Cornsay.

Further east a definite north–south scarp of Permian limestones (as in neighbouring North Yorkshire) wriggles over the landscape from south of Shildon to South Shields via Sunderland and Houghton-le-Spring – the underlying unconformity marking both the eastern limit to exposed Coal Measures rocks and the northernmost tip to Mesozoic lowlands of eastern England. Here, in East Durham were the last great pits to mine the concealed Coal Measures, thousands migrating into the growing settlements of Blackhall, Horden, Peterlee, Easington and Seaham during the late-nineteenth to mid-twentieth centuries. These are nowadays stranded and neglected urban enclaves in sweeping green countryside of gentle undulations with abandoned limestone and dolomite quarries providing oases of lime-rich flora and nesting places, all protected as nature reserves.

To the west the Carboniferous bedrock is notably intruded by the Whin Sill along Cross Fell and its inliers of Lower Palaeozoic rocks, the resistant dolerite also popping up, due to folding and faulting, to form High Force waterfall and Cauldron Snout's cataracts in Upper Teesdale. Buried deep below the Alston Plateau, the Weardale granite was revealed by a borehole in the 1960s (more on this below). Above its weathered surface are interstratified limestone, sandstone and mudrock members of the Yoredale Group. This is the locus of the North Pennine orefield centred on Weardale, Upper Teesdale and the Allendales. Sturdy stone-built (sandstone and limestone) villages and the market towns of Wolsingham, Stanhope and Alston (England's highest) grew up, the orefield prominent in lead and then zinc production well into the early twentieth century, and subsequently into the 1990s for its non-metallic output of fluorite and barite.

The west-to-east pattern of North Pennine drainage determined that north–south travel and communication through the coalfield depended upon sturdy bridges to cross the wide and fast-flowing rivers and their sea-fringe of long, wide estuaries. So it is that modern road and rail visitors to the North East arrive via bridges and viaducts that offer splendid urban views. Given the close bonding brought about by the coal trade it is interesting to reflect on the strong similarities between Newcastle-on-Tyne and London in this respect. Both have Roman origins, each nucleated along a large estuarine river with clusters of iconic buildings joined by distinctive and much-loved bridges. The big difference today is in the quality of life.

To the south, medieval Chester-Le-Street, Durham City and Bishop Auckland became nuclei for scores of surrounding pit villages. The Royal County Hotel in Durham City was the venue for visiting socialist and union dignitaries who viewed the parades of the Durham Miners' Gala from its balcony, an annual celebration of the might of coal and its miners. In the mid-nineteenth century, shipyards like Jarrow grew up along Tyneside's south bank and on Wearside. They used iron and steel plate produced in new industrial towns, notably Consett on the western edge of the coalfield – then somewhat later came Middlesbrough on Teesside.

The First Swamp Forests

Early Carboniferous rift tectonics were particularly vigorous here, with extremely thick sedimentary accumulations, as much as five kilometres in north Cumberland and west-central Northumberland. The hyperactive tectonics that produced the subsidence necessary to preserve this pile of shallow-water sediment were accompanied by volcanic activity. This occurred initially along the north and south margins to the Northumberland–Solway rift (Cockermouth, Birrenswark and Kelso lavas) as a minor accompaniment but building up over time and in space northwards into midland Scotland. As important was the influx of sandy detrital sediment brought into lake, lagoon and bay from seasonal rainfall feeding locally sourced rivers and by longer-travelled discharges sourced to the north-east. These Early Carboniferous riverine and deltaic swamps were populated by forest stands of early Lepidodendrales capable of generating metres of peats. Their branches overlooked watery and aerial habitats full of crustaceans, fish, amphibian and insect pioneers that were to take over the whole equatorial world by the Late Carboniferous. It was this heritage, the first economically viable forest swamp coals, that enabled the early mining communities of Spittal and Scremerston in north Northumberland and Berwickshire.

Sea-level oscillations caused deltaic swamps to advance and retreat across the subsiding former rift margins in the Middle Carboniferous. The repeated process gave rise to stratal repetitions of marine limestone and detrital deltaic sandy and muddy sediment of the aforementioned 'Yoredales' (cf. old dialect for Wensleydale – 'Uredale'). The west Durham and Northumberland fells strongly feature such cycles with their numerous coals, usually thin, but sometimes up to half a metre thick. They accumulated as the vigour of rifting declined and the effects of thermal sagging took over across the whole region during deposition of the Coal Measures in the Late Carboniferous.

Taking Coals to London: Why Tyneside?

The success since medieval times of Tyneside coal in providing warmth for the capital's citizens 250 miles away hinged on the ease and economy of both extraction and transport. The efficacy of extraction was courtesy of the geology – thick, horizontal to gently dipping and unfaulted coal seams close to surface were available in large acreage, single-lease swathes. These coal-bearing acres were owned by Tudor, Stuart and Restoration landed gentry with a common mind to entrepreneurship.[1] Transport was easy and cheap from the pitheads down onto keelboats (lighters) and then into the sheltered outer estuary to be loaded onto sturdy collier brigs for a week-long voyage along a generally lee shore (though not always so in winter, as Defoe reminded us) to the Thames estuary.

The details need a little fleshing out. The estuary is one of Britain's longest, tidal from Tynemouth upstream for over twenty-five kilometres as the crow flies to its limit at Wylham a few kilometres west of Newcastle. It is also the business end of a large upland catchment whose peak spring discharges include snowmelt from such faraway ranges as Cross Fell (via the South Tyne) and the south-western extensions of the Cheviots (via the North Tyne). Strong tidal currents together with such periodic floods combine to keep the estuarine channel flushed of sediment, and therefore largely navigable along the parts that mattered. The main Tyne channel itself incises into the Coal Measures along its whole tidal length, with the thick coal seams of the Lower and Middle Coal Measures (especially the legendary High Main, never less than 1.5 m thick) but also a dozen or so significant other seams. These all outcrop or are in shallow subcrop over long stretches of its length, particularly west of Newcastle on the north bank and around Gateshead on the south bank. Further east, almost to Tynemouth, the outcropping measures are dominated by thick sandstone with few coals. These were uneconomic until deeper mines eventually reached through them into the riches beneath during the mid-nineteenth century.

The incised Tyne valley slopes steeply below the surrounds of the adjacent coal-rich catchment, descending to the river banks below. Coal was brought downhill at first by pack animal, as Celia Fiennes encountered. In the bigger concerns animal-led wagons ran on wooden, then iron rails. The impediment to ordered loading of this coal along the narrow upstream channel frontages to the main coal outcrops and subcrops west of Newcastle/Gateshead was overcome by the use of hundreds of keels. These transported it to the wide (but still sheltered) outer estuary between North and South Shields, where substantial numbers of colliers could safely wait out seasonal gales if necessary. The arrangement is dramatically illustrated by the Early Modern map previously featured in Figure 1.1. In the nineteenth century, steam locomotives took the coal along steel tracks direct to the outer estuary.

From the thirteenth century to the late eighteenth century there was simply nowhere else in Britain (or elsewhere in Europe) that could compete with all these favourable factors for coal extraction and transport. Sunderland (on Wearside) could have rivalled Newcastle, but serious portage facilities on its exposed coast were absent prior to the hard engineering of harbour and dockyard complexes in late-Georgian/early Victorian times.[2] The Lothian coalfields east of Edinburgh were also coastal and on a lee-shore for much of the year, but Scotland was an independent country, often a deadly rival and, latterly, an economic rival until the early 1700s, by which time the Tyneside trade was well established. Yorkshire coal was too far from the sea, mostly on unnavigable rivers. South Wales around Swansea was too far round the gale- and privateer-ridden south-west peninsula. Bristol had the same disadvantage but did supply some tonnage to London though no major, long-lasting, coherent trade developed, in contrast to its vast international dealings in other commodities.

Of Keelmen

The colossal nature of the Tyneside coal trade provided employment for thousands, and its increasing expansion in the seventeenth century required more and more wage labourers. The trade was controlled by middlemen in Newcastle with almost medieval guild-style patronage and protectionism, the so-called 'hostmen' who bought the coal from the mine captains and landowners before contracting for transport with the keelmen, and then in their turn to the owners of the sea-going collier fleets to transport it down to the London coal merchants. It was quite an involved loop – in the words of historian Andy Burn it depended upon:

> a huge workforce to move the coal before mechanization was introduced much later, which the hostmen contracted through sub-merchant employees known as 'fitters'. Typically four keelmen would shift 21 tons of coal from a riverside staithe onto a keel, sailing or rowing it out to colliers, which they loaded through portholes before returning to Newcastle hours later, ideally with help from the rising tide. In line with the rapid expansion of

coal shipments, keelmen grew from fewer than one in ten fathers recorded on baptisms at the beginning of the seventeenth century to more than a quarter by the end, the largest single group by some distance. Newcastle, in the process of precocious industrialization brought about by the demand for coal, had become a heavily proletarian town.[3]

A problem in enlisting the numbers of keelmen required for efficient down-estuary transport of coal was the seasonal nature of the trade, with no winter sailings, as the North Sea turned from lee shore to windward without shelter along most of its length during violent northerly and easterly storms. Hearsay in 1708 had a servant explain to a visitor that:

> I have heard good saylors say, they had rather run the hazard of an East-Indie voyage, then be obliged to sail all the winter between London and Newcastle.[3]

In the early days many of the keelmen were Scottish Borderers who might return to their shielings and valleys in the winter, but as time went on many settled permanently in Newcastle, concentrated in the Sandgate district down by the river's left bank in All Saints parish (*see* heading to Chapter 6).

We are fortunate in having some firm knowledge of the way of life of these early industrial workers and their families in the seventeenth century.[3] Their organized recruitment and working practices were critical in enabling the efficient workings of an industry that dealt with the muscle-powered transfer of vast tonnages of bulky and dusty raw material to waiting sea-going vessels. From probate records we know by name of several individuals amongst the keeling community, both working keelmen and their employers, though only ten per cent or so left estates that demanded a probate inventory. Similarly the records of scrivener-written wills for the unfortunates caught up in the great Newcastle plague of 1636 included keelboat owner and worker, Thomas Holmes.[4] From estimates of the number of seasonal journeys up and down the estuary their wages through the seventeenth century were thought to have been sufficient to sustain a basic family on a single wage (though less so by the troubled 1690s). Perks of the job might include provision of beer, the right to keep coal 'sweepings' (leftovers from a keel load) and, from the evidence of a few criminal prosecutions, the occasional private sale of coal unknown to the all-powerful hostmen.

Yet the majority of experienced working keelmen were seasonally poor and frequently near-destitute. These were tough, skilled men whose knowledge of river lore and tide gave them a special place in the coal trade. In the interests of business efficiency (and, one hopes, common decency) seasonal welfare was provided in due course by employers, hostmen and the town's parish administration. This included the provision of winter credit from hostmen and fitters as part of their contract of employment and parish relief (increasing throughout the century) for casual winter labouring work in and about the town in a variety of guises.

Joseph Turner's aforementioned 'Keelmen' painting of 1835 that adorns this book's cover (*see* also Chapter 15) can also be seen with the benefit of hindsight to be an elegy on the imminent destruction of a whole way of life and culture of manual work. A year later in 1836 we have the following description of gravity-driven railway coal wagons at work on elevated jetty-like staithes, the exact scene probably located along the eastern Tyne right bank:

> to see a railway carried from the high bank and supported on tall piles, horizontally above the surface of the river, and to some distance into it, as if to allow these vagabond trains of wagons to run right off, and dash themselves down into the river…as they draw near the river…a pair of gigantic arms separate themselves from the end of the railway. They catch the wagon; they hold it suspended in the air; they let it softly and gently descend…[to] a ship [that] already lies below the end of the railway. The wagon descends to it; a man standing there strike a bolt – the bottom falls, and the coals which it contains are nicely deposited in the hold of the vessel! Up again soars the empty wagon in that pair of gigantic arms. It reaches the railway; it glides like a black swan into its native lake, upon it, and away it goes as if of its own accord, to a distance to await its brethren, who successively perform the same exploit, and then joining it, all scamper back again as hard as they can over the plain to the distant pit.[5]

The scene so-described depicts direct loading independent of tide, wind and keels. By the 1840s raised staithes received steam locomotive-driven coal wagons from the pits – all mines then had their own gravity-free access to the riverside and the waiting collier vessels. Eventually, long-distance rail transport from any number of inland coalfields would destroy the Tyneside sea coal trade entirely. By then the burgeoning industrialization and heavy engineering concerns of the north-east were able to take over the livelihoods of the keeling community.

Of Glasshouses

There is a link between Tyneside keelmen and the glass working that was prominent[6], perhaps pre-eminent, in Newcastle (beginning at Lemington, on the Tyne just to the west; Fig. 23.2) and South Shields from the eighteenth century, both until just a decade or so ago. It developed a little later at Sunderland on Wearside, with James Hartley's Wear Glass Works producing much of the country's sheet glass from the nineteenth century, again until little over a decade ago. The link comes from the extremely fuel-intensive process of glass-making that took place in the pioneering, all-English development of tall, conical, brick-built 'glass cones' where the glassblowing process was carried on around central coal-fired furnaces.

The chief raw mineral constituent in the recipe to make the molten 'metal' (the term used by glassmakers to describe their fused mixture) was pure quartz sand, required

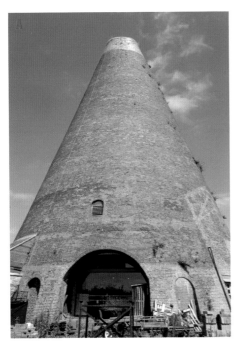

Figure 23.2 **A–C** The last surviving glass cone in north-east England at Lemington, a few miles west of central Newcastle-on-Tyne, at Newburn (54.583166, 1.425277). **A** The cone, built around 1787, was originally 40.6 metres high but the top 6.5 metres has been removed – still a fine sight on the northern slopes to the Tyne valley. **B** The ground interior (at June 2019) is below a handsome timber roof and is the salesroom for a firm of stove manufacturers who have inventively used the original glass-blowing alcoves that surrounded the original central furnace set-up (now removed) where the blowers would come to collect their molten 'metal'. **C** The lower brickwork shows the centimetre-by-centimetre horizontal offlap of successive brick rows necessary to build up smoothly and safely from the base to the cone apex. Source: Author.

as loose and free of clay and iron oxides as possible. Such glass sands were needed in bulk but were almost unknown in the North Pennines, though several localities existed amongst the younger Mesozoic sediment outcrops of southern and eastern England. Some entrepreneur, possibly Sir Robert Mansell,[7] who in 1623 had been given a crown monopoly on glass manufacture in Newcastle, realized that the empty holds of northbound returning colliers could be filled with glass sand. In this reverse process to coal export the sand would have been unloaded in the Shields anchorages for the keelmen to take back up-river to the early Tyneside glass houses, or later in the nineteenth century to plate glass works in South Shields itself. The port of King's Lynn is one that supplied Lower Cretaceous Leziate sands to the Wearside glassmakers:

> Glass had first been made on Wearside in the time of Bede, and the industry
> revived in the late Middle Ages with the availability of sand ballast. Later,

specialist glassmaking sand was imported, especially white sand from King's Lynn, and soapers' ashes from Yarmouth and London, serving as ballast and sometimes bartered for Sunderland coal.[8]

In this way a new take on the old adage was enacted. 'Taking Sand to Sunderland' proved perfect business sense, the lower-cost item travelling to meet the chief *in situ* energy source of the day.

The Transition to Making: Local Iron and Steel for Ships, Guns and Bridges

The high-grade coking coals of the southern part of the coalfield played a central role in the continuing Industrial Revolution of the mid-to-late nineteenth century when the primary regional economy (now including Middlesbrough, on Teesside) of coal exporting was replaced by iron and steel manufacturing – great ironclad ships, massive armaments and steel bridges. The cheap transport of raw materials around the region (and to adjacent areas, like coke to West Cumbria) was made possible by early steam locomotion pioneered for the purpose by the Stephensons of Tyneside, as in the Stockton and Darlington Railway of 1825.

By the 1840s such transport had changed the game in the iron and steel industry, enabling bulk transport of ore, coke, iron and, eventually, steel products far and wide. Early ironworks and cementation steelworks using imported Swedish bar iron (Fig. 23.3) along the Tyne and swift-flowing tributaries like the Derwent gave way to vast production in West Durham from the early 1840s. The Consett Iron Company works[9] was the first

Figure 23.3 **A**, **B** Two views of the beautiful sandstone-built Derwentcote cementation steel furnace (the dovecote-like central building) and adjoining auxiliary buildings at Hamsterley, County Durham (54.541128, 1.475315), now safely preserved and reconstructed by English Heritage (since the 1990s). Source: Author.

giant integrated iron (and, eventually, steel) works to be sited in the region, mining its raw materials locally, especially the coking coals of several seams (notably the Harvey Coal) and three seams of clay ironstone in the Lower Coal Measures found together at outcrop in the western limits to the coalfield. These were the 3-metre thick Number 1 Ironstone above the Harvey Coal (mined together), the Ten Band Ironstone and the German Bands Ironstones. Interestingly from the point of view of the history of metallurgy, the latter took their name from a group of seventeenth-century German Protestant master-swordmakers and their families who had fled persecution in their native Solingen – the Sheffield of Germany – to set up their industry using imported iron around the small town of Shotley Bridge on the River Derwent near Consett – a sort of Sheffield-in-minuature. The ores around Consett were no richer or purer than any other clay ironstone ores from the Coal Measures, but in the deeply embayed scarplines of the Derwent valley and its tributaries around the town the gently dipping coal seams and iron ores could be easily mined at multiple localities by adits and relatively shallow shafts without the need for expensive deep mines, much to their economic advantage. The geological and logistic situation was much like that found in north-east Glamorgan at the beginning of the ironworks there a hundred years before (Chapter 16).

Outlets for Consett iron plate came both with the railways and the new-fangled ironclad ships that had begun to be built as a speciality of both Tyneside and Wearside shipyards.[10] After periods in the economic doldrums during the difficult years of the 1850–70s, the advent of the Gilchrist-Thomas process by the 1870s saw scores of both Bessemer and open-hearth furnaces with the basic lining set up at Consett and its subsequent rise as one of the chief steel-makers to the industrialized world, especially for the now all-powerful shipyards. Additional manufacturing came with the growth of armament works like that of Armstrong of Newcastle, whose presiding mechanical genius was the self-taught solicitor William George Armstrong[11], inventor of hydraulic machinery, notably dockyard cranes. He later designed the first breech-loading cannon, donated the patent to the government and set up the independent Elswick Ordnance Company to manufacture it and other new weapons of war. However, his longest-standing and most passionate interest was in the field of hydraulics, as visitors to his Cheviot-side retreat, Cragside (now a National Trust property), at Rothbury will discover. He is right to be remembered as the mechanical inventor who forecast the demise of coal power earlier than foreseen by contemporary pundits and its replacement by renewable sources like his beloved water power (a beautiful water-wheel of his features later in this chapter), but also including solar power.

By the 1870s a major new source of British iron ore was exploited amongst the Jurassic (Middle Liassic) sedimentary deposits of the Cleveland Hills in North Yorkshire. The discovery eventually introduced industrial Teesside and the new 'Ironopolis' of Middlesbrough to the planet[12]. Nascent iron smelting using Durham coke was already in place along the Tees from the 1850s, at first using imported high-grade magnetite ores from Goatland in the North York Moors. With the rapid exhaustion of this source, Basque hematite was imported. It was not until the Gilchrist-Thomas process was

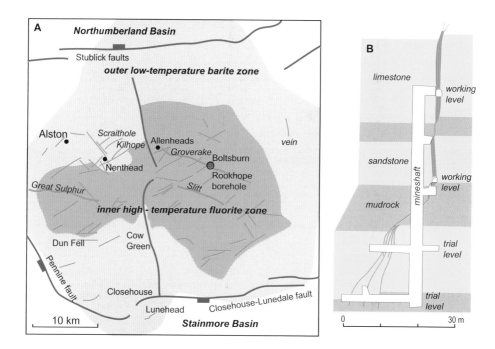

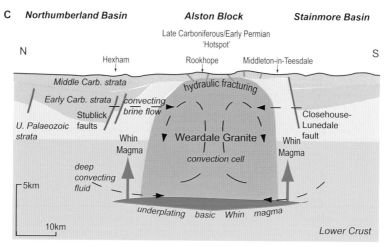

Figure 23.4 **A–C** North Pennine Mineralization. **A** The extent of mineralization, chief faults and the extent (shaded) of the inner fluorite and outer barite mineralized zones. The location of this map is shown on the smaller-scale map of Figure 23.1. **B** Section to show a typical North Pennine lead/zinc vein deposit, the Killhopehead Vein between Nenthead and Allenheads. The lower (failed) trial levels show the serendipity of geological prediction in mine work. **C** Schematic model for the most recent and authoritative version of the origin of the North Pennine mineralization. Sources: A, B material redrawn, recrafted and recoloured from: Dunham (1990) C ditto from Bott and Smith (2018).

finalized in Middlesbrough some twenty years later that the high-phosphorus Cleveland ores could be exploited – the whole Teesside 'iron-rush' began its unstoppable course.

Metal Mineralization in the North Pennines

The North Pennine orefield (Fig. 23.4) yielded over three million tonnes of lead concentrates, as well as plentiful zinc, fluorite, barite and the world's only commercial source of witherite. It became perhaps the geologically best-understood orefield in Britain through the captivating story of its geological evolution, as discovered and developed by successive research initiatives at the University of Durham involving K.C. Dunham that began in the 1930s. Further, during its heyday the orefield was partly owned, leased and managed by the aforementioned London Lead Company (1692–1905), whose efficient mining practices were accompanied by an advanced system of miners' welfare in those isolated and remote upland fell villages of the Allendales that formed the human nucleus to the mining fields.[13]

The orefield lies in what is known geologically as the Alston Block, an up-faulted plateau of Early and Middle Carboniferous strata to the east of the Coal Measures of the Durham coalfield. It is largely a vein orefield with some thick strata-like 'flats' of rich ore where the limestone, under a claystone seal, has been replaced by ore-bearing minerals. The vein mineral deposits are concentrated along minor fault lines that periodically widen into substantial (≤10 m) ore-filled cavities where they cut through brittle limestone and sandstone strata. The faulted orebodies trend predominantly north-east to south-west with subsidiary north-west to south-east orientations.

Figure 23.5 Pale purple cubic fluorite crystals with two octahedral galena crystals. The specimen is from a former working mine in Upper Weardale (it may have come from reworked spoil tips at Boltsburn). I had written as a schoolboy to the mine manager in 1965 of my wish for a specimen of purple fluorite to add to my growing mineral collection, and a few weeks later I was astonished to receive this carefully wrapped specimen in a small box – an act of kindness I have never forgotten.

During the 1930s, detailed field mapping by Dunham, then a research student under Arthur Holmes at Durham, described a marked ovoid concentration of mineralization located mostly east of Alston and centred a few kilometres south of Allenheads.[14] Within the orefield, lead ores were mined everywhere, but with a marked zonation of other minerals: random concentrations of zinc ore were mostly to the north-west and south; there was a central zone of fluorite (Fig. 23.5) and an outer zone of barite. Such zoned mineralization existed elsewhere in England, notably in Cornwall, and so Dunham hypothesized that the mineralization was similarly due to a Variscan granite that had intruded and crystallized under the Alston Block. Arthur Holmes himself made clear his scepticism about Dunham's guess at a Variscan age for any Alston granite, as did others, preferring an older Acadian age around 400 Ma, as in granite bodies in the neighbouring Lake District and Southern Uplands.

Dunham's hunch that a granitic source might occur below surface was confirmed in the 1950s by geophysicist M.H.P. Bott, newly hired by Dunham as a lecturer in geophysics at Durham and an expert innovator in the interpretation of Earth's microgravity as influenced by buried rock masses of contrasting density. Following earlier pioneering gravity measurements in south-west and northern England and southern Scotland, Bott and Geological Survey colleagues undertook further surveying and measurements that indicated tell-tale gravity gradients centred on the ovoid of the orefield. These indicated the presence of an outward-sloping, low-density granite-like body less than a kilometre below the surface at its highest crest in Upper Weardale.[15]

Dunham and Bott subsequently prepared the detailed case for the UK's most ambitious onshore deep-proving borehole since the discovery in 1890 of the concealed Kent coalfield. Drilling at the cusp of the suspected granite at Rookhope in Upper Weardale they penetrated and cored deep below the Early Carboniferous Yoredale group, directly proving the existence of a buried granite.[16] Holmes was proven right, however, for the granite had an obviously weathered top several metres thick, upon which overlying Carboniferous sedimentary strata rested with unconformity. Radiometric dates from borehole core samples established that the granite intrusion was, indeed, of Acadian age, as Holmes and others had predicted. The mineralization therefore post-dated the arrival of the granite magma by almost a hundred million years, by which time the intrusion was well-solid.

At this time, Dunham and his post-doctoral and research students were engaged in original and fundamental studies of base metal mineralization by migrating formation waters (deep thermal aquifer flow) in the North Pennines using the novel technique of 'fluid inclusions' – the analysis of minute fluid drops preserved in mineral crystals during their crystallization by cooling. These revealed higher temperatures for the inner fluorite mineralisation (≤180°C) compared to that of the outer barite (*c.*50°C), with the composition of the mineralizing fluids being chloride- and sulphate-rich brines.[17]

The very latest ideas for the onset of North Pennine mineralization appeal to the intrusion of basic alkaline magma below the Weardale granite towards the end of the Carboniferous period.[18] This is thought to have caused convection currents that brought in mineralizing brines from deeply buried sediments in the adjacent Northumberland basin (Fig. 23.4C). These scavenged lead, zinc and barium ions both from the magma itself and the older granite above – the magma eventually intruding upwards via dykes into the now-inverted, tilted and faulted upper crustal Carboniferous strata to form the Whin Sill and its associated dykes. Just a few million years later, renewed alkali basaltic magma arrived under the North Pennines (and under midland Scotland). It was this final burst of convectional heating that added fluorine and rare earth metals to the mineralizing fluids and caused them to shoot upwards, hydraulically fracturing the Yoredale limestones and sandstones of the Alston Block by faulting to form the ore veins and zoned mineral assemblage of the orefield.

Figure 23.6 **A–C** The Killhope (North of England) Lead Mining Museum in Upper Weardale (54.465790, 2.162121) is a mineworks preserved as an industrial museum complex with restored hydraulic-driven machinery. **A** View of the site: the functioning overshot water-wheel fed from uphill reservoirs powered the crushing and washing plant buildings to the left. The wheel drove four sets of crushing rollers and was manufactured by William Armstrong at his Elswick Works on Tyneside, the oldest and largest surviving wheel of its kind. In the foreground is the open-air washing plant where final treatment of the mined, crushed and sorted ore took place. **B** The 10.5 metre diameter wheel rotates three times per minute. It was manufactured around 1860 and installed here in 1877, quite late in the development of the North Pennine ore field – lead prices were soon to drop with the onset of fierce competition from New World and Colonial mines. **C** Walnut-sized and smaller fragments of crushed ore and associated rock and non-metallic minerals were processed in this 'jigger' shed, where belt drives from the main wheel powered the 'jiggers' – agitated water-filled tanks where the denser lead ore was partly separated from waste rock and mineral. Further separation took place next door in the 'buddle' shed and subsequently down on the open-air dressing floor.

Something of the grandeur of the metal mines that captured the imaginations of both the young W.H. Auden and K.C. Dunham high in the North Pennines is captured in the preserved and restored mine complex at the visitor centre of Killhope in Upper Weardale (Fig. 23.6). It is a place for reflection on the efforts of humans to exploit natural resources making use of renewable water power – its adits, the magnificent overshot water-wheel (William Armstrong of Newcastle's contribution) and the remains of extensive ore-dressing sheds and floors are set in a timeless landscape of deep valleys, high pasture and rough fellside. Here the curlew's legato disyllabic cry echoes through the air on misty late-summer mornings, filling visiting walkers (or returning Durham alumni) with a profound sense of awe at the permanence of nature amidst lovingly preserved industrial remains of long ago; all there for future generations to get to know.

24

Scottish Midlands

The Lady is the last of all her kind
Headframe in the clouds, these pulley-whorls
Change with the light, a beacon to remind
Who fuelled Scotland, lit us, kept us warm.
.....
Keep her headgear. The Lady burns our minds.
Without her wheels we could never know
How in the miner's eye, a coal glows.
The Lady is the last of all her kind.

First and last verses of *The Lady Victoria Colliery* by Valerie Gillies, the Edinburgh
Makar: a poem commissioned by the Royal Commission on the Ancient and Historical
Monuments of Scotland (2008)

Perspectives

Like the South Wales coalfield whose exploitation defined an economic nucleus that
became the core of a modern nation, the Scottish Midland coalfields (Fig. 24.1) can be
seen in a similar light. Yet there was, and is, an all-important difference. Midland Scotland
had always been the economic, demographic and political nexus of the country. It was
the concentrated seat of landed power (still so) for successive cultures, repeated waves
of invaders and tattered refugees from famine and dispossession in the neighbouring
Gàidhealtachd, Galloway and the Southern Uplands for hundreds of years.[1] For it is
here that Scotland's chief agriculturally rich lowlands are situated, the greatest area
below 200 metres altitude north of the Borders. Their location is due to large areas of
relatively soft and erodible Carboniferous sedimentary bedrock – albeit separated by
rugged intrabasinal uplands founded largely upon contemporaneous and older volcanic
rocks. Most importantly, the Carboniferous strata are often overlain by unconsolidated
glacial-age mixed sedimentary detritus, potentially fertile substrates ice-scoured out of

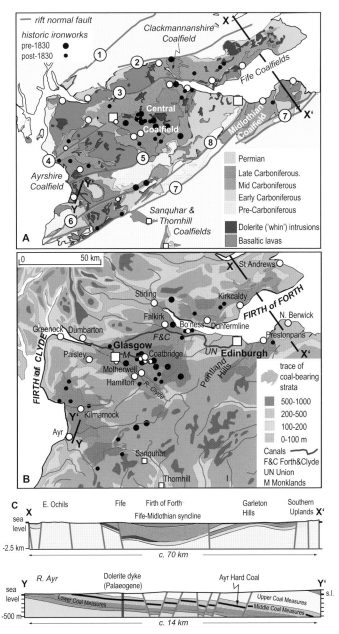

Figure 24.1 **A–C** Midland Scotland. **A** Solid geology. The labelled faults considered active during Early Carboniferous rifting are : **1** Highland Boundary Fault, **2** Ochil Fault, **3** Campsie Fault, **4** Dusk Water Fault, **5** Inchgotrick Fault, **6** Kerse Loch Fault, **7** Southern Uplands Faults, **8** Pentland Fault. **B** Topography. **C** Geological sections along the lines X–X' and Y–Y' shown on A. Sources: material redrawn, recrafted and recoloured from: A British Geological Survey (2007a) and Duncan (2005) for the location of blast furnace plants. B *Times Atlas of the World* (1987), C British Geological Survey (2007b); British Geological Survey 1:50 000 Scotland (Ayr) Sheet 14W.

surrounding mountains to the north and south and brought towards the coastlines to form reclaimed and immensely fertile estuarine 'carseland'.

So it is that today many of the industrialized and heavily populated former coalfields are closely bordered by breathtakingly beautiful hill and semi-mountainous country: the Midlothian coalfield by the wriggly silhouette of the rugged Pentlands to the west with the Moorfoots of the Southern Uplands opposite; Ayrshire by the Carrick Hills and the stepped lava landscapes of the Renfrewshire Hills; Fife and Stirling by the more distant Ochils and Sidlaws; the Central Coalfield enveloped by the Kirkpatrick, Campsie and Gargunnock Hills and the Southern Uplands. Suitably fertilized, limed and ploughed by Norfolk-style (one man/two horse) ploughs, the produce of oats and neeps together with sheep, dairy herds and livestock continued to nurture the new centres of population that grew up during the eighteenth and nineteenth centuries. The agricultural revolution began a couple of decades before the mineral riches of the region became the key to its industrial strength in the 1750s.

It remains a paradoxical fact of history that 1707, the year of Union with England and Wales, was almost coincident with Abraham Darby I's use of coke in iron smelting and just five years before Thomas Newcomen's final success with the invention of his revolutionary steam engine. Before 1707 a bankrupt and agriculturally enfeebled Scotland had been shaken by a failed colonial adventure in Panama (the Darien scheme[2]) and by the severe climatic vicissitudes and near-starvation engendered by a succession of disastrous harvests during the worst decades of the Little Ice Age.[3] Had it waited out its misery for a few decades more, perhaps employment might have boomed without English collateral and stimulus – the national coffers refilling from home-produced iron giving a material basis to its nascent race of original and pioneering entrepreneurial scientists, engineers and industrialists – one of the many 'Ifs' of history!

Carboniferous Midland Scotland: Forest Swamps in a Volcanic Rift Basin

Long-lived basic volcanic activity from numerous eruptive centres[4] makes the Midland Valley stand unique amongst Carboniferous rift basins across Britain, Ireland and Maritime Canada. The basaltic volcanism was probably initiated by the rifting process itself with the possible later connivance of a regional mantle plume. It lasted in aggregate for almost the full sixty-million-year span of Carboniferous time. At any one time there would have been active eruptive centres at or just below sea level, releasing surface spreads of lava and ash. Coastlines with river deltas sourced from the north-east (beyond Fife) and the all-enveloping forest swamps in the rift had to adjust their margins continuously with respect to these piles of black newly-erupted basalt that would have stood island-like in a surround of green swamp forest. The lavas are seen most spectacularly today as the rugged upstanding landscapes founded on outcrops of the Clyde Plateau Lavas that half-circle the central coalfield. Coastal Fife, on the other hand, is peppered by the eroded stubs and surface scars of explosive volcanic vents, some of them maars whose low craters were ring-faulted, with central downthrown eruptive deposits still preserved.

Tectonic subsidence took place over several areas undergoing active crustal stretching – in the Lothians, Clackmannan/Fife, the Central Belt between the Clyde and Forth, and in Ayrshire. In all these places thick sequences of coal- and ironstone-bearing strata accumulated, at first during the Middle Carboniferous as part of the succession of Yoredale-like cycles in the Limestone Coal Group – familiar from northern England, but here with much thicker coals and important blackband ironstones. There are also indications in eastern areas of oblique fault slippage and fold growth due to the kind of right-handed transpressive tectonics outlined in Part 2. During Late Carboniferous times the eruption of extensive lava fields had largely ceased but explosive volcanism was still ongoing. This is seen around the periphery of the eastern forest swamps that grew mightily during periods of low sea level. Their own depositional cores were built up from the deltaic effluent brought in from the great rivers of the Caledonian and Precambrian hinterlands of modern Scandinavia, Greenland and Maritime Canada. Subsequently there were bursts of magmatic activity and renewed extensional tectonics around the Carboniferous/Permian boundary. A plethora of intrusive igneous dykes (first noted by George Sinclair) and great sills, notably the Forth Sill, were emplaced within the sedimentary successions that had accumulated over the previous fifty million years.

All of this mix of coal-bearing strata, lavas, ashes, vents, dykes and sills were then subject to the tectonic inversion familiar across the other British coal basins. This produced broad folds in whose eroded synclinal remains we recognize the individual coal-bearing basins seen today. This mix of igneous and tectonic goings-on made coal mining in midland Scotland considerably more difficult than elsewhere. Hence the reasons for the repeated mention by George Sinclair[5] of the perils to straightforward mining in coastal Midlothian brought on by the presence of hard and unyielding bodies of 'Gaes' and 'Whin'.

Hearths of Midlothian – Early Years

By 1672 when Sinclair published his opus discussed in Chapter 2, the Lothian coalfields and their continuation under the Forth to outcrop in the coastal parts of East Fife, Clackmannan and Stirling were the locus of the majority of production from midland Scotland, the main mines dotted around the coast from Aberlady in East Lothian to Elie in Fife via Bo'ness, Stenhousemuir, Alloa, Dunfermline and Kirkcaldy.[6] Most was mined for domestic use, the majority destined for the many firegrates of Edinburgh's old town – hence the epithet for Edinburgh as 'Auld Reekie' and its homely blessing for good health – 'Lang may yere lum [chimney] reek' – and for other urban centres like Falkirk, Stirling and St Andrews. It was also used for salt production in the numerous pans around the Forth coastline (as at Prestonpans). In the decades leading up to Union the Scottish government had taken firm hold of their indigenous coal and salt supplies, periodically prohibiting export and heavily taxing incoming English coals and salt (chiefly from Newcastle-on-Tyne). This despite the export success and income accruing

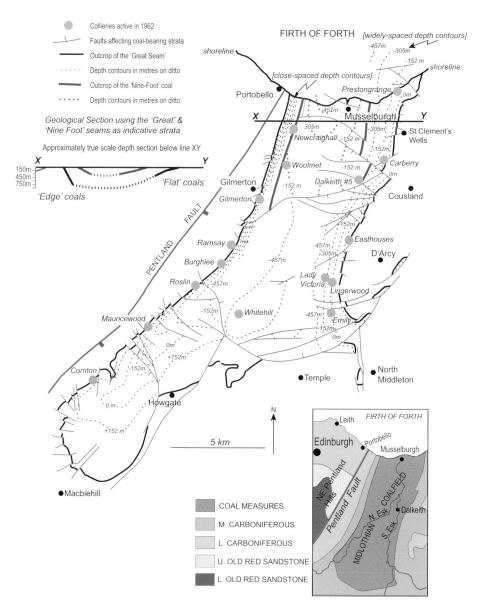

Figure 24.2 George Sinclair first determined the synclinal (downfolded) geological structure of both the Midlothian and East Lothian coalfields in the seventeenth century. This modern-era geological map and section of the former is based upon almost 300 subsequent years of geological and mining investigations by generations of practising geologists – Sinclair's numerous professional successors. Redrawn, modified and crafted by the author from Mitchell *et al.* (1962).

to the Scottish Crown from the larger Lothian and Fife mines when they stood in for the London sea coal trade during the Tyneside blockade enacted by Parliament during the Civil War – the *milieu* that George Sinclair lived in prior to his dismissal from the University of Glasgow in 1666.

Sinclair's account of the morphology of the Midlothian coalfield contains a litany of place names and aristocratic lairds and is the first description of a whole, structurally defined sedimentary basin and coalfield ever made. It follows examples of opposite dip across tracts of land where the coal runs too deep to be directly visible. As the dip changes through 180 degrees it defines what we today call a syncline or downfold (Figs 2.3, 24.2). In what follows, Sinclair's spelling and grammar are retained with a few notes by the present author in square brackets:

> The greatest field I know wherein this is conspicuous, is in Mid-lothian where is to be found, the cropping of a Coal of a considerable thickness, which is termed their great-seam, or Main-Coal [Great Seam], and the other Coals lying below it, which may be traced in the order following. At Preston-Grange these Coals are found dipping to the NW, and rising to the SE, which have been wrought up to Wallifoord [Wallyford]: from that along the foot of Fauside Hill [Falside Hill; 144 m], the dipp lying in the Lands of Inneresk [the Wedderburn estates around Inveresk], which marches therewith on the North.

In the absence of direct evidence, Sinclair's essential observations came from the nature of the body of rock surrounding the 'Great Coal' (later named the 'Great Seam') and of the coal itself, being exactly the same thickness and nature across the land, the rock lying above and below the coal having the same particulars. The narrative continues through Carberry Tower, Smeaton, Cowden and so on till he comes to Carrington where:

> the course of the Coal turns…thence taking its Dipp, quite contrary to what it had before, the other Dipping N and NW, or NE, according to the turn of the Streek, it Dipps there S, SE, etc…to Magdalen Pans [W of Fisherrow Harbour], where the turn of the cropp being within the Sea, is not seen, till it be found at Preston-Grange, where we began to remark its course. The parcel of ground, under which this great body of Coal lyes, is of considerable extent, it being eight miles in length, and five or six in breadth; in regard whereof many other Coals are found lying above the Great Coal, the cropps whereof doth not come near the Cropp of it, by a considerable distance.

The reader will notice that Sinclair's description of contrary dips involves use of the phrase 'the turn of the cropp…', an exact description of the curved trace to a fold's termination.

The Serf Status of Certain Scottish Miners till 1795

When Edmund Burke wrote his protestations quoted earlier in Chapter 2 on the lot of miners who were 'without the least Prospect of being delivered from it…' he probably meant that alternative employment to working in the mining fields was difficult to obtain or simply not available to the vast majority – as in upland Glamorgan or West Yorkshire. Yet in Scotland his words expressed the literal truth, for from the early 1600s to 1795 Scottish miners (and saltworkers) included a substantial number (probably a thousand or so at the early date and up to several thousand later) who were basically serfs – bound together with their families as the legal (indentured) property of landowners in whose mines they could be inveigled to labour all their working lives, and into their children's lives too, in many cases. We have to be careful here about these people. They were not slaves as such – they could not be sold on and could live where they chose around the mineworks, and were paid at a prescribed rate. The majority were recruited widely from the population – as the social historian of Scottish mining, Robert Duncan, puts it:

> the harsh rigour of the Scots poor law [the Act of 1606] included the punishment of able bodied unemployed men and vagabonds by sentencing them to a lifetime of enforced labour in coal-works…In the first instance the landowning coal-masters… sought to use state law to secure their existing workforce. By this means, they aimed to avoid a labour shortage brought about by workers leaving for other employment or being poached by other coal-masters who were prepared to offer better earnings and conditions.[7]

This was done by 'securement', the right to recapture any absconding mining families who left without permission and a written testimonial within the year and, if thought necessary by them, to apply an arbitrary level of corporal punishment. The life sentence was implicit rather than explicit, for annual contracts could be given repeatedly and a written reference for work elsewhere (allowed under the law) refused. That these and other such practices in later Acts (1641, 1661) could be sanctioned by Enlightenment Scotland up until 1795, when the Acts were finally and completely repealed (preliminary reform came in the 1770s), may seem unbelievable to us today, even at this length of time. Yet in his codification of Scottish law in 1681, Sir James Dalrymple, the Viscount Stair, a former Lord President of the Court of Session and Scotland's most senior judge, justified enserfment of miners and salt workers in the national (financial) interest:

> …these services being so necessary for the kingdom, where the fuel of coal is in most parts necessary at home and very profitable abroad…so that colliers and salters, while they live, must continue in these services.[8]

Such 'Binding Acts' meant that any right of free choice of labour by Scottish miners was nugatory – a far cry from the liberties and rights long-owned by the English working class through the (relatively) benign Elizabethan Poor Laws. There were also no hereditary rights from Royal Assent like those English mine courts in the Forest of Dean and Peak District, which held rights and special powers of governance and protection. Elsewhere in Europe, J.U. Nef informs us that miners enlisted in special guilds, doubtless a legacy from the days of Agricola or earlier when metal mining began to have special status in society.[9] These organizations could be suffocatingly protectionist when confronted by instruments of economic change, but they protected miners' rights and those of their families from abusements by mine owners – such protection took an inordinate time to come to Scotland.

Industrial Developments

By the middle of the eighteenth century the Scottish economy had started to revive under the stimulus brought about by Union. The country's more northern latitude and the direct nature of the Clyde approaches from the west also gave it some advantage for the North Atlantic trade over western English ports like Bristol and Liverpool. Most notable was the tremendous increase in revenues coming into the growing portage along the Clyde from the trade in sugar and tobacco[10] – likewise, the growing textile trade with westwards export of cotton and linen from the new mills of Lanarkshire and burgeoning Glasgow. In this atmosphere of financial opportunities an accommodating and sympathetic banking system made capital freely available to investors. Merchants and entrepreneurs turned to the the exploitation of native coal and blackband/claystone iron ores to develop an English-style industrial base. But there were few in Scotland itself who could bring the necessary technical and managerial experience to bear.

So it was that in the late-1750s William Cadell, a Scottish merchant in the iron and timber trade, met a Birmingham doctor, John Roebuck, Sheffield-born with eclectic industrial chemical interests (*see* also Chapter 4) and Samuel Garbett, a Birmingham businessman.[11] The Englishmen had set up an analytical laboratory in Birmingham and a sulphuric acid plant at Prestonpans in Lothian. The men joined forces and travelled as a group around midland Scotland, primarily along the shores of the Forth, seeking localities where the most judicious combination of factors might accommodate an ambitious venture to smelt iron from coke. They needed the usual suspects: ample local reserves of coking coal, ironstone and limestone; a reliable water supply to run mechanical bellows, rolling mills and slit mills; good transport links to national and international markets. They did fieldwork to determine the course and extent of ironstone strata and tested various iron ores by assay for their chemical make-up, those from Limestone Coal Group outcrops near Bo'ness proving best. Just as important, they hired in key skilled workers from the English Midlands and imported necessary raw materials needed to build, maintain and operate coke-fired blast furnaces and to cast and forge the output of iron.

The locality the trio eventually chose in 1759 was a site east of Falkirk on Carron Water near Bo'ness on the Forth shore. Here, after a shaky initial couple of decades, grew the successful Carron Ironworks, Scotland's answer to England's Coalbrookdale. Similarly, they manufactured domestic ironware (a notably fine kitchen range was modelled on a design by Benjamin Franklin) but in addition an increasing variety of armaments (the Coalbrookdale Quakers would not latterly touch such stuff) to answer the great demands placed first by the Seven Years War and later by the Napoleonic Wars. After some disasters with quality control (perhaps related to choice of high-phosphorus ironstone) and the reliability of their naval cannon, an improved production line culminated in the invention of that most brutal of short-range naval cannon, the carronade or 'smasher' (*see* Patrick O'Brien's novels for details), which they manufactured for over forty years.

By the late eighteenth century the pioneers of the Carron Ironworks were situated close to the newly opened Forth and Clyde Canal that revolutionized mass transport of materials and goods across their narrow isthmus. The canal seemed blessed by the talents that engineered it (John Smeaton) and surveyed variants of its final course (James Hutton, James Watt), but lack of capital led to a very long gestation, over twenty years. By the time the canal opened in 1790 Carron Ironworks had been joined by nine others, mostly at inland sites in Lanarkshire and Ayrshire (Fig. 24.1A). Over subsequent decades there was a huge expansion in output and quality of Scottish pig-iron as new techniques (the all-Scottish invention of hot blast in the 1830s;[12] the Gilchrist-Thomas process) and the increased use of coaly blackband ironstones (almost self-calcining in the ovens) led to economies in scale and fuel consumption. Increased mobility across central and western areas, in particular the rise of Ayrshire engineering in Kilmarnock (Fig. 24.3) and the Monkland canal in Lanarkshire, and a later myriad of railway branches linked mines, furnaces, forges and manufacturing plant. Major concerns now dominated the industry, with the Bairds of Coatbridge[12] often at the forefront of aggressive acquisitions in both coal and ironstone mining. The recruitment problems of old were solved by huge influxes of desperate dispossessed Lowland, Highland and Irish families from the Highland and Lowland Clearances and the 1840s Irish Famine.[1] Once-rural Coatbridge grew to resemble other coal, iron and steel industrial centres of Britain – witness this extract from a visitor in 1869:

> Though Coatbridge is a most interesting seat of industry, it is anything but beautiful. Dense clouds of smoke roll over it incessantly, and impart to all the buildings a peculiarly dingy aspect. A coat of black dust overlies everything, and in a few hours the visitor finds his complexion considerably deteriorated by the flakes of soot which fill the air, and settle on his face.…From the steeple of the parish church, which stands on a considerable eminence, the flames of no fewer than fifty blast furnaces may be seen.…The flames then have a positively fascinating effect. Their form is ever changing, and the variety of their movements is endless. Now they shoot far upward, and breaking short off, expire among the smoke; again spreading outward, they curl over the lips of the furnace, and dart through the doorways, as if determined to annihilate the bounds within which they are confined.[12]

The growth of Clydeside as the premier British location for construction of mercantile ironclad ships and the technical innovations that accompanied their evolution (screw

Figure 24.3 A General view of the horizontal winding engine (originally steam-powered) and drum of Lady Victoria Colliery, Lothian, preserved and reconditioned to run electrically in the National Mining Museum of Scotland at Newtongrange, Lothian (55.514452, 3.040008). The 2400 horse-power engine was manufactured by Grant, Ritchie and Company of Kilmarnock in the 1890s. Former Lothian NCB engineer, now museum guide, Dave McDougal, is instructing Joy Lawlor and Kate Johnston in its operation. **B** Close-up to show the drive linkage between the cylinder and winding drum with the piston at the top of its stroke. Source: Author.

Figure 24.4 A Union banner hanging in the National Mining Museum of Scotland at Newtongrange, Lothian, promoting the need for pit-head baths. **B** The preserved exterior of the baths at Prestongrange Industrial Heritage Museum, Lothian, built in the 1950s. Source: Author.

propellers; advanced engine and boiler design) led to further demand for quality iron and steel products from Midland ironmasters and steel plants throughout the nineteenth and well into the twentieth century. After the Second World War the Scottish mining industry went into long decline as reserves of machine-mineable coal shrank. At least by the 1950s many miners were able to at least shower before going home from their shifts – the result of a union-inspired campaign that sought to have pit-head baths at every colliery (Fig. 24.4).

Figure 24.5 The community-funded memorial sculpture to the Prestonpans miners of Midlothian, who worked in the local pits of Prestongrange and Prestonlinks for over one hundred years. Two massive yellowish sandstone blocks split along quarry drillholes are separated by 'pit props' framing two bent miners shovelling mined coal in an eerie, dream-like recreation of the claustrophobic dark and dusty conditions of a deep mine. The site is within a modern housing development, just off a busy local road (*c*.55.571811, 2.580205). In the local *argot* it's called 'The Shrine' since it is perched on the site of a former colliery ventilation shaft around whose perimeter the locals once collected to place (illegal) bets on the horses. The sculpture was designed, created and installed by Gardner Molloy. Photo: Author.

Coda: Collapse of West Lothian Oil Shale and Coal Mining

As noted in Chapter 10, James Young's original cannel coal source for petroleum distillation in West Lothian was soon worked out. Recent research[13] shows that he was pointed in the oil shale direction by consulting by the young Archibald Geikie (later Director of the Geological Survey and often a controversial character) who as a rookie Geological Survey geologist had recently recorded exposures of these in the West Lothian area where he was mapping. The oil shale industry existed for just over a hundred years, during which time Young's retorting technique was endlessly refined and made more efficient, not to mention the clever marketing of the industrial products.[14] During its halcyon years over 1000 miners hewed out the shales at scores of mine and retorting sites in West Lothian. They provided paraffin, high-class smokeless candles and other organic distillates far and wide – the author's paternal grandparents had paraffin lighting downstairs and candles upstairs in their Norfolk house well into the 1950s. Indeed, it was a miracle that the industry survived for so long in the face of the arrival of cheap imports of crude oil from the Middle East in the 1950s to 1960s. Though minuscule in relation to later twentieth-century offshore oil production from Scottish waters, its forlorn 'bings' (spoil tips of retorted oil-shale residue) dotted about West Lothian leave an important industrial heritage, both in terms of landscape and industrial, social and economic history. As D. Hallett and other ex-Britoil colleagues put it in a review written at the peak of offshore oil production in 1985:

> Scotland for all practical purposes can be regarded as the birthplace of the oil industry….Oil shale exploration was never subject to Government control by licensing, and was specifically exempted from the provisions of successive Petroleum Production Acts. Furthermore it was exempted from both royalties and the levy on imported oil which the Government imposed in an effort to encourage domestic production. It was for these reasons, plus progressive improvements in technology, and the extensive utilization of by-products which allowed production to continue for so long.[15]

The industry collapsed once its subsidy was withdrawn in 1962.

A similar fate befell midland Scotland's deep-mine collieries forty years later when the last, the giant Longannet Complex in Fife, was shut down as inoperable after serious flooding. Yet the Scottish mining heritage lives on in industrial museums, preserved workings (for instance at Newtongrange) and the proud memories of local communities who still remember the hard-working lives of their mining ancestors courtesy of artwork[16] and community-funded sculpture (Fig. 24.5).

Notes and Sources

Preface

1 Allen, R.C. (2009)

2 This book is not encyclopaedic so I have to apologize to readers whose native lands include the smaller Scottish (Sanquhar, Thornhill, Canonbie) and English (Ingleton, Kent) coalfields; the latter, anyway, a late (1860s) and completely concealed incomer (though with four deep, brave collieries) and with no links to early industrialization. A more substantial omission has been the metal mining province of Cornwall and Devon, where modified Newcomen and Trevithick engines reigned supreme, many made locally. As documented in the text, the mines supplied ores to the great coal-based smelters of South Wales and Bristol. Their geology and socio-economic history includes highly deformed Devonian bedrock and early Permian mineralization whose non-Carboniferous complexities must await some future biographer.

3 Mazisi Kunene (1970)

Part One: Economy in Motion

General

A varied sourcebook of quotes from contemporary observers concerning the impact of the coming of the machine/industrial/factory age is *Pandæmonium* by Jennings, H. (2012). Such material was turned into a brilliantly executed tableau by Danny Boyle for the 2012 Olympic Opening Ceremony. Several of the featured Midland English and midland Scottish eighteenth-century persona are assuredly placed within their *milieu* by Uglow, J. (2002), a rich and rewarding source. The first comprehensive book that dealt with the then-new (since the early 1960s) science of industrial archaeology was Raistrick, A. (1972). Cossons, N. (1993) is a worthy successor, a magnificent and richly illustrated achievement. Falconer, K. (1980) is still a useful illustrated gazetteer for English historical industrial sites, though out of date with respect to subsequent preservation initiatives.

Chapter 1 Travellers' Tales 1: Celia Fiennes and Daniel Defoe in Pre-Industrial Britain

1 Hatcher, J. (1993); Nef, J.U. (1932, 1972)

2 Fagan, B. (2002)

3 Fiennes, C. (1888; 2016)

4 Defoe, D. (1724–6; 1971); I have used extracts from the edited and abridged Penguin edition of 1971

5 Priestley, J.B. (1934; 1977)

Chapter 2 Beginning of Coallery: George Sinclair, 'Scoto-Lothiani'

1 Sinclair, G. (1672) A copy can be downloaded from the *Wikipedia* entry for 'George Sinclair (mathematician)', courtesy of Boston Public Library (USA). The *Short History of Coal* is Item 24, pp. 258–302 of his *Miscellaneous Observations*

2 Skempton, A.W. (1996)

3 A new account of Sinclair's life, his contemporary and posthumous scientific reputation and his various works on physics (but not in geology; that is in this book) is given by Craik, A.D.D. (2018). It was Sir Edward Bailey (1952) who first drew attention to Sinclair's geological work, though his account is thin and matter-of-fact.

4 Agricola, G. (1556)

5 Steno, N. (1669); *see* popular biography by Cutler, A. (2003)

6 Sinclair was mostly mapping and recording the Great Seam of the Limestone Coal Group in the Midlothian coalfield syncline. This and adjacent seams occur interbedded with sandstone strata, not limestone – *see* Mitchell, G.H. *et al.* (1962)

7 Dudley, D. (1665)
8 McManus, J. (2017) A geologist reconstructs the microhistory of Fife mining in its broader context, with vivid original artwork by Derek Slater in Ch. 8
9 Burke, E. (1756)

Chapter 3 New Reductions: Iron, Darby and 'Charking Coles'

1 Defoe, D. op. cit.
2 Much of the material consulted and quoted in this chapter comes from Raistrick, A. (1989) to which the interested reader is referred. Additional background material concerning Coalbrookdale in general may be found in Hayman, R. & Horton, W. (2009) and for the geology in particular in Hamblin, R.J.O. & Coppack, B.C. (1995)
3 Raistrick op. cit. p. 18
4 Raistrick op. cit. p. 20
5 Raistrick op. cit. p. 38
6 Hamblin, op. cit.
7 The smelting abilities of coke, like those of charcoal, depend upon the reducing action of carbon monoxide acting on iron oxides in the (calcined) ore, the gas newly formed from carbon dioxide in the intense heat towards the base of the blast furnace charge.
8 Hopkins, W. 1954
9 Evans, C. *et al.* (2002)
10 Allen, R.C. (2009)

Chapter 4 Steam Engine Works: Newcomen to Watt and Boulton

1 Papin, D. *Wikipedia* entry for Denis Papin
2 Clear accounts of the technical and economic background to Newcomen's and later steam engine variants are given by Allen, op. cit. Ch. 7 and Cossons, op. cit. Ch. 4
3 Letter from Hutton, J. to Watt, J., late summer 1774. *See* Jones, J. *et al.* (1994) for the complete letter and for three other letters from Hutton to other individuals at this time of intensive fieldwork on his part.
4 Quoted by Uglow, J. (2002) Ch. 9 p. 101 who notes 'Recounted by JW in 1817 to the Glasgow engineer Robert Hart (*see* 1859; the wording varies in different accounts)'.
5 Letter from Hutton, J. to Clerk-Maxwell, G., August 1774. *See* Jones, J. *et al.* op. cit. but their editorial reference to Watt's engine is *not* of his subsequent horizontal-motion machine.
6 Uglow, op. cit. Ch. 21 p. 255
7 Uglow, op. cit. Ch. 21 p. 257

Chapter 5 Still Waters Run Shallow: Canal Mania

1 There is much to be gained from studying the historical work of a modern civil engineer, the late Sir A.W. Skempton (op. cit.) whose numerous papers on the subject were collected in his book on early coastal, river and canal engineering projects and the engineers responsible.
2 Defoe, D. op. cit.
3 See *Wikipedia* entries for Newry Canal; other information kindly provided by Goggins, B. (pers. comm.)
4 *See* Boucher, C. (1968) and Malet, H. (1990). In what follows I have tended to side with the balanced version of events outlined by historian Hugh Malet concerning the primacy of Bridgewater and his agent John Gilbert in championing the Bridgewater Canal project, rather than with the efforts of James Brindley as championed by others, including his biographer, Cyril Boucher.
5 The Duke's correspondence during his Grand Tour is held in the University of Salford's *Duke of Bridgewater Archive* as part of Box 1 and comprises eight letters. Reference DBA/1/7-18
6 Uglow, J. (op. cit.) has detailed sections in her biography dealing with the involvement of several of the Lunar men with canal 'mania'.
7 *See* Barton, R.M. (1966) for a comprehensive account of the history of the Cornish china-clay industry and of the logistics involved in its export.
8 I have lost the source for this letter
9 Uglow, J. op. cit. Chs 13, 14, 25
10 According to the *Wikipedia* entry for James Brindley this poem appeared in the *Chester Courant* of 1 December 1772.

Chapter 6 Travellers' Tales 2: Louis Simond Under Tyneside, 1811

1 Nef, J.U. op. cit.
2 Abbé le Blanc (1745)

3 Simond, L. (1968)
4 Mills, D.A.C. & Holliday, D.W. (1998)
5 *See* the introductory chapters to Moore, P. (2018)

Part Two: Redress of Time: Carboniferous Worlds Reconstructed

General

Beerling, D. (2007) is a fine, illuminating, semi-popular account of plant hegemony over the rest of us. I am very fond of the late Nigel Trewin's (2013) book *Scottish Fossils* which contains some lovely images of classic Devonian and Carboniferous fossil plants and creatures, many of them unique in world terms (e.g. the Rhynie and Granton *Lagerstätten*). Toghill, P. (2000) serves as a clearly-written and well-illustrated introduction to British geology for the layperson. Davies, S.J. *et al.* (2012) is a more advanced account of British Carboniferous geology. We are lucky indeed to have Torsvik, T.H. and Cocks, L.R.M. (2017) superb collection of past world reconstructions to guide global thoughts about Devonian and Carboniferous climate and tectonics – without their labours the efforts herein would be even more broad-brush and imprecise.

Chapter 7 Devonian Prequel: Scintillas, then Splashes of Green

1 *See* Kenrick, P. & Crane, P.R. (1997); Edwards, D. & Wellman, C.H. (2001); Wellman, C. *et al.* (2003); Steemans, P. *et al.* (2009); Wellman, C.H. (2010)
2 Graham, J.R. (1983)
3 Retallack, G.J. (1997)
4 Burgess, I.C. (1960) is a pioneering contribution
5 Hernick, L.V. *et al.* (2008)
6 Edwards, D. & Feenan, J. (1980)
7 Edwards, D. *et al.* (1992)
8 Glasspool, I.J. *et al.* (2004)
9 Mackie, W. (1913); Kidston, R. & Lang, W.H. (1916–21); Rice, C.M. & Trewin, N.H. (1988); Trewin, N.H. (2001)
10 Neatly summarized by Trewin, N.H. & Thirlwall, M.F. (2002)
11 Matsunaga, K.K.S. & Tomescu, A.M.F. (2016)
12 Goldring, W. (1924, 1927)
13 Stein, W.E. *et al.* (2007, 2012)
14 Stein, W.E. *et al.* (2019)
15 Bridge, J.S. (2000)
16 Volkova, I.B. (1994)
17 Wang, D. & Liu, L. (2015)
18 Berry, C.M. & Marshall, J.E.A. (2015)
19 Wang, D. *et al.* (2019)
20 Algeo, T.J. & Scheckler, S.E. (1998), with updates on root architecture by Meyer-Berthaud, B. *et al.* (2013) and, in a major way, by Stein, W.E. *et al.* (2019)
21 Lakin, J.A. *et al.* (2016)

Chapter 8 Carboniferous Tectonic Geography and Climate

1 Leeder, M.R. (1982); Leeder, M.R. & McMahon, A.H. (1988); Gawthorpe, R.L. *et al.* (1989); Fraser, A.J. & Gawthorpe, R.L. (2003)
2 For a general introduction to British Carboniferous geology *see* Toghill, P. (2000) Chs 7, 8. A more advanced summary is by Davies, S.J. *et al.* (2012) Ch. 14
3 *See* faunal discoveries by Smithson, T.R. *et al.* (2012)
4 Gilligan, A. (1920)
5 Glover, B.W. *et al.* (1996); Hallsworth, C.R. & Chisholm, J.I. (2017)
6 Hallsworth, C.R. & Chisholm, J.I. (2000, 2008)
7 Drewery, S.E. *et al.* (1987)
8 Cliff, R.A. *et al.* (1991)
9 Lancaster, P.J. *et al.* (2017)
10 Blanford, W.T. *et al.* (1859)
11 Wegener, A. (1924) p. 98
12 Du Toit, A.L. (1937) p. 71
13 Wanless, H.R. & Shepard, F.P. (1936)
14 Hays, J.D. *et al.* (1976)

Chapter 9 Carboniferous Equatorial Swamp Forests

1 I have used Rydin, H. & Jeglum, J.K. (2013) as my chief source for much of this and the subsequent section.
2 Dargie, G.R. *et al.* (2017)
3 *See* John Gaudet's (2014) brilliant interdisciplinary book on papyrus swamplands.
4 Falcon-Lang, H.J. (2004a)
5 Lyell, C. (1838)
6 Phillips, T.L. & DiMichele, W.A. (1992)
7 See *inter alia* Berner, R.A. & Canfield, D. E. (1989); Dudley, R. (1998); Beerling, D.J. *et al.* (2002) and the fine popular accounts by Lane, N. (2002) and Beerling, D.J. (2007)
8 *See* Scott, A.C. (1989); Scott, A.C. & Jones, T.P. (1994)
9 *See* Falcon-Lang, H.J. & Scott, A.C. (2000); Falcon-Lang, H.J. & Bashforth, A.R. (2004); Dolby, G. *et al.* (2011)
10 Falcon-Lang, H.J. *et al.* (2011; 2015a)
11 Falcon-Lang, H.J. (2015)
12 *See* the pioneering work in Britain of A.C. Scott (e.g. 1977)
13 Robinson, J.M. (1990)
14 *See* https://www.weforum.org/agenda/2017/08/how-trees-in-the-amazon-make-their-own-rain/ I inserted this section after conversations in an Irish forest with Sam Lawlor who pointed out to me the startling results of the recent work on self-seeding aerosols produced by Amazonian trees and the pre-monsoonal generation of rain in the forest's own convectional boundary layer.
15 Wright, J.S. *et al.* (2017)
16 Pöhiker, C. *et al.* (2012)
17 Wang, X. *et al.* (2017)
18 Bisat, W.S. (1924, 1928)
19 Maynard, J.R. & Leeder, M.R. (1992); *see* also the exhaustive review of Rygel, M.C. *et al.* (2008)
20 Wells, M.R. *et al.* (2005)
21 Besly, B.M. & Fielding, C.R. (1989)
22 DiMichele, W.A. *et al.* (1996); Falcon-Lang, H.J. (2004b)
23 Falcon-Lang, H.J. *et al.* (2018)

Chapter 10 Carbon Accumulation and Mineral Additions

1 The classic work on sediment compaction is by Baldwin, B. (1971)
2 A pioneering text on coalification and hydrocarbon generation is by Tissot, B.P. & Welte, D.H. (1978)
3 Stopes, M. (1919; 1935)
4 Sagan, D. & Margulis, L. (1995)
5 Childers, S.E. *et al.* (2002)
6 *See* the incisive review by Fisher, Q.J. *et al.* (1998)
7 Boardman, E.L. (1989) is the outstanding study of blackband ironstones.
8 *See* Thelemann, M. *et al.* (2017) for a recent well-documented geo-archaeological study.
9 Mitchell, G.H. *et al.* (1947) p. 46 *et seq.*
10 Quoted by Duncan, R. (2005) Ch. 3, p. 78
11 Stopes, M.C. & Watson D.M.S. (1909)
12 Made crystal clear by Berner, R.A. & Raiswell, R. (1984); Raiswell, R. *et al.* (1988)
13 Jones, N.S. (2007)
14 *See* Parnell, J. (1988)
15 Harvey, A. *et al.* (2018)

Chapter 11 Tectonic Inversion: Preservation of Coal Basins

1 Corfield, S. *et al.* (1996)
2 For western Cenozoic uplift mechanisms *see* the enlightening papers by Bott, M.H.P. & Bott, J.D.J. (2004); Arrowsmith, S.J. *et al.* (2005) and Davis M. W. *et al.* (2012)

Part Three: Legacies: Carbon Cycling, Chimneys and Creativity

General

For economic history *see* R.C. Allen (2009) – concise, authoritative and sufficiently technical. The chemistry of the carbon cycle is similarly analysed by the late R.A. Berner (2004). I found A. Malm's (2016) polemical account of 'Fossil Capital' unputdownable and immensely important (it seems to me) for a true understanding of British industrial history and the possibilities for non-carbon economies in the future.

Concerning the natural sublime, it is still worth reading Edmund Burke's fresh and original 1757 opus, readily available in modern paperback.

Chapter 12 Atmospheres, Carbon Cycling and Glaciations

1 Babbage, C. (1835) p. 18
2 Ébelmen, J.-J. (1845)
3 Accounts of Ébelmen's work in the light of modern ideas on the carbon cycle may be found in Berner, R.A. & Maasch, K.A. (1996) and Berner, R.A. (2012)
4 Tyndall, J. (1859); for a clear and enjoyable account of Tyndall's work in this field, *see* Hulme, M. (2009)
5 Arrhenius, S. (1896)
6 Hutton, J. (1795) Volume 2, p. 180
7 Gregor, C.B. *et al.* (1988)
8 Berner, R.A. (2004). Berner dedicates this book, which summarizes an important part of his life's work, to the memory of J-J Ébelmen.
9 Lovelock, J. (1979)
10 Raymo, M.E. & Ruddiman, W.F. (1992)
11 *See* Algeo, T.J. & Scheckler, S.E. (op. cit.); Morris, J.L. *et al.* (2015); Berry, C.M. & Marshall, J.E.A. (2015)
12 For biogeochemical stable isotope trends for an Early Carboniferous onset of severe Gondwanan glaciation *see* Barham, M. *et al.* (2012). For the North African story *see* Le Heron, D.P. (2018), especially his Fig. 5, but note that the dating of these features and deposits does not seem particularly exact.
13 *See* the very interesting results of Chen, J. *et al.* (2018)
14 Robinson, J.M. (1991)
15 Obtaining pCO$_2$ from stable isotope composition of ancient carbonate soils requires *a priori* knowledge of soil respiration rates at the time of soil formation which, by definition, cannot be obtained.
16 Franks, P. *et al.* (2014). Leaf stomata CO$_2$ exchange models using stable isotopes do not support the existence of large swings (>1000 ppm) in pCO$_2$ over the interval of post-Devonian time.
17 The calculations that follow update those in Leeder, M.R. (2007), with the computation of the minimum area for Euro-American coal basins obtained by integrating the enveloping area containing coalfields shown in Falcon-Lang *et al.* (2006, their Figure 8).
18 Hartmann, J. *et al.* (2013) is an interesting proposed geo-engineering initiative to use the natural weathering of crushed mineral silicate minerals like olivine in order to reduce atmospheric CO$_2$, though the practicality of such schemes seems suspect.
19 Gerlach, T. (2011) is an essential review of this topic, with many telling statements and important primary references.
20 Watson A. *et al.* (1978)
21 Glasspool, I.J. *et al.* (2015). An attempt to correlate global Phanerozoic fossil charcoal abundance (as measured in thin sections of coal) with the partial pressure of oxygen, but without convincing deviation statistics or error magnitudes.
22 Dudley, R. (1998)

Chapter 13 Britain: First 'Chimney of the World'

1 As referenced by Brimblecombe, P. (1987; 2011) pp. 7, 19
2 *op cit*. pp. 47, 61, from Evelyn, J. (1659)
3 From *Fumifugium* by Evelyn, J. (1661)
4 Simond, L. op. cit.
5 Keats, J. (1818) Letter 67 of Thursday 4th June to the Misses M. and S. Jeffrey. In: Forman, M.B. (1935)
6 Allen, R.C. (2009)
7 Hobsbawm, E.J. (1969)
8 Malm, A. (2016)
9 Mokyr, J. (2017)
10 *See* Uglow, J.S. *passim*, op. cit.
11 Foreman, A. (2001)
12 From an undated letter of Josiah Wedgwood to Thomas Bentley, printed in Eliza Meteyard's *The Life of Wedgwood*, 1865
13 Hawkes, J. (1950) pp. 206–207
14 *See* Yallop, J. (2015) for a delightful personal tour of many of these towns and villages.
15 Hobsbawm, E.J. & Rudé, G. (1972)
16 Devine, T.M. (2018)

17 I am straying a little from the main themes of this book here, but nevertheless add this short section on political and union developments, as they do seem to be consequences of the Industrial Revolution, if not of its geology. Thompson, E.P (1980) is a key reference, his p. 206 the source of my quotation. Malm, A. (2016) also has good stuff on the textile industry. Kirby, M.W. (1977), his Ch. 1, is succinct on the late-Victorian/Edwardian coal industry. Duncan, R. (2005) is a valuable source for Scottish experiences.

18 From Hodgson, Rev. J. (1813); *see* also online: Hodgson, John (1999) [1812], *Felling Colliery 1812: An Account of the Accident* (pdf), Picks Publishing.

19 I cannot recommend too highly the brilliant account of Davy's invention and his fieldwork underground in Durham, to be found in Richard Holmes' fine biographical account of Georgian artistic and scientific persons, *The Age of Wonder* (2008)

20 From the *Journal of Lord Shaftesbury* quoted by Hodder, E. (1888)

21 *See* Cleeland, J. & Burt, S. (1995)

22 *See* Seaton, A. (2018) for a highly informed and readable non-specialist account by an expert practitioner and researcher of lung disease and particulate matter, together with his further and wider reasons for the celebration of the death (at least in Britain) of King Coal.

23 I have written this section partly from memory, living at the time in West Yorkshire with friends supporting the miners (regardless of political persuasion or doubts about NUM strategy) and partly from diverse online sources. Seaton, A. (op. cit.) outlines a particularly clear history of those difficult times.

24 This was the procedure followed in Germany, shutting down its whole deep-mined bituminous coal industry by 2018, and her future plans to shut down the huge brown coal industry by 2038. *See* article in *Sydney Morning Herald*, 14 July 2019 'How Germany closed its coal industry without sacking a single miner'.

25 From Tegner, H. (1969)

Chapter 14 Chimneys of the Modern World

1 Malm, A. (2016) op. cit.

2 Tharoor, S. (2017) is a vigorous and coruscating take on the history of exploitation by the British of the once-independent regional economies of the former Indian states that were to form the 'Jewel in the Crown' of its vast Empire.

3 Oldham, T. (1859)

4 *See* the comprehensive *Wikipedia* entry: https://en.wikipedia.org/wiki/Song_dynasty. Needham, J. (1986) is a wonderful multivolume primary reference source on all matters Chinese in terms of technology, inventions and pre-industrial applications.

5 Polo, M. (*c.*1270) as translated by Yule, H. (1903)

6 *See* https://en.wikipedia.org/wiki/List_of_countries_by_coal_production; based on data from British Petroleum.

7 *See* the views of some of the chief players skilfully and informally laid out by journalist Nicola Davison in: The Anthropocene epoch_ have we entered a new phase of planetary history_ _ Environment _ The Guardian.html. You can then chase up individual contributions online.

Chapter 15 Industrial Sublime and Other Creative Legacies

1 Quoted by Raistrick, A. (1989) p. 71. Hannah Darby was a dutiful correspondent to her aunt, and three years later she wrote a long letter describing severe food riots at Coalbrookdale (Raistrick op. cit.78–79). The following year (1757) she married Richard Reynolds, who played an important role in Coalbrookdale industrial affairs over the next fifteen years.

2 Burke, E. (1757)

3 Oxford English Dictionary (1972)

4 Smith, A. (2013)

5 Arthur Young (1776)

6 To view image *see* Wikimedia Commons File: An Iron Forge1772.jpg

7 Hayman & Horton (op. cit.) point out that de Loutherbourg's title is misleading since the scene depicts the Bedlam Furnaces in nearby Madeley Wood with the highlighted house, Bedlam Hall, and the hearth as part of the smithy. This author would concur with their statement that 'the substance of the picture is actually rather shallow', especially the showy litter of valuable castings left out to rust in the elements!

8 Cited with no further details in Jennings, H. (2012), item 113

9 Gilchrist, A. (1863)

10 Nicholson, N. (1981)

11 For these suggestions and a full list of critical sources see *Mont Blanc (poem)* file in *Wikipedia*. I drew on many of these with a geologist's eye in my own reading.

12 Ingpen, R. (1915)

13 To view image *see* William Holman Hunt – Our English Coasts ('Strayed Sheep') 1852 – Google Art Project.jpg

14 For another reading of this picture and for the reminiscences of Elaine Webster *see*: https://eyrecrowe. com/pictures/1870s/the-dinner-hour-wigan/

15 Lawrence, D.H. (1965) the quotes are from this Penguin edition, successively from pp. 7, 8, 26–27, 389

16 In: Stallworthy, J. (2013)

Part Four: Landscapes of Industrial Revolution

General

In addition to the many cited sources in each chapter, my simplifications of industrial history and geology have made extensive use of three key 'overview' volumes. Cossons, N. (1993) volume on industrial archaeology has already been highlighted with respect to Part 1, but I also recommend it here for both locations of surviving relics and explanations of industrial processes. J.U. Nef (1932, reprinted 1972) is the classic account of the rise of the British coal industry into its prime, with much primary and secondary material. Hatcher, J. (1993) is a modern and meticulous version of the background and history of the pre-1700 coal industry at the very beginning of the Industrial Revolution.

Chapter 16 South Wales

1 *See* Tacitus (AD 98) The Agricola Book 2/17. *...subiit sustinuitque molem Iulius Frontinus, vir magnus, quantum licebat, validamque et pugnacem Silurum gentem armis subegit, super virtutem hostium locorum quoque difficultates eluctatus.* Translation (by Harold Mattingly): 'But Julius Frontinus was equal to shouldering the heavy burden, and rose as high as a man then could rise. He subdued by force of arms the strong and warlike nation of the Silures, after a hard struggle, not only against the valour of his enemy, but against the difficulties of the terrain.'

2 Wales is lucky indeed to possess what is perhaps the best of all British Geological Survey Regional Guides, by the late Malcolm Howells (2007) from which I have taken much inspiration and pertinent information.

3 Simond, L. op. cit. p. 66

4 *See* the very full and detailed account of the South Wales iron industry by Atkinson, M. and Baber, C. (1987)

5 The Glamorgan-Gwent Archaeological Trust: http://www.ggat.org.uk/cadw/historic_landscape/blaenavon/english/Blaenavon_Features.htm

6 Borrow, G. (1862, 2009) Ch. CIII

7 ibid. Ch. CIV

8 The reader is referred to a recent coruscating BBC documentary account that leaves no doubt as to the culpability for the disaster by the negligence at all levels within the National Coal Board and its insouciant boss Lord 'Alf' Robens. *See*: Aberfan_The mistake that cost a village its children – BBC News.html

9 Simond, L. op. cit. p. 67

10 *See* pp. 316–318 of Darwin, C. (1840; 2009). He also gives accounts of local mountain scenery and employment details of the Chilean copper miners.

11 Cadw (2016): a rich archive of historico-industrial history that gives an idea of the inheritance provided by the Lower Swansea valley landscape.

12 Davies, I. quoted in http://www.ggat.org.uk/index.html

Chapter 17 England's West Country: Somerset and Gloucestershire

1 Williams, G.D. & Chapman, T.J. (1986) give a technical summary

2 Green, G.W. (1992), especially Figure 35

3 The following site has much information on the history of coal mining in the area: http://www.cems.uwe.ac.uk/~rstephen/livingeaston/index.html

4 The two papers by Joan M. Day (1975, 1988) have proved invaluable in writing the main part of this section.

5 McFarlane, D.A. *et al.* (2014)

6 Evans, C. *et al.* (2002)

7 First brought to serious light in modern times by Fuller, J.G.C.M. (2007), but originally by Sir Edward Bailey (1952, Fig. 1) in his history of the British Geological Survey.

8 Strachey, J. (1717)

9 Phillips, J. (1844); modern edition with editorial and biographical material by Torrens, H. (2000; 2003) who seems to have been unaware of Strachey's contributions revealed by J.G.C.M. Fuller (op. cit.)

Chapter 18 North Wales

1 *See* Hatcher, J. (1993) pp.129–135; Nef, J.U. (1932) pp. 55–56

2 An excellent source is: Engineering Timelines-Bersham Ironworks.html and the references quoted therein

3 *See* unattributed quote in: https://www.wrexham.gov.uk/english/heritage/bersham_ironworks/early_years.htm

4 Cotton Twist Company.html

5 Raistrick, A. (1938, 1977) is the historian of the Company, a Quaker himself and pioneering Coal Measures geologist and palaeontologist

6 Burt, R. (1977)

7 *See* John Taylor's report on the Mold Mines, 1827, held at Clwyd Record Office and Figure 10 of Burt, R. (op. cit.)

8 Ebbs, C. (2008)

9 Appleton, P. (2013)

Chapter 19 English South Midlands

1 *See* Toghill, P. (2006) Ch. 8 for an introductory account to Shropshire's Carboniferous geological story and the fine accounts of historic coal and iron ore mining and industrial archaeology in Hayman, R. and Horton, W. (2009)

2 Hatcher, J. (1993)

3 Raistrick, A. (1989)

4 I believe the artist Warington Smythe to be one and the same as the economic geologist from the British Geological Survey who wrote the memoir *The Iron Ores of Britain* (1856; *see* Coda of Chapter 22) and who probably created the field sketch of the Cumberland iron ore mine featured in that section. His artistic talents go unmentioned in various biographical entries.

5 Young A. (1776)

6 From a letter of Thomas Carlyle to Alexander Carlyle, printed in Norton, C.E. (1886)

7 King, P.W. (1999) is a very full account of the litigious nature of this early industry

8 Belford, P. (2010) combines field industrial archaeology with a deep sense of respect and tradition

Chapter 20 East of the South Pennines (Yorks/Notts/Derbys)

1 The Pennine region is lucky to possess an outstanding British Geological Survey Regional Guide by the late Neil Aitkenhead *et al.* (2002)

2 Bristow, C.S. (1987, 1988)

3 Kendall, P.F. & Wroot, H.E. (1924) p. 194

4 ibid., pp. 194–197

5 the words of C.E.N. Bromehead in Mitchell, G.H. *et al.* (1947) p. 46

6 Kendall, P.F. & Wroot, H.E. op. cit. p. 199

7 ibid. pp. 199–200

8 Cobbett, W. (1832)

9 Hey, D. G. (2005) p. 91

10 Hey, D. G. (1969)

11 *See* Smythe, W. W. (1856), Strahan, A, *et al.* (1920). The latter has numerous analyses of sideritic clay ironstones from Late Carboniferous strata across the UK showing phosphorus (calculated as phosphoric acid) usually in the range 0.3–1.5 weight %.

12 Hey, D. G. (2005) pp. 91–92

13 An informally written but highly informative account is by *Steel Revolution. 1. The Kelly–Bessemer Process*.html *See* also others in that series on steel.

14 Anstis, R. (1997) and the *Wikipedia* entry for 'Robert Forester Mushet'

15 Percy Carlyle Gilchrist, 1851–1935. *Obituary Notices of Fellows of the Royal Society*, Vol. 2, No. 5 (Dec., 1936), pp. 19–24

16 Defoe, D. (op.cit.)

Chapter 21 West of the South Pennines (Lancs/North Staffs)

1 Malm, A. (2016)

2 *See* Aitkenhead *et al.* (op. cit.) for much geological background

3 Malet, H. (1990)

4 Egerton, Francis, eighth Earl of Bridgewater (1812); available as Google ebook but *see* also *Wikipedia* entry: 'Worsley Navigable Levels' for an image of the extent of the levels in plan view used by the author in Figure 21.2

5 Letter by J. Wedgwood to T. Bentley March 2 1767; cited by Uglow, J. (2002) p. 525

6 Boucher, C.T.G. (1968) pp. 84–86

7 *See*: www.historyofparliamentonline.org/volume/1820–1832/member/kinnersley-william-1780–1823

8 Anderson, D. (1975)

9 Wright, W.B. *et al.* (1927) p. 3

10 ibid. p. 111

11 *See* the landmark papers by Collinson, J.D. (1969, 1970)

Chapter 22 West and South Cumberland

1 Atkinson, M. and Baber, C. (1987)

2 Hatcher, J. (1993)

3 http://www.cumbria-industries.org.uk/a-z-of-industries/iron-and-steel/#content

4 Kendall J. D. (1884–85)

5 http://www.lakedistrict.gov.uk/learning/archaeologyhistory/archaeologydiscoveryzone/
   archaeologyindepth/archaeologycunseybeckforge

6 https://www.gracesguide.co.uk/Barrow_Hematite_Steel_Co

7 https://en.wikipedia.org/wiki/Barrow_Hematite_Steel_Company

8 The source for this section is the magisterial memoir authored by Rose, W.C.C. and Dunham, K.C. (1977)

9 Bailey, Sir Edward (1952) p. 46 et seq.

10 *See* Smythe, W. (1856); pp. 15–28 concerns Cumberland

Chapter 23 Northumberland and Durham

1 Hatcher, J. (1989)

2 Skempton, A.W. (op. cit.) Ch. 2; the initial funds for beginning engineering works to improve Sunderland
   harbour in 1717 were funded by a levy on all coal shipped down the Wear below Newbridge (Chester-le-
   Street)

3 Burn, A. (2017)

4 Wrightson, K. (2011); a painstakingly collected microhistory of the 1636 Newcastle plague from scrivener
   Ralph Tailor's records, with much mention of keelmen. Fascinating and moving in turn.

5 From Howitt, W. (1842)

6 *See*: Chemical Industry and Glass Making in the North East.html; Lemington Glass Works – Wikipedia.
   html; Tyne and Wear industries lime_glass_pottery_paper.pdf

7 https://en.wikipedia.org/wiki/Robert_Mansell

8 *See* https://www.victoriacountyhistory.ac.uk/sites/default/files/work-in-progress/lime_glass_pottery_
   paper.pdf. This glass sand trade does not seem to have attracted any defining research, though port
   records, keelmen archives and glasshouse accounts material must surely exist.

9 Consett Iron Co.html

10 Industrial Period (1750–1950) _ sitelines.newcastle.gov.uk.html

11 William Armstrong, 1st Baron Armstrong – Wikipedia.html

12 For timelines *see*: Origins of Teesside iron and steel .html; For an early historic account *see* Frey, J.W.
    (1929); For a lovingly crafted photo-history *see* Heggie, J.F.K. (2013)

13 *See* Raistrick, A. (1977) for much of interest on the Weardale and Allendale mines and mining
    communities

14 Dunham, K.C. (1934)

15 Bott, M.H.P. & Masson-Smith, D. (1957)

16 Dunham, K.C. (1961, 1965)

17 Sawkins, F.J. (1966)

18 Bott, M.H.P. & Smith, F.W. (2018)

Chapter 24 Midland Scotland

1 Devine, T.M. (2018) is comprehensively authoritative on the Scottish Clearances (Highland and Lowland)
   that 'released' manpower into industrial midland Scotland and the world at large.

2 Prebble, J. (2000); https://en.wikipedia.org/wiki/Darien_scheme

3 Fagan, B. (2000); https://en.wikipedia.org/wiki/Seven_ill_years

4 *See* Upton, B. (2015) on Scottish volcanoes through geological time

5 Sinclair, G. (1672)

6 Hatcher, J. (1989)

7 Duncan, R. (2005) is a fine, well-illustrated source for the human and social side of the history of mining
   in Scotland. McManus, J. (2017) is a microhistorical source for development of the East Fife coalfield.

8 *See*: James Dalrymple, 1st Viscount of Stair – Wikipedia.html

9 Nef, J.U. (1932) p. 400 et seq.

10 A succinct popular review is: The Scottish Industrial Revolution, or The Scottish First Industrial Miracle

1700–1800.html

11 *See*: Hamilton, H. (1928) and Carron_ Historical perspective for Carron.html. Relations between the three founders and economic aspects are by Garbett, S. and Gascoigne, C. (1958)

12 For an account of James Neilson, the inventor of the 'hot blast' *see* https: Livingston, A. //www.academia.edu/12645097/Some_Dumfries_and_Galloway_Pioneers_of_the_Industrial_Revolution; For archive records of the Bairds *see*: Records of William Baird & Co Ltd, coal and iron masters, Coatbridge, North Lanarkshire, Scotland - Archives Hub.html; For the account of Coatbridge in 1869, *see* also Livingstone, A. op. cit.

13 Underhill, J.R. & Craig, J. (2019)

14 McKay, J. (2012) is the fundamental economic account; *see* also Duncan op. cit. who has recorded some social aspects

15 Hallett, D. *et al.* (1985)

16 *See* McManus, J. (2017) for artwork by ex-miner Derek Slater of Ferryhill, County Durham

Bibliography

Agricola, G. (1556) *De Re Metallica*. Frieburg: (Transl. H.C. and L.H. Hoover, New York, 1950)

Aitkenhead, N. *et al.* (2002) *British Regional Geology: The Pennines and Adjacent Areas*. Nottingham: British Geological Survey

Algeo, T.J. & Scheckler, S.E. (1998) Terrestrial-marine teleconnections in the Devonian: links between the evolution of land plants, weathering processes, and marine anoxic events. *Philosophical Transactions of the Royal Society of London* **B353**, 113–130

Allen, R.C. (2009) *The British Industrial Revolution in Global Perspective*. Cambridge: Cambridge University Press

Anderson, D. (1975) *The Orrell Coalfield, Lancashire 1740–1850*. Buxton: Moorland Publishing Company

Anstis, R. (1997) *Man of Iron – Man of Steel: The Lives of David and Robert Mushet*. London: Albion House

Appleton, P. (2013) Mineralisation and mining at Minera, North Wales. *Open University Geological Society Journal* **27**, 23–26

Arrhenius, S. (1896) On the influence of carbonic acid in the air upon the temperature of the ground. *Philosophical Magazine and Journal of Science* **41**, 237–276

Arrowsmith, S.J. (2005) Seismic imaging of a hot upwelling beneath the British Isles. *Geology* **33**, 345–348

Atkinson, M. & Baber, C. (1987) *The Growth and Decline of the South Wales Iron Industry 1760–1880*. Cardiff: University of Wales Press

Babbage, C. (1835) *On the Economy of Machinery and Manufactures*. London: John Murray

Bailey, Sir Edward (1952) *Geological Survey of Great Britain*. London: Thomas Murby & Co

Baldwin, B. (1971) Ways of deciphering compacted sediments. *Journal of Sedimentary Petrology* **41**, 293–301

Barham, M. *et al.* (2012) The onset of the Permo-Carboniferous glaciation: reconciling global stratigraphic evidence with biogenic apatite delta O^{18} records in the late Visean. *Journal of the Geological Society of London* **169**, 119–122

Barton, R.M. (1966) *A History of the Cornish China-Clay Industry*. Truro: D. Bradford Barton Ltd

Bayona, G. & Lawton, T.F. (2003) Fault-proximal stratigraphic record of episodic extension and oblique inversion, Bisbee basin, southwestern New Mexico, USA. *Basin Research* **15**, 251–270

Beerling, D. (2007) *The Emerald Planet: How Plants changed Earth History*. Oxford: Oxford University Press

Beerling, D.J. *et al.* (2002) Carbon isotope evidence implying high O_2/CO_2 ratios in the Permo-Carboniferous atmosphere. *Geochimica et Cosmochimica Acta* **66**, 3757–3767

Belford, P. (2010) Five centuries of iron working: excavations at Wednesbury Forge. *Post-Medieval Archaeology* **44**, 1–53

Berner, R.A. (2004) *The Phanerozoic Carbon Cycle*. Oxford: Oxford University Press

Berner, R.A. (2012) Jacques-Joseph Ébelmen, the founder of earth system science. *Comptes Rendus Geoscience* **344**, 544–548

Berner, R.A. & Raiswell, R. (1984) C/S method for distinguishing freshwater from marine sedimentary rocks. *Geology* **12**, 855–862

Berner, R.A. & Canfield, D. E. (1989) A new model for atmospheric oxygen over Phanerozoic time. *American Journal of Science* **289**, 333–361

Berner, R.A. & Maasch, K.A. (1996) Chemical weathering and controls on atmospheric O_2 and CO_2: Fundamental principles were enunciated by J. J. Ebelmen in 1845. *Geochimica et Cosmochimica Acta* **60**, 1633–1637

Berry, C.M. & Marshall, J.E.A. (2015) Lepidodendrales forests in the early late Devonian paleoequatorial zone of Svalbard. *Geology* **43**, 1043–1046

Besly, B.M. & Fielding, C.R. (1989) Palaeosols in Westphalian coal-bearing and red-bed sequences, Central and Northern England. *Palaeogeography, Palaeoclimatology, Palaeoecology* **70**, 303–330

Bisat, W.S. (1924) The Carboniferous goniatites of the north of England and their zones. *Proceedings of the Yorkshire Geological Society* **20**, 40–124

Bisat, W.S. (1928) The Carboniferous goniatites of England and their continental equivalents. *Comte Rendue de Congress Internationale de Stratigraphie Carbonifère 1927*, 117–133

Blanford, W.T. *et al.* (1859) On the geological structure and relations of the Talcheer Coal Field, in the District of Cuttack. *Memoirs of the Geological Survey of India* **1**, 33–89

Boardman, E.L. (1989) Coal Measures (Namurian and Westphalian) Blackband Iron Formations: fossil bog iron ores. *Sedimentology* **36**, 621–633

Borrow, G. (1862, 2009) *Wild Wales: Its People, Language and Scenery*. London: John Murray (2009 reprint by Bridge Books, Wrexham)

Bott, M.H.P. & Masson-Smith, D. (1957) The geological interpretation of a gravity survey of the Alston Block and the Durham Coalfield. *Quarterly Journal of the Geological Society of London* **113**, 93–117

Bott, M.H.P., & Bott, J.D.J. (2004) The Cenozoic uplift and earthquake belt of mainland Britain as a response to an underlying hot, low density upper mantle. *Journal of the Geological Society of London* **161**, 19–29

Bott, M.H.P. & Smith, F.W. (2018) The role of the Devonian Weardale Granite in the emplacement of the North Pennine mineralisation. *Proceedings of the Yorkshire Geological Society* **62**, 1–15

Boucher, C.T.G. (1968) *James Brindley Engineer 1716–1772*. Norwich: Goose and Sons

Bridge, J.S. (2000) The geometry, flow patterns and sedimentary processes of Devonian rivers and coasts, New York and Pennsylvania, USA. *Geological Society of London Special Publication* **180**, 85

Brimblecombe, P. (2011) *The Big Smoke: A history of air pollution in London since medieval times*. Abingdon: Routledge

Bristow, C.S. (1987) *Sedimentology of large braided rivers ancient and modern*. Ph.D. thesis, University of Leeds

Bristow, C.S. (1988) Controls on sedimentation in the Rough Rock Group. In: Besley, B. & Kelling, G. (eds) *Sedimentation in a Synorogenic Basin Complex – the Upper Carboniferous of NW Europe*, 114–131. Glasgow: Blackie

British Geological Survey (2007a) *Bedrock Geology: UK South*. 1:625 000 Scale. Keyworth

British Geological Survey (2007b) *Bedrock Geology: UK North*. 1:625 000 Scale. Keyworth

British Museum (Natural History Museum) (1964) *British Palaeozoic Fossils*. London: Trustees of the British Museum, 3rd Edn

Burgess, I.C. (1960) Fossil soils of the Upper Old Red Sandstone of south Ayrshire. *Transactions of the Geological Society of Glasgow* **24**, 138–153

Burke, E. (1756) *A Vindication of Natural Society* (Liberty Fund 1982 https://oll.libertyfund.org/titles/850)

Burke, E. (1757, 2008) *A Philosophical Enquiry into the Origin of our Ideas of the Sublime and Beautiful*. Oxford: Oxford World's Classics

Burn, A. (2017) Seasonal work and welfare in an early industrial town: Newcastle-upon-Tyne 1600-1700. *Continuity and Change* **32**, 157–182

Burt, R. (1977) *John Taylor: Mining Entrepreneur and Engineer 1779–1863*. Buxton: Moorland Publishing Company

Cadw (2016) *Hafod and the Lower Swansea Valley: Understanding Urban Character*. Llywodraeth Cymru/ Welsh Government

Chaloner, W.G. (2005) The palaeobotanical work of Marie Stopes. In: *Geological Society of London Special Publication* **241**, 127–135

Chen, J. *et al.* (2018) Strontium and carbon isotopic evidence for decoupling of pCO_2 from continental weathering at the apex of the late Palaeozoic glaciation. *Geology*, **46**, 395–398

Childers S.E. *et al.* (2002) *Geobacter metallireducens* accesses insoluble Fe(III) oxide by chemotaxis. *Nature* **416**, 767–9

Cleeland, J. & Burt, S. (1995) Charles Turner Thackrah: a pioneer in the field of occupational health *Occupational Medicine*, **45**, 285–297

Cliff, R.A. *et al.* (1991) Sourcelands for the Carboniferous Pennine river system: constraints from sedimentary evidence and U-Pb geochronology using zircon and monazite. In: *Geological Society of London Special Publication* **57**, 137–159

Cobbett, W. (1832) *Scotland and Northern Tour*. London. (modern edition by *Zenticula*)

Collinson, J.D. (1969) The sedimentology of the Grindslow Shales and the Kinderscout Grit: a deltaic complex in the Namurian of northern England. *Journal of Sedimentary Petrology* **39**, 194–221

Collinson, J.D. (1970) Deep channels, massive beds and turbidity current genesis in the central Pennine Basin. *Proceedings of the Yorkshire Geological Society* **37**, 495–519

Corfield, S. *et al.* (1996) Inversion tectonics of the Variscan foreland of the British Isles. *Journal of the Geological Society of London* **152**, 17–32

Cossons, N. (1993) *The BP Book of Industrial Archaeology*. Newton Abbot: David and Charles

Craik, A.D.D. (2018) The hydrostatical works of George Sinclair (*c*.1630–1696): their neglect and criticism. *Notes and Records of the Royal Society of London*: doi:10.1098/rsnr.2017.0044

Cutler, A. (2003) *The Seashell On the Mountaintop*. New York: Dutton

Dargie, G.R. *et al.* (2017) Age, extent and carbon storage of the central Congo basin peatland complex. *Nature* **542**, 86–90

Darwin, C. (1840) *Journal of Researches into the Geology and Natural History of the Various Countries Visited by H.M.S. Beagle*. London: H. Colburn. (Cambridge University Press facsimile edition, 2009)

Davies, J.R. *et al.* (1999) *British Geological Survey Flint Memoir*. England and Wales Sheet 108, Solid Geology at 1:50 000

Davies, S.J. *et al.* (2012) Carboniferous Sedimentation and Volcanism on the Laurussian Margin. In: Woodcock, N. and Strachan, R. (eds) *Geological History of Britain and Ireland*, 233–273. Chichester: John Wiley and Sons Ltd

Davis M.W. *et al.* (2012) Crustal structure of the British Isles and its epeirogenic consequences. *Geophysical. Journal International,* **190**, 705–725

Day, J.M. (1975) The Costers: copper-smelters and manufacturers. *Transactions of the Newcomen Society* **47**, 47–58

Day, J.M. (1988) The Bristol brass industry: Furnace structures and their associated remains. *Journal of the Historical Metallurgy Society* **22**, 24–41

Defoe, D. (1724–26; 1971 edition) *A Tour Through the Whole Island of Great Britain*. Harmondsworth: Penguin English Library

Devine, T.M. (2018) *The Scottish Clearances*. London: Allen Lane

DiMichele, W.A. *et al.* (1996) Persistence of Late Carboniferous tropical vegetation during glacially driven climatic and sea-level fluctuations. *Palaeogeography, Palaeoclimatology, Palaeoecology* **70**, 303–330

Dolby, G. *et al.* (2011) A conifer-dominated palynological assemblage from Pennsylvanian (late Moscovian) alluvial drylands in Atlantic Canada: implications for the vegetation of tropical lowlands during glacial phases. *Journal of the Geological Society of London* **168**, 571–584

Drewery, S.E. *et al.* (1987) Provenance of Carboniferous sandstones from U-Pb dating of detrital zircons. Nature **325**, 50–53

Du Toit, A.L. (1937) *Our Wandering Continents: An Hypothesis of Continental Drifting*. Edinburgh: Oliver and Boyd

Dudley, D. (1665) *Mettallum Martis*. (2004 reprint Dudley: Kates Hill Press)

Dudley, R. (1998) Atmospheric oxygen, giant Palaeozoic insects and the evolution of aerial locomotor performance. *Journal of Experimental Biology* **201**, 1043–1050

Duncan, R. (2005) *The Mine Workers*. Edinburgh: Birlinn Ltd

Dunham, K.C. (1934) The genesis of the North Pennine ore deposits. *Quarterly Journal of the Geological Society of London* **90**, 689–720

Dunham, K.C. (1990) *Geology of the North Pennine Orefield*. Memoir of the British Geological Survey

Dunham, K.C. *et al.* (1961) Granite beneath the northern Pennines. *Nature,* **190**, 899–900

Dunham, K.C. *et al.* (1965) Granite beneath Visean sediments with mineralisation at Rookhope, northern Pennines. *Quarterly Journal of the Geological Society of London* **121**, 383–414

Ebbs, C. (2008) *The Milwr Tunnel*. 2nd edn www.lulu.com/crisebbs Also: The Bookshop, High Street, Mold

Ébelmen, J-.J. (1845) Sur les produits de la décomposition des espèces minérales de la famile des silicates. *Annales des Mines* **7**, 3–66

Edwards, D. & Feenan, J. (1980) records of *Cooksonia*-type sporangia from late Wenlock strata in Ireland. *Nature* **247**, 41–42

Edwards, D. *et al.* (1992) A vascular conducting strand in the early land plant *Cooksonia*. *Nature* **357**, 683–685

Edwards, D. and Wellman, C.H. (2001) In: *Plants Invade the Land*. Gensel, P.G. and Edwards, D. (eds), Ch 2. New York: Columbia University Press

Edwards, W. (1954) The Yorkshire/Nottinghamshire Coalfield. In: Trueman, A.E. (ed.) *The Coalfields of Great Britain*. London: Edward Arnold

Egerton, F. (1812) *Description du Plan Incliné Souterrain*. Bureau des Annales des Arts et Manufactures. Google eBook

Engels, F. (1845) *The Condition of the Working-Class in England*. English Transl. & Edition 1892 (1969 paperback edition, ed: Eric Hobsbawm, 301–302. London: Granada)

Evans, C. *et al.* (2002) Baltic iron and the British iron industry in the eighteenth century. *Economic History Review* **60/4**, 642–665

Evelyn, J. (1659) *A Character of England*. London

Evelyn, J. (1661) *Fumifugium*. London

Fagan, B. (2002) *The Little Ice Age*. New York: Basic Books

Falcon-Lang, H.J. (2004a) Early Mississippian Lepidodendrales forests in a delta-plain setting at Norton, near Sussex, New Brunswick, Canada. *Journal of the Geological Society of London* **161**, 969–981

Falcon-Lang, H.J. (2004b) Pennsylvanian tropical rain forests responded to glacial-interglacial rhythms. *Geology* **32**, 689–692

Falcon-Lang, H.J. (2015) A calamitalean forest preserved in growth position. *Review of Palaeobotany and Palynology* **214**, 51–67

Falcon-Lang, H.J. & Scott, A.C. (2000) Upland ecology of some late Carboniferous cordaitalean trees from Nova Scotia and England. *Palaeogeography, Palaeoclimatology, Palaeoecology* **156**, 225–242

Falcon-Lang, H.J. & Bashforth, A.R. (2004) Pennsylvanian uplands were forested by giant cordaitalean trees. *Geology* **32**, 417–420

Falcon-Lang, H.J. *et al.* (2006) The Pennsylvanian tropical biome reconstructed from the Joggins Formation of Nova Scotia, Canada. *Journal of the Geological Society of London* **163**, 561–576

Falcon-Lang, H.J. *et al.* (2011) Pennsylvanian coniferopsid forests in sabkha facies reveal the nature of seasonal tropical biome. *Geology* **39**, 371–374

Falcon-Lang, H.J. *et al.* (2015a) A Late Pennsylvanian coniferopsid forest in growth position, near Socorro, New Mexico, U.S.A.: Tree systematics and palaeoclimatic significance. *Review of Palaeobotany and Palynology* **225**, 67–83

Falcon-Lang, H.J. *et al.* (2018) New insights on the stepwise collapse of the Carboniferous Coal Forests. *Palaeogeography, Palaeoclimatology, Palaeoecology* **490**, 375–392

Falconer, K. (1980) *Guide to England's Industrial Heritage*. London: B.T. Batsford Ltd

Fielding, C.R. *et al.* (2008) The Late Paleozoic Ice Age – a review of current understanding and synthesis of global climate patterns. In: Geological Society of America Special Publication **441**, 343–354

Fielding, C.R. & Frank, T.D. (2015) Onset of the glacioeustatic signal recording late Palaeozoic Gondwanan ice growth: New data from palaeotropical East Fife, Scotland. *Palaeogeography, Palaeoclimatology, Palaeoecology* **426**, 121–138

Fiennes, C. (1888) *Through England on a Side Saddle in the Time of William and Mary*. (2016 reprint www.folkcustoms.com.uk)

Fisher, Q.J. *et al.* (1998) Siderite concretions from nonmarine shales (Westphalian A) of the Pennines, England: Controls on their growth and composition. *Journal of Sedimentary Research* **68**, 1034–1045

Foreman, A. (2001) *Georgiana, Duchess of Devonshire*. London: Random House

Forman, M.B. (1935) *The Letters of John Keats*. Oxford: Oxford University Press

Franks, P.J. *et al.* (2014) New constraints on atmospheric CO_2 concentration in the Phanerozoic. *Geophysical Research Letters* **41**, 4685–4694

Fraser, A.J. & Gawthorpe, R.L. (2003) *An Atlas of Carboniferous Basin Evolution in Northern England*. Geological Society of London Memoir **28**

Frey, J.W. (1929) Iron and Steel Industry of the Middlesbrough District. *Economic Geography* **5**, 176–182

Fuller, J.G.C.M. (2007) Smith's other debt: John Strachey, William Smith and the strata of England 1719–1801. *Geoscientist* 17/7, 5–12

Garbett, S. and Gascoigne, C. (1958) The Struggle for Carron. *Scottish History Review* **37**, 136–145

Gaudet, J. (2014) *Papyrus: The plant that changed the World – from Ancient Egypt to today's Water Wars*. New York: Pegasus

Gawthorpe, R.L. *et al.* (1989) Late Devonian and Dinantian basin evolution in northern England and North Wales. In: Yorkshire Geological Society Occasional. Publication **6**, 1–24

Gelsthorpe, D. (2007) Marie Stopes the palaeobotanist: Manchester and her adventures in Japan. *The Geological Curator* **8(8)**: 375–380

Gerlach, T. (2011) Volcanic versus anthropogenic carbon dioxide. *EOS, Transactions of the American Geophysical Union* **92**, 201–208

Gibling, M.R. *et al.* (1992) Late Carboniferous and early Permian drainage patterns in Atlantic Canada. *Canadian Journal of Earth Sciences* **29**, 338–352

Gilchrist, A. (1863) *Life of William Blake*. (publisher details unknown)

Gilligan, A. (1920) The petrography of the Millstone Grit of Yorkshire. *Quarterly Journal of the Geological Society of London* **75**, 251–293

Glasspool, I.J. *et al.* (2004) Charcoal in the Silurian as evidence for the earliest wildfire. *Geology* **32**, 381–383

Glasspool, I.J. *et al.* (2015) The impact of fire on the Late Paleozoic Earth system. *Frontiers in Plant Science*, **6**, DOI: 10.3389/fpls.2015.00756

Glover, B.W. *et al.* (1996) A second major fluvial sourceland for the Silesian Pennine basin of northern England. *Journal of the Geological Society of London* **153**, 901–906

Goldring, W. (1924) The Upper Devonian forest of seed ferns. *New York State Museum Bulletin* **521**, 50–72

Goldring, W. (1927) The oldest known petrified forest. *Science Monthly* **24**, 514–529

Gough, P. *et al.* (2013) *Graham Sutherland: From Darkness into Light.* Bristol: Sansom & Company Ltd

Graham, J.R. (1983) Analysis of the Upper Devonian Munster basin, an example of a fluvial distributary system. *Special Publication International Association of Sedimentologists* **6**, 473–483

Green, G.W. (1992) *British Regional Geology: Bristol and Gloucester region.* British Geological Survey

Gregor, C.B. *et al.* (1988) *Chemical Cycles in the Evolution of the Earth.* New York: John Wiley and Sons

Grigson, G. (2017) *Selected Poems.* (Ed. J.Greening) London: Greenwich Exchange

Hains, B.A. & Horton, A. (1969) *British Regional Geology: Central England.* British Geological Survey

Hallett, D. *et al.* (1985) Oil exploration and production in Scotland. *Scottish Journal of Geology* **21**, 547–570

Hallsworth, C.R. & Chisholm, J.I. (2000) Stratigraphic evolution of provenance characteristics in Westphalian sandstones of the Yorkshire coalfield. *Proceedings of the Yorkshire Geological Society* **53**, 43–72

Hallsworth, C.R. & Chisholm, J.I. (2008) Provenance of late Carboniferous sandstones in the Pennine Basin (UK) from combined heavy mineral, garnet geochemistry and palaeocurrent studies. *Sedimentary Geology* **203**, 196–212

Hallsworth, C.R. & Chisholm, J.I. (2017) Interplay of mid-Carboniferous sediment sources on the northern margin of the Wales-Brabant High. *Proceedings of the Yorkshire Geological Society* **61**, 285–310

Hamblin, R.J.O. & Coppack, B.C. (1995) *Geology of Telford and the Coalbrookdale Coalfield.* Memoir of the British Geological Survey, parts of Sheets 152 and 153

Hamilton, H. (1928) The Founding of the Carron Ironworks. *The Scottish Historical Review* **25**, 185–193

Harrison, Tony (2005) *Under the Clock.* London: Penguin

Harrison, Tony (2015) Polygons. *London Review of Books* **37**, 4

Hart, R. (1859) 'Reminiscences of James Watt'. *Transactions of the Glasgow Archaeological Society*, 1859

Hartmann, J. *et al.* (2013) Enhanced chemical weathering as a geoengineering strategy to reduce atmospheric carbon dioxide, supply nutrients, and mitigate ocean acidification. *Reviews of Geophysics* **51**, 113–149

Harvey, A. *et al.* (2018) Shale prospectivity onshore Britain. In: *Petroleum Geology of NW Europe: 50 Years of Learning – Proceedings of the 8th Petroleum Geology Conference*, 571–584. Geological Society of London

Hatcher, J. (1993) *The History of the British Coal Industry. Volume 1. Before 1700: Towards the Age of Coal.* Oxford: Oxford University Press

Hawkes, J. (1951) *A Land.* London: The Cresset Press

Hayman, R. & Horton, W. (2009) *Ironbridge: History and Guide.* Stroud: The History Press

Hays, J.D. *et al.* (1976) Variations in the Earth's orbit: Pacemakers of the ice ages. *Science* **194**, 1121–1131

Heggie, J.F.K. (2013) *Middlesbrough's Iron and Steel Industry.* Stroud: Amberley Publishing

Hernick, L.V. *et al.* (2008) Earth's oldest liverworts – *Metzgeriothallus sharonae* sp. Nov. from the Middle Devonian (Givetian) of eastern New York, USA. *Review of Palaeobotany and Palynology* **148**, 154–162

Hey, D. (1969) A dual economy in South Yorkshire. *The Agricultural History Review* **17**, 108–119

Hey, D. (2005) The South Yorkshire Steel Industry and the Industrial Revolution *Northern History* **42**, 91–96

Hibbard, J. and Waldron, J.W.F. (2009) Truncation and translation of Appalachian promontories: Mid-Palaeozoic strike-slip tectonics and basin initiation. *Geology* **37**, 487–490

Hill, G. (1998) *The Triumph of Love.* Harmondsworth: Penguin Books

Hobsbawm, E.J. (1969) *Industry and Empire.* Harmondsworth: Pelican Books

Hobsbawm, E.J. and Rudé, G. (1972) *Captain Swing.* Harmondsworth: Penguin Books

Hodder, E. (1888) *The Life and Work of the Seventh Earl of Shaftesbury, K.G.* London: Cassell & Co

Hodgson, Rev. J. (1813) *The Funeral Sermon of the Felling Colliery Sufferers.* Newcastle.

Holmes, R. (2008) *The Age of Wonder.* London: Harper

Hopkins, W. (1954) The Coalfields of Northumberland and Durham. In: Trueman, A.E. (ed.) *The Coalfields of Great Britain* London: Edward Arnold

Howells, M.F. (2007) *British Regional Guide: Wales.* Keyworth, Nottingham: British Geological Survey

Howitt, W. (1842) *Visits to Remarkable Places.* 1907 edition London: Longmans, Green

Hulme, M. (2009) On the origin of the 'greenhouse effect': John Tyndall's 1859 interrogation of nature. *Weather* **64**, 121–123

Hutton, J. (1795) *Theory of the Earth with Proofs and Illustrations.* Edinburgh (Electronic Edition at www. Qontro.com)

Ingpen, R. (1915) *The Letters of Percy Bysshe Shelley*. London: Pitman

International Commission on Stratigraphy (2012) *International Chronostratigraphic Chart*

Jennings, H. (1985, 2012) *Pandæmonium 1660–1886*. London: Icon Books

Jones, J. *et al.* (1994) The correspondence between James Hutton (1726–1797) and James Watt (1736–1819) with two letters from Hutton to George Clerk-Maxwell (1715–1784): Part 1. *Annals of Science* **51**, 637–653

Jones, N.S. (2007) The West Lothian Oil-Shale Formation: Results of a sedimentological study. *Geology and Landscape Northern Britain Programme Internal Report IR/05/046*. Keyworth: British Geological Survey

Jukes, J.B. (1859) *The South Staffordshire Coalfield*. Memoir of the Geological Survey of Great Britain 2nd Edn

Kendall J.D. (1884–85) Notes on the History of Mining in Cumberland and North Lancashire. *Transactions of the North of England Institute of Mining and Mechanical Engineers* **34**, 83–124

Kendall, P.F. & Wroot, H.E. (1924) *Geology of Yorkshire*. Privately printed.

Kenrick, P. & Crane, P.R. (1997) The origin and early evolution of plants on land. *Nature* **389**, 33–39

Kidston, R. & Lang, W.H. (1916–21) On Old Red sandstone plants showing structure, from the Rhynie chert bed, Aberdeenshire: Parts I–V. *Transactions of the Royal Society of Edinburgh* **51** & **52**

King, P.W. (1999) The development of the Iron Industry in South Staffordshire in the 17th Century: History and Myth. *Transactions of the Staffordshire Archaeological and Historical Society* **38**, 59–76

Kirby, M.W. (1977) *The British Coalmining Industry, 1870–1946*. London: Macmillan Press Ltd

Kunene, M. (1970) 'The Gold-miners' from *Zulu Poems*. London: André Deutsch

Lakin, J.A. *et al.* (2016) Greenhouse to icehouse: a biostratigraphic review of latest Devonian-Mississippian glaciations and their global effects. In: *Geological Society of London Special Publication* **423**, 439–464

Lancaster, P.J. *et al.* (2017) Interrogating the provenance of large river systems: multi-proxy in situ analyses in the Millstone Grit, Yorkshire. *Journal of the Geological Society of London* **174**, 75–87

Lane, N. (2002) *Oxygen: the Molecule that made the World*. Oxford: Oxford University Press

Larkin, P. (1974) *High Windows*. London: Faber and Faber

Lawrence, D.H. (1965) *Sons and Lovers*. Harmondsworth: Penguin

LeBlanc, Abbé (1745) *Lettres d'un François*

Le Heron, D.P. (2018) An exhumed Paleozoic glacial landscape in Chad. *Geology* **46**, 91–94

Leeder, M.R. (1982) Upper Palaeozoic basins of the British Isles – Caledonide inheritance versus Hercynian plate margin processes. *Journal of the Geological Society of London* **139**, 479–491

Leeder, M.R. & McMahon, A.H. (1988) Upper Carboniferous (Silesian) basin subsidence in northern Britain. In: *Sedimentation in a syn-orogenic basin complex: the Carboniferous of northwest Europe*. Glasgow: Blackie and Son

Leeder, M.R. (2007) Cybertectonic Earth and Gaia's weak hand: sedimentary geology, sediment cycling and the Earth system. *Journal of the Geological Society of London* **164**, 277–296

Livingstone, A. www.academia.edu/12645097/Some_Dumfries_and_Galloway_Pioneers_of_the_Industrial_Revolution

Lovelock, J.E. (1979) *Gaia: A New Look at Life on Earth*. Oxford: Oxford University Press

Lyell, C. (1838) *Elements of Geology*. London: John Murray

Mackie, W. (1913) The rock-series of Craigbeg and Ord Hill, Rhynie, Aberdeenshire. *Transactions of the Edinburgh Geological Society* **10**, 205–236

MacNeice, L. (1979) *Collected Poems*. London: Faber and Faber

Malet, H. (1990) *Coal, Cotton and Canals*. Manchester: Neil Richardson

Malm, A. (2016) *Fossil Capital: The Rise of Steam Power and the Roots of Global Warming*. London: Verso

Matsunaga, K.K.S. & Tomescu, A.M.F. (2016) Root evolution at the base of the lycophyte clade: insights from an early Devonian lycophyte. *Annals of Botany* **117**, 585–598

Maynard, J.R. & Leeder, M.R. (1992) On the periodicity and magnitude of Late Carboniferous glacio-eustatic sea-level changes. *Journal of the Geological Society of London* **149**, 303–311

Maynard, J.R. *et al.* (1997) The Carboniferous of western Europe: the development of a petroleum system. *Petroleum Geology*, **3**, 97–115

McCluskey, S. *et al.* (2000) Global positioning system constraints on plate kinematics and dynamics in the eastern Mediterranean and Caucasus. *Journal of Geophysical Research* **105**, 5695–5719

McFarlane, D.A. *et al.* (2014) A speleothem record of early British and Roman mining at Charterhouse, Mendip, England. Archaeometry **56**, 431–443

McKay, J. H. (2012) *Scotland's First Oil Boom: The Scottish Oil-Shale Industry, 1851–1914*. Edinburgh: John Donald

McManus, J. (2017) *Coal Mining in the East Neuk of Fife*. Edinburgh & London: Dunedin Academic Press

Meteyard, M. (1865) *The Life of Wedgwood*. London: Hurst & Blackett

Meyer-Berthaud, B. & Decombeix, A-L. (2007) A tree without leaves. *Nature*, **446**, 861–862

Meyer-Berthaud, B. & Decombeix, A-L. (2012) In the shade of the oldest forest. *Nature* **483**, 41–42

Meyer-Berthaud, B. *et al.* (2013) Archaeopterid root anatomy and architecture: new information from permineralized specimens of Fammenian age from the Anti-Atlas, Morocco. *International Journal of Plant Sciences* **174**, 364–381

Mills, D.A.C. & Holliday, D.W. (1998) *Geology of the district around Newcastle-on-Tyne, Gateshead and Consett*. Memoir of the British Geological Survey, Sheet 20 (England and Wales)

Mitchell, G.H. *et al.* (1947) *Geology of the Country around Barnsley*. Memoir of the Geological Survey of Great Britain Sheet 87 London: His Majesty's Stationary Office

Mitchell, G.H. (1954) The Coalfields of the South Midlands. In: Trueman, A.E. (ed.) *The Coalfields of Great Britain* London: Edward Arnold

Mitchell, G.H. *et al.* (1962) *The Geology of the Neighbourhood of Edinburgh*. Memoir of the Geological Survey of Scotland. Edinburgh: Her Majesty's Stationary Office

Mokyr, J. (2017) *Culture of Growth: The Origins of the Modern Economy*. Princeton: Princeton University Press

Moore, P. (2018) *Endeavour: The Ship and the Attitude that Changed the World*. London: Chatto and Windus

Morris, J.L. *et al.* (2015) Investigating Devonian trees as geo-engineers of past climates: Linking palaeosols to palaeobotany and experimental geobiology. *Palaeontology*, **58**, 787–801

Needham, J. (1986) *Science and Civilisation in China* (5 Vols). Cambridge: Cambridge University Press

Nef, J.U. (1932, 1972) *The Rise of the British Coal Industry*. New York: Books for Libraries Press, Freeport

Nicholson, N. (1981) *Sea to the West*. London: Faber & Faber

Norton, C.E. (1886) *Early Letters of Thomas Carlyle*. London: Macmillan Ltd

Oakley, K.P. & Muir-Wood, H.M. 1962 *The Succession of Life through Geological Time*. London: British Museum

Oldham, T. (1859) Preliminary Notice on the Coal and Iron of Talcheer, in the Tributary Mehals of Cuttack. *Memoirs of the Geological Survey of India* **1**, 1–32

Parnell, J. (1988) Lacustrine petroleum source rocks in the Dinantian Oil Shale group, Scotland: a review. *Geological Society of London Special Publication* **40**, 235–246

Phillips, J. (1844) *Memoirs of William Smith LL.D.* London: John Murray. Available online and for download at https://archive.org/details/memoirsofwilliam00philrich

Phillips, T.L. & DiMichele, W.A. (1992) Comparative ecology and life-history biology of arborescent Lepidodendrales in Late Carboniferous swamps of EuroAmerica. *Annals of the Missouri Botanical Gardens* **79**, 560–588

Pöhiker, C. *et al.* (2012) Biogenic Potassium Salt Particles as Seeds for Secondary Aerosol in the Amazon. *Science* **337**, 1075–1078

Polo, M. (*c.*1270) The Travels of Marco Polo: The Complete Yule-Cordier Edition of 1903. Reprint by Courier Corporation 1993

Prebble, J. (2000) *Darien: the Scottish Dream of Empire*. Edinburgh: Birlinn

Priestley, J.B. (1934; 1977) *English Journey*. London: Heinemann and Harmondsworth: Penguin

Raistrick, A. (1972) *Industrial Archaeology*. London: Eyre Methuen Ltd (paperback edition by Granada Publishing, 1973)

Raistrick, A. (1977) *Two Centuries of Industrial Welfare: The London (Quaker) Lead Company 1692–1905*. Buxton: Moorland Publishing Company, revised 2nd Edition

Raistrick, A. (1989) *Dynasty of Iron Founders*. Ironbridge Gorge Museum Trust, revised 2nd Edition

Raiswell, R. *et al.* (1988) Degree of pyritisation of iron as a paleoenvironmental indicator of bottom water oxygenation. *Journal of Sedimentary Petrology* **58**, 812–819

Ramsbottom, W.H.C. (1978) Carboniferous. In McKerrow, W.S. (ed.) *The Ecology of Fossils* London: Duckworth

Raymo, M.E. & Ruddiman, W.F. (1992) Tectonic forcing of late Cenozoic climate. *Nature* **359**, 117–122

Retallack, G.J. (1997) Early Forest Soils and their role in Devonian Global Change. *Science*, **276**, 583–585

Rice, C.M. and Trewin, N.H. (1988) Lower Devonian gold-bearing hot spring system near Rhynie, Scotland. *Transactions of the Institution of Mining and Metallurgy* **97**, B141–144

Robinson, J.M. (1990) Lignin, land plants, and fungi: Biological evolution affecting Phanerozoic oxygen balance. *Geology* **15**, 607–610

Robinson, J.M. (1991) Phanerozoic atmospheric reconstructions: a terrestrial perspective. *Palaeogeography, Palaeoclimatology, Palaeoecology* **97**, 51–62

Rose, W.C.C. & Dunham, K.C. (1977) *Geology and hematite deposits of South Cumbria*. Geological Survey of Great Britain. London: Her Majesty's Stationary Office

Rotherham, J. (2013) *The Lost Fens: England's greatest ecological disaster*. Stroud: The History Press

Rydin, H. & Jeglum, J.K. (2013) *The Biology of Peatlands*. Oxford: Oxford University Press

Rygel, M.C. *et al*. (2008) The magnitude of late Paleozoic glacioeustatic fluctuations: a synthesis. *Journal of Sedimentary Research* **78**, 500–511

Sagan, D. & Margulis, L. (1995) *Garden of Microbial Delights: A Practical Guide to the Subvisible World*. Dubuque: Kendall/Hunt Publishing Company

Sahney, S. *et al*. (2010) Rainforest collapse triggered Carboniferous tetrapod diversification in Euroamerica. *Geology*, **38**, 1079–1082

Sawkins, F.J. (1966) Ore genesis in the North Pennine orefield in the light of fluid inclusion studies. *Economic Geology* **61**, 385–401

Scotese, C. R. (2000) Palaeomap Project:http://www.scotese.com/scotesepubs.htm

Scott, A.C. (1977) A review of the ecology of Upper Carboniferous plant assemblages with new data from Strathclyde. *Palaeontology* **20**, 447–473

Scott, A.C. (1989) Observations on the nature and origin of fusain. *International Journal of Coal Geology* **12**, 443–475

Scott, A.C. & Jones, T.P. (1994) The nature and influence of fire in Carboniferous ecosystems. *Palaeogeography, Palaeoclimatology, Palaeoecology* **106**, 96–112

Seaton, A. (2018) *Farewell, King Coal: From industrial triumph to climatic disaster*. Edinburgh & London: Dunedin Academic Press

Simond, L. (1968) *An American in Regency England*. (ed. C. Hibbert) London: Pergamon.Press Ltd

Sinclair, G. (1672) *The Hydrostaticks; or The Weight, Force or Pressure of Fluid Bodies, Made evident by Physical, and Sensible Experiments. Together With some Miscellany Observations, the last whereof is a short History of Coal, and of all the Common, and proper Accidents thereof; a Subject never treated of before*. Edinburgh: Swinton, Glen and Brown. Online at: *www.electricscotland.com* courtesy of Boston Public Library

Skempton, A.W. (1996) *Civil Engineers and Engineering in Britain*, 1600–1830. Aldershot: Variorum Collected Studies

Smiles, S. (1863) *Industrial Biography: Iron Workers and Tool makers*. http://www.gutenberg.org/ebooks/404

Smith, A. (2013) *The Sublime in Crisis: Landscape Painting after Turner*. Tate Gallery, London, Research Publications: https://www.tate.org.uk/art/research-publications/the-sublime/alison-smith-the-sublime-in-crisis-landscape-painting-after-turner-r1109220

Smithson, T.R. *et al*. (2012) Earliest Carboniferous tetrapod and arthropod faunas from Scotland populate Romer's Gap. *Proceedings of the National Academy of Sciences of the United States of America*, **109**, 4532–4537

Smythe, W. W. (1856) *The Iron Ores of Britain Part 1*. Memoir of the Geological Survey of Great Britain

Stallworthy, J (ed.) (1990) *The Poems of Wilfred Owen*. London: Chatto and Windus

Stallworthy, J. (2013) *Wilfred Owen*. London: Pimlico

Steemans, P. *et al*. (2009) Origin and radiation of the Earliest Vascular Land Plants. *Science* **324**, 353

Stein, W.E. *et al*. (2007) Giant cladoxylopsid trees resolve the enigma of the earth's earliest forest stumps at Gilboa. *Nature,* **446**, 904–907

Stein, W.E. *et al*. (2012) Surprisingly complex community discovered in the mid-Devonian fossil forest at Gilboa. *Nature,* **483**, 78–81

Stein, W.E. *et al*. (2019) Mid-Devonian *Archaeopteris* Roots Signal Revolutionary Change in Earliest Fossil Forests. *Current Biology*, **30**, 1–11

Steno, N. (1669) *Prodromus* (or *De Solido*). Florence (Transl. H. Oldenburg, London, 1671). Online at: https://archive.org/stream/prodromusnicola00stengoog/prodromusnicola00st

Stewart, W.N. & Rothwell, G.W. (1993) *Paleobotany and the Evolution of Plants*. Cambridge: Cambridge University Press, 2nd Edn

Stopes, M.C. (1919) On the Four Visible Ingredients in Banded Bituminous Coal: Studies in the Composition of Coal, No. 1. *Proceedings of the Royal Society of London* **B90**, 470–487

Stopes, M. C. (1935) *On the petrology of banded bituminous coal. Fuel in Science and Practice* **14,** *4–13.*

Stopes M.C. & Watson D.M.S. (1909) On the Present Distribution and Origin of the Calcareous Concretions in Coal Seams, known as 'Coal Balls'. *Philosophical Transactions of the Royal Society of London* **B200**, 167–218

Strachey, J. (1717) A Curious Description of the Strata Observ'd in the Coal Mines of Mendip in Somersetshire. *Philosophical Transactions of the Royal Society of London* **30**, 968–73

Strahan, A. *et al*. (1920) Special Reports on the Mineral Resources of Great Britain: XIII – Iron Ores. *Memoirs of the Geological Survey*

Tacitus (98, 1970) *The Agricola and the Germania*. Harmondsworth: Penguin Books

Tegner, H. (1969) *Ghosts on the road*. In: A Gleaming Landscape. Wainwright, M. (ed.) London: Guardian Country Diary for February 1969

Tharoor, S. (2017) *Inglorious Empire: What the British Did to India*. London: Penguin

Thelemann, M. *et al.* (2017) Bog iron ore as a resource for prehistoric iron production in Central Europe – A case study of the Widawa catchment area in eastern Silesia, Poland. *Catena* **140**, 474–490

Times Atlas of the World. (1987) London: Times Books

Thompson, E.P (1980) *The Making of the English Working Class*. Harmondsworth: Pelican

Tissot, B.P. & Welte, D.H. (1978) *Petroleum Formation and Occurrence*. Berlin: Springer

Toghill, P. (2000) *The Geology of Britain: An Introduction*. Marlborough: Airlife Publications

Toghill, P. (2006) *Geology of Shropshire*. Marlborough: The Crowood Press 2nd Edn

Torrens, H. (2000) The William Smith Lecture 2000. Bath: The Bath Royal Literary and Scientific Institution, 153–193

Torrens, H. (2003) An Introduction to the Life and Times of William Smith (1769–1839). Ibid. xi-xxxviii

Torsvik, T.H. & Cocks, L.R.M. (2017) *Earth History and Palaeogeography*. Cambridge: Cambridge University Press

Trewin, N.H. (2001) The Rhynie chert. In: *Palaeobiology II*. Oxford: Blackwell Science Ltd

Trewin, N.H. (2013) Scottish Fossils. Edinburgh and London: Dunedin Academic Press Ltd

Trewin, N.H. & Thirlwall, M.F. (2002) Old Red Sandstone. In: *Geology of Scotland*, N.H. Trewin (ed.), Geological Society of London

Trueman, A.E. (1938) *The Scenery of England and Wales*. London: Gollancz, reprinted in 1949 as: *Geology and Scenery in England and Wales*. Harmondsworth: Pelican Books

Tyndall, J. (1859) On the transmission of heat of different qualities through gases of different kinds. *Proceedings of the Royal Institution* **3**, 155–158.

Uglow, J. (2002) *The Lunar Men: Five Friends Whose Curiosity Changed the World*. New York: Farrar, Straus and Giroux

Underhill, J.R. & Craig, J. (2019) Archibald Geikie and the establishment of the Scottish Oil Shale Industry. *Special Publication of the Geological Society of London,* **480**, 379–399

Upton, B. (2015) *Volcanoes and the Making of Scotland*. Edinburgh: Dunedin Academic Press

Volkova, I.B. (1994) Nature and Composition of the Devonian Coals of Russia. *Energy and Fuels* **8**, 1489–1493

Waldron, J.W.F. *et al.* (2015) Late Paleozoic strike-slip faults in Maritime Canada and their role in the reconfiguration of the northern Appalachian orogeny. *Tectonics* **34**, 1661–1684

Wang, D. & Liu, L. (2015) A new Late Devonian genus with seed plant affinities. *BioMed Central Evolutionary Biology* doi: 10.1186/s12862-015-0292-6

Wang, D. *et al.* (2019) The Most Extensive Devonian Fossil Forest with Small Lepidodendrales Trees bearing the Earliest Stigmarian Roots. *Current Biology* **29**, 2604–2615

Wang, X. *et al.* (2017) Hydroclimate changes across the Amazon lowlands over the past 45 000 years. *Nature* **541**, 204–207

Wanless, H.R. & Shepard, F.P. (1936) Sea level climatic changes related to late Palaeozoic cycles. *Bulletin of the Geological Society of America* **47**, 1177–1206

Waters, C.N. (ed.) (2011) *A Revised Correlation of Carboniferous Rocks in the British Isles*. Special Report **26**, Geological Society of London

Watson A. *et al.* (1978) Methanogenesis, fires, and the regulation of atmospheric oxygen. *Biosystems* **10**, 293–298

Wegener, A. (1924) *The Origin of Continents and Oceans*. English Translation of the 3rd German Edition London: Methuen and Co Ltd

Wellman, C.H. (2010) The invasion of the land by plants: when and where? *The New Phytologist* **188**, 306–309

Wellman, C.H. *et al.* (2003) Fragments of the earliest land plants. *Nature* **425**, 282–285

Williams, G.D. & Chapman, T.J. (1986) The Bristol–Mendip foreland thrust belt. *Journal of the Geological Society* **143**, 63–73

Wells, M.R. *et al.* (2005) Large sea, small tides: the late Carboniferous seaway of NW Europe. *Journal of the Geological Society of London* **162**, 417–420

Woodcock, N.H. & Strachan, R.A. (eds) (2012) *Geological History of Britain and Ireland*. Oxford: Blackwell Science Ltd

Wright, J.S. *et al.* (2017) Rainforest-initiated wet season onset over the southern Amazon. *Proceedings of the National Academy of Sciences of the United States of America* **114**, 8481–8486

Wright, W.B. *et al.* (1927) *The Geology of the Rossendale Anticline.* His Majesty's Stationary Office: Geological Memoirs of England

Wrightson, K. (2011) *Ralph Tailor's Summer: a Scrivenor, his City, and the Plague.* New Haven and London: Yale University Press

Xu, H-H *et al.* (2017) Unique growth strategy in the earth's first trees revealed in silicified fossil trunks from China. *Proceedings of the National Academy of Sciences* doi: 10.1073/pnas.1708241114

Yallop, J. (2015) *Dreamstreets.* London: Jonathan Cape

Young, A. (1776) *Tours in England and Wales 3. Tour from Essex to Shropshire.* (online as: 'A Vision of Britain Through Time'; see visionofbritain.org.uk)

Yule, H. (1903) *The travels of Marco Polo: the complete Yule-Cordier edition: including the unabridged third edition (1903) of Henry Yule's annotated translation, as revised by Henri Cordier, together with Cordier's later volume of notes and addenda (1920).* Hathi Trust Digital Library

Zeigler, P. (1989) *Evolution of Laurussia.* Dordrecht: Kluwer Academic

Glossary

A

Acadian Mid-Devonian episode of mountain-building and tectonics in northern Britain; named from previously recognized events in former French Maritime Canada (Acadie, via the Greek, Arcadia)

Adit Tunnel cut and driven horizontally or at slight incline into a hillside to intersect ore-bearing veins or coal seams; allows outward drainage and transport of persons, animals and materials

Algae Large group of diverse photosynthesizing aqueous plants with cellular structure having chlorophyll and enclosed nuclei

Alkalinity The concentration of carbonate (CO_3^{--}) and bicarbonate (HCO_3^-) in natural waters; chief control ('buffering') of acidity

Alluvial Pertaining to rivers and their transported sediment

Alpine-Himalayan Early Cenozoic mountain-building episode from the Pyrenees to Sichuan by closure of the equatorial Tethys ocean and its adjacent seas and straits

Anaerobic Oxygen-free conditions in a water body or in the pore waters of bottom sediment

Anthracite High-rank, bright-burning ashless coal with high carbon content (>92%) and little water

Anthropocene Proposed period of historical time when Earth surface and atmospheric processes were strongly influenced by the enhanced economic activity of *Homo Sapiens'* exploitation of fossil fuels, the onset conveniently marked by radiation fallout from the US Army's Trinity nuclear test of 1945

Anticline Stratal up-bend; a fold with a crest and two outward-tilted limbs

Archaeopteris Extinct tall tree-like fern (woody bark, spore-bearing), a major Middle Devonian to Early Carboniferous forest-former with deep-penetrating and widely radiating woody roots

Archaean Eon of time 4000–2500 Ma

Ash Fine volcanic deposit sedimented from eruptive plumes or as hot, ground-hugging flows (*nuée ardentes*)

Ashlar Frontwork of a smoothly masoned freestone block

Asthenosphere Upper mantle in convection-driven flow on which the rigid lithosphere plates move

Atmospheric steam engine (aka **reciprocating beam engine/Newcomen engine**) First coal-burning, steam-driven pumping machine (1712) marking the onset of the Industrial Revolution

Autocompaction Self-driven compaction by deposition and gravity; primarily evident in water-saturated organic (peaty) and fine-grained (muddy) sediment

B

Bacteria Primitive microscopic organic forms without cell nuclei; they mediate microbial anaerobic alteration of (particularly) organic-rich sediments in the shallow subsurface

Barite (aka barytes) Mineral form of barium sulphate ($BaSO_4$); dense, pinky-cream, platy

Basalt Common, dark, finely crystalline, silica-poor lava

Basic-lined smelting (aka Gilchrist-Thomas Process) Natural carbonate removes phosphorus from coke-smelted iron ore as calcium phosphate slag

Basin Tectonically subsiding crust that slowly accumulates deposited sediment

Basin inversion When a sedimentary basin is shortened, uplifted and eroded

Basset Outcrop; an old term

Batholith A large three-dimensional mass of intruded plutonic igneous rock

Battery ware Metallic ware (copper, brass, pewter, etc.) produced by mechanical percussion, originally by water-powered tilt-hammers

Bed Layer (stratum) of sedimentary rock bounded by bedding planes

Bedding plane A surface in strata marking a depositional pause or a different sediment influx

Bedding The depositional geometry or course of sedimentary beds

Bedrock Indurated outcrop or subcrop below soil, alluvium or glacial deposits

Bellpit Bell-shaped excavation of ore or coal below a vertical shaft

Bessemer converter Pivoting, squat, open-topped furnace, the molten iron blown by hot air from below to first decarbonize and then mix in known quantities of manganese, carbon, etc. to make mild steel

Bicarbonate ion Normally the chief constituent that determines the alkalinity of natural waters; acts as a buffering agent to control pH at (normally) slightly alkaline values

Bing Reddened surface spoil tip of mined and retorted oil-shale waste in West Lothian

Bioclastic Describes limestone comprising calcitic fossil shell fragments

Biogeochemistry Organically driven chemical processes of the biosphere that determine the elemental cycling of carbon, oxygen, hydrogen, nitrogen, phosphorus and sulphur

Bitumen Black, highly viscous, liquid-to-vitreous, semi-solid form of petroleum

Blackband ironstone Mixed sedimentary ironstone with alternating laminae rich in siderite and carbonaceous materials, the latter enabling self-fuelled calcining

Blast furnace Originally from Song Dynasty, China; a top-charged, high-temperature furnace using charcoal (or coke in modern times) in a hearthstone-lined base ventilated by bellows, the molten metal let out at tapping points for direct casting as pig-iron

Bloomery Updraught-fed, low-temperature furnace using charcoal to reduce iron ore below its melting point; its 'bloom' of spongy, bulbous slag and metal mix is then hammered hot to low-carbon wrought (bar) iron in a smithy hammer forge

Bog iron ore Ochreous iron oxide sourced from reducing and acidic wetlands; the soluble iron later oxidizes and precipitates in permeable river terrace deposits

Bone china 'English porcelain' pioneered by Josiah Spode: a fused, finely milled mix of china clay (kaolinite), Cornish stone (feldspar-rich granite) and bone ash; produces a strong crystalline mix of calcic-feldspar, tricalcic phosphate and natural glass

Boulder Clay An older name for glacial till

Brachiopod Largely extinct marine two-shelled bilaterally symmetrical creatures attached to a substrate via a holdfast. Abundant in shallow Carboniferous seas

Brass Alloy of copper and 35–50% zinc

Breccia Coarse-grained sedimentary rock or fault-rock comprising angular fragments

Bronze Age Division of archaeological time; in Britain *c*.4–2.8 ka, getting older eastwards to Anatolia

Bryophyte Non-vascular, moisture-loving, spore-bearing plants without true roots, though some with holdfasts; liverworts, hornworts and mosses

C

Caking Property of certain coals to form larger aggregates through surface crusting on burning; a property valued by smiths and sought after for slow-burn coking

Calamine Mineral form of zinc carbonate ($ZnCO_3$); nowadays known as smithsonite

Calamites Extinct tree-like horsetail growing up to ten metres; mostly as understorey in Carboniferous forest swamps and as pioneers on rapidly depositing deltaic margins; distinctive bamboo-like segmented, hollow and striated stems; reproduction via spores, also by rapid clonal growth from rhizomes

Calcine To drive off carbon dioxide from carbonate ores (e.g. siderite) by intense heat to leave iron oxide

Calcite Mineral form of calcium carbonate ($CaCO_3$); common, light-coloured, softish, with two perfect cleavage planes

Calcrete Calcareous lower soil horizon (calcisol) formed in semi-arid climates

Caledonian Lower Palaeozoic mountain-building episode associated with closure of the Iapetus ocean

Carbon burial/sequestration The part of the geological (long-term) carbon cycle when tectonic subsidence buries organic carbon for tens to hundreds of millions of years

Carbon capture/drawdown Attempts to remove atmospheric carbon dioxide by various artificial methods

Carbon cycle The various throughputs of carbon dioxide that control its short, medium and long-term abundance in the global atmosphere

Carbon isotopes There are three: two are stable, C^{12} and C^{13}; their ratio with respect to a universal standard is used in assessing biogeochemical reactions; one is unstable, C^{14}; its decay constant used in radiocarbon dating

Carboniferous Period of time 359–299 Ma

Carse(land) (Scots) Drained and reclaimed former estuary-margin wetlands

Cassiterite Mineral form of tin oxide, SnO_2; stable, hard and chemically inert

Cathaysian Pertaining to Cathay (China); the country as a distinctive Permo-Carboniferous floral province

Celtic Catch-all referring to Iron Age cultures and peoples in Europe

Cenozoic Era from 66 Ma to present

Chalcopyrite Mineral form of copper-iron sulphide ($CuFeS_2$); commonest copper ore, easily oxidized, soft with brownish streak

Charcoal Brittle, pure carbon: product of wood burnt very slowly under oxygen-poor conditions in 'clamps'

Charophyte Freshwater green algae extant since Devonian times: sometimes calcifies in hard waters

Chemical gradient A spatial change in chemical composition of porewater in sediments that leads to diffusion of ions along the maximum rate of change of concentration

Chemical rock weathering The effects of attack by rain and groundwater upon bedrock by dissolution, alteration, oxidation and reduction

China clay The mineral kaolinite, an alumium silicate

Chrome spinel Dense magnesium aluminium oxide mineral from upper mantle rocks such as peridotite; common in ophiolites

Civil War (largely English) – fought 1642–1651 between Parliament and the Royalist forces loyal to the absolute monarchical principles insisted upon by the Stuart king, Charles I

Clarain (aka Huminite) A maceral component of bituminous coal; 'Bright' or 'glance' coal; cellular structure unfilled by humic breakdown products (like bitumens); shows relatively low light reflectance on the microscope stage

Clastic Describes detrital sediment

Clay (aka Mud) The finest grade of clastic sediment, grains 0.0039–0.0625 mm diameter

Clay ironstone Low-grade iron ore as siderite, iron carbonate, usually as decimetric nodules in mudrock; product of bacterial mediation under shallow, often non-marine burial conditions

Clay minerals Various flaky alumino-silicate minerals with sheet-like lattices

Cleavage (mineral) Splitting plane(s) along atomic weakness in a mineral

Cleavage (rock) Splitting plane parallel to orientated flaky silicate minerals in folded rocks

Club moss Living, mostly diminutive, members of Lepidodendrales spore-bearing plants with simple scaly leaves whose tree-like ancestors included the giant scale-trees of Carboniferous swamp forests

Coal Carbonaceous accumulation of variously-compacted and partly-to-wholly deconstructed vegetation

Coal ball Aggregate of three-dimensionally preserved plant remains held together by mineral precipitate, usually calcitic in composition

Coal basin Part or all of a sedimentary basin containing coal-bearing strata

Coal (boghead/parrot) Coal rich in algae, spores and waxy plant remains (liptinite); sometimes retortable to yield hydrocarbon

Coal (cannel) Bright- and clean-burning dull black homogenous coal (from north British dialect for 'candle')

Coal rank Hierarchy of coal types ranging from fibrous brown coal to anthracite via bituminous coals; increasing rank has increased proportions of carbon and decreased volatile components

Coallery The practice of coal-seeking mineral engineers and mining engineers introduced by George Sinclair in 1672

Coaxial drilling Drilling within cylindrical casing to avoid wall collapse; first use by Drake in Pennsylvania, 1859

Cog vessel Single-masted, flattish-bottom, square-sail, clinker-built ships with raised bow and stern areas, the latter bearing a central external rudder

Coke Cinder-like, hard and porous slow-burnt bituminous coal; cakes into fissured aggregates as the airless burning cycle is concluded

Collier vessel Traditional sturdy, shallow-draught, flat-bottomed brig with open central hatches for loading and unloading; Cook's *Endeavour* was an ex-Whitby collier

Colliery Coal mine

Compaction Of sediment; largely by water expulsion under gravitational force due to overlying sediment

Conifer Vascular woody plants, mostly trees, as gymnosperms – cone-bearing seed plants. Evolved from stout upland Carboniferous Cordaites, with cone-like fertile structures. Root: Latin for *conus* (cone), as in ice-cream.

Conglomerate Coarse-grained sedimentary rock comprising boulders to granules > 2 mm diameter

Conodont Microscopic phosphatic tooth-like structure in the otherwise soft-bodied, worm-like marine creature *Clydagnathus*. World's first complete specimen described in 1983 from the Granton Shrimp Bed *Lagerstätte* (Early Carboniferous), Lothian

Continental collision Contiguous formerly separate continents juxtaposed along faults after plate subduction

Continental drift Casual (but poetic) English for Alfred Wegener's original more serious 'Kontinente Verschiebung' – 'continental displacement'

Cooksonia The oldest vascular plant dating back to the Middle Silurian; has trumpet-shaped sporangia with a blocking 'mute' which eventually disintegrated to release spores

Cordaitales Extinct vascular woody plants as stout and tall upland Carboniferous trees with cone-like fertile structures; ancestors to conifers

Cornish stone Feldspar-rich mèlange of altered granite used in manufacture of porcelain

Country rock Ambient rock of any surrounding district into which molten magma intrudes to form plutonic igneous rock as a batholith or pluton

Cross bedding Inclined stratification formed by granular avalanche deposition on the leeside slopes of ripples, dunes, and bars due to current flow of water or wind

Crust Outer solid part of the lithosphere down to the Moho boundary with the upper mantle

Crustal recycling Remelting of crust due to deep burial during mountain building

Crustal shearing Transverse sliding along major crustal-depth shear zones

Crustal shortening Thickening of the crust due to thrust faulting; forms mountain ranges during continental collision

Crustal stretching Thinning of the crust due to normal faulting; forms active rift basins, often during plate separation to form new oceans

Cryptospore Primitive spores from Lower Palaeozoic strata

Cupilo Cupola-like furnace for metal smelting (Bristol area)

Cuticle In botany as the hydrophobic waxy and resinous outer (epidermal) layers of certain leaves, stems and seeds; prevents desiccation and undue wetting

Cwm (Cymru) Scoop-shaped excavation into a mountainside at the head of a former valley glacier

Cyanobacteria Formerly blue-green algae; non-nucleic bacterial fibrous forms developing mat-like aggregates (stromatolites) as far back as the early Proterozoic

D

Deltaic Pertaining to deltas

Desulfovibrio vulgaris Anaerobic bacteria which reduce seawater sulphate (the electron acceptor) to sulphur to obtain metabolic energy; in so-doing enables ferrous iron to precipitate as pyrites

Devonian Period of time 419–359 Ma

Diagenesis Physical, chemical and mineral changes to deposited sediment during progressive burial

Dip Maximum angle and direction of the tilt of strata

Dolerite Dark, basic igneous rock; moderately crystalline, characteristic of dykes and sills

Dolomite Double magnesium/calcium mineral carbonate

Dropstone Pebble- or larger-sized rock fragments released by iceberg melting that plop into sea or lake floor sediment, deforming it characteristically by piercing bedding or lamination

Durain A maceral component of bituminous coal: 'dull' hard coal

Dyke Linear vertical to sub-vertical intrusive sheet of igneous rock, commonly dolerite

E

Engine house Usually chimneyed brick or stone housing for steam engine and boiler; the best multi-storeyed and definitely Georgian intrusions into often wild moorland countryside

Enlightenment 18th century (*c.*1715–1789) optimistic secular and republican philosophical movement emphasizing order and reason, brought to an end by the French Revolution

Eruptive Subaerial ejection of lava and/or volcanic ash

Estuarine Pertaining to estuaries

Euroterrane Informal term used herein for numerous 'platelet' terranes that rifted away from northern Gondwana and travelled westwards in the Late Devonian/Early Carboniferous

Eustatic A global change to sea level brought about either by ice cap meltout or mid-ocean ridge uplift

Evaporite Mineral salt precipitated from brine by evaporation

Exponential Mathematical expression for a rate of change in some quantity, y (like temperature), that changes with time according to the function $y = e^x$ where e is the base of natural logarithms

Extension Tectonic process whereby stretching of the lithosphere by attenuation and normal faulting forms rift basins

F

Far-field In geology (borrowed somewhat clumsily from electromagnetism) a tele-effect arising as a consequence of some far-distant action, e.g. tsunami, eustatic sea-level rise

Fault A fractured slip-plane in rock that has undergone 3D sliding displacement

Fault plane Plane of displacement on a fault

Fault (normal) Fracture where rock is displaced down a tilted fault plane

Fault (thrust) Fracture where rock is displaced up a tilted fault plane

Fault (wrench) Fracture where rock is displaced along a fault plane

Fault-bounded Sedimentary basins whose limits are thus defined

Fault footwall Lower side of a faulted rock mass below the fault plane

Fault hangingwall Upper side of a faulted rock mass above the fault plane

Fault line Hidden fault plane traced by juxtaposed rocks

Fault rock Broken-up breccia, recrystallized (mylonite) or cemented rock along a fault plane

Faunal province A region (once) containing distinctive life forms (as fossils in coeval strata)

Feldspar Common hardish silicate mineral with three perfect cleavages; contains variable amounts of alkaline earth elements calcium, sodium and potassium. Chemical weathering breaks it down to china clay (kaolinite)

Fen Freshwater wetland, usually on river floodplains, founded on mineral sediment or on thin peats resting on such materials. With permanent or seasonal inundations generally nutrient-rich flood water of neutral acidity; supports reed (*Phragmites*) beds and brushwood stands

Fermentation Enzyme-mediated anaerobic reactions with carbohydrate; contributes to diagenesis in deposited sediment by creating carbon dioxide, hydrogen, formate, acetate, carboxylic acids; elsewhere vital fluids like alcohol

Ferns Ancient vascular plants that reproduce by free-living spores abundantly provided on branching foliage with complex leaves (megaphylls)

Finery (forge) Ancient processes producing malleable wrought (bar) iron from brittle pig (cast) iron by decarburization, the carbon removed by oxidation, the iron then mercilessly hammered to develop a fibrous microstructure. Replaced by the puddling process and the roller mill due to Henry Cort in 1783–4

Fissility The rock property of easy-splitting along sedimentary layers or along slaty cleavage

Flagstone Sandstone splittable into centimetric-thick slabs

Flat Horizontal to shallow-dipping ore-seam more-or-less continuous along strike; often in limestones formed by replacement of the rock by ore during mineralization

Flint White-rinded, grey-black lustrous and amorphous form of chalcedony; as lumps (nodules) in chalk

Fluid inclusion Discovered microscopically by H.C. Sorby in the nineteenth century. Minute fluid drops preserved in mineral crystals during their crystallization by cooling; amenable to modern chemical microanalysis and the determination of the temperature of crystallization

Fluorite (aka fluorspar) Mineral form of calcium fluoride (CaF_2)

Fold Up- or down-fold of strata; undulose forms often produced by stratal shortening under conditions of compressional crustal tectonics

Forest boundary layer Zone of marked retardation of wind flow over forest stands leading to sheltering inside and increased velocity gradient and turbulence above

Forest swamp Well-wooded freshwater marshes populated by standing water-loving species with aerated surface layers of neutral acidity above anaerobic peat or mineral-rich substrates.

Forth Sill Thick quartz-dolerite sill of large areal extent in Lothian, Firth of Forth and Fife; contemporaneous (Late Carboniferous/Early Permian) with Whin Sill of NE England fame

Fossil capital Fossil fuel reserves

Fossil economy A system dependant upon burning fossil fuel

Fossil fuel Fuel originating from fossilized organic carbon or hydrocarbon

Fracking Natural gas forced from 'tight' reservoirs: the last resort, as from November 2019 illegal in the UK

Freestone Stone with no significant internal splitting planes that can fracture in any guided way; massive structureless stone

Free trade Without duties or tariffs

Fulvic acids From breakdown of organic humus; similar to humic acids, but with contrasting carbon and oxygen contents, acidity, degree of polymerization, molecular weight, and colour (yellowy)

Fusain A charcoal maceral in bituminous coals and elsewhere

Furnace (reverberatory) Tall-chimneyed furnace, the ore or metal intensely heated to melting point by indirect application of heat as lateral blast (usually from above)

G

Gaels Q-Celtic (Irish Gaelic) speaking inhabitants of Ireland's Gaeltachts and Scotland's Dál Riata (greater Argyll), the latter since at least the fifth century, subsequently throughout Highland Scotland; extant in both countries

Gae A fault in the seventeenth-century Lothian vernacular as used by George Sinclair (1672); perhaps because faults underground cause coal seams to vanish or go (gae) away

Gaian The hypothesis of James Lovelock that conditions at the planetary surface are due to self-regulatory mechanisms inherent in the evolution of the biosphere

Galena Mineral form of lead sulphide, PbS

Galvanize The process of 'hot-dip' coating of steel plate or sheet with a thin film of zinc-iron alloy to counter rusting

Gannister A tough massive sandstone overwhelmingly composed of silica grains, usually also cemented (lithified) by silica cement

Geobacter metallireducens Bacteria attracted by chemical gradients (Chemotactic) with the metabolic ability to reduce ferric iron to ferrous iron whilst digesting organic matter in deposited sediment; the ferric iron accepts an electron produced during the metabolism

Geo-engineering An approach to solving the problem of increased carbon dioxide and global warming by enhanced natural silicate weathering of finely crushed olivine

Geological map Topographic map with projection of mapped distribution of geological features and formations

Geological map (drift) Geological map of all surface outcrops and all Quaternary sediment

Geological map (solid) Geological map that ignores Quaternary sediment and extrapolates solid bedrock between observed outcrops

Geological section A vertical section drawn down through part of the upper crust to show the disposition of outcropping and buried rocks as proven from field mapping and inferred by extrapolation

Georgian Period of the Hanoverian (German) Protestant monarchs almost synchronous with the early Industrial Revolution as pursued in this book, 1714-1837

Geothermal gradient The rate of increase of temperature with depth through the crust, typically around 25 degrees centigrade per kilometre

Gigantism Abnormally large insect life forms that evolved in response to heightened oxygen content and enhanced air density of the Carboniferous atmosphere

Glacial erratic Often far-travelled rock or mineral brought by glacier ice

Glacial outwash Sediment deposited by glacial meltwater at a glacier front

Glass cone Tall, conical, stone or brick-built structure with ground-level coal-fired furnaces and glassblowing alcoves

Glossopteris Extinct fossil seed fern leaves (no complete shrub or tree exists) restricted to the middle and high latitudes of Permian Gondwana. Woody fragments show marked growth rings with thick spring/summer wood layers. Abundant fossil accumulations represent autumnal leaf drifts.

Goethite Mineral form of hydrated iron (ferric) oxide (oxyhydroxide); source of brown ochre pigment, commonly found as bog iron ore

Gondwana Palaeozoic southern hemisphere continental mass finally joined with Laurussia by Variscan mountain building: subsequently split by Atlantic and Indian oceans opening

Goniatite Extinct, small, coiled and chambered pelagic marine cephalopods; ammonite-like creatures; widely used to correlate Middle to Late Carboniferous strata

Graded bedding Gradual change in sediment size upwards through a bed; usually fining-upwards

Grainstone Limestone whose larger constituents are packed against each other, the interstices occupied by younger cementing calcite crystals

Granite Light-coloured, coarsely crystalline, slowly cooled (plutonic) igneous rock comprising quartz, feldspar and mica silicate minerals

Greenhouse effect (on Earth) The atmospheric trapping of re-radiated (infrared) sunlight by diatomic gases like carbon dioxide, water vapour and methane

Gritstone Granule-grade (2–4 mm diameter) sandstone with scattered pebbles

Groundwater Water held below surface in porous and permeable sediment and rock; saturated conditions pertain below the water table where flow occurs along pressure gradients towards rivers, streams and the oceans

Growth fault Fault actively displacing surface sediment with deposition taking place preferentially on the downthrown side

Gypsum Mineral form of hydrated calcium sulphate, $CaSO_4.2H_2O$; an evaporite salt

H

Halite Mineral form of sodium chloride, NaCl; an evaporite salt

Hearth tax An English Early Modern wealth tax on the number of hearths in a single family dwelling;

records are a source of information on relative wealth of different occupations in certain areas

Hearthstone Large fire-resistant sandstone pieces making up a furnace hearth; usually of gannister freestone

Hematite Mineral form of iron (ferric) oxide (Fe_2O_3); 'red ore' from colour of powder (streak), source of red ochre

Hercynian (aka Variscan) Late-Carboniferous mountain-building episode associated with closure of the Rheic Ocean and the welding together of the supercontinent Pangea along the line of the mountainous Hercynides (Variscides) from Cantabria to Donetz

Highstand Advent of highest stable post-glacial sea level; latest was around 5–6 ka

Holocene Epoch of time from 11.7 ka to present

Hostmen Powerful middlemen of the Tyneside coal trade acting between collieries, collier vessels and London merchants

Humic acids Derived from microbial breakdown of organic humus; similar to fulvic acids, but with contrasting carbon and oxygen contents, acidity, degree of polymerization, molecular weight, and colour

Hushing Crude mineral exploration technique whereby ponded upslope reservoirs release their water suddenly downhill to erode soil and unconsolidated sediment in an attempt to expose mineral veins cutting bedrock

I

Ice age Periods in Earth history when large polar ice caps on continental crust expanded equatorwards to cover significant land and shelf sea areas of the mid-latitudes

Igneous rock Formed by crystallization during cooling of intruded magma

Inlier Older strata surrounded by younger; often due to partial erosion of an anticlinal fold or of an unconformable cover of younger strata

Inertinite (aka fusain, charcoal) Type of resistant residual carbon maceral left after anaerobic combustion

Intrusion Body of rock formed by cooling and crystallization of magma injected into country rock

Intrusive Describes rock body formed by magma injecting country rock

Inversion tectonics Tectonics acting in a sedimentary basin as strata are shortened, thrust, uplifted and eroded

Ironstone Sedimentary rock largely or wholly comprising iron-bearing minerals (siderite, hematite, etc.)

Iron Age Division of archaeological time; in Britain 2.8–2 ka, earlier elsewhere

Iron (bar/wrought/rod) Cast/pig-iron reworked in a finery furnace, and then whilst red-hot by hammering or rolling

Iron (ferric) Ions in the ferric state have three unfilled electron vacancies, hence Fe^{3+}

Iron (ferrous) Ions in the ferrous state have two unfilled electron vacancies, hence Fe^{2+}

Iron (öregrund) Purest iron smelted in Central Sweden from magnetite and widely considered the best for historic artisanal cementation and crucible steel making

Iron (cast/pig) Newly smelted molten iron run from furnace tapping into sand moulds

Iron smelting The reduction of iron ore to metallic iron in a furnace with heat provided by either charcoal or coke and with limestone as a flux to form calcium silicate slag from siliceous impurities

J

Jasperware Original Wedgwood pottery from the 1770s, its composition approximately 56% barium sulphate (barite), 30% ball clay, 10% ground flint, 4% barium carbonate (witherite). The fired body is naturally white; initially it was stained throughout with striking metallic oxide colors, later (for economy) just surface-coloured or 'dipped'.

Joint More-or-less planar or gently curved dilational fracture in a rock mass

K

Kaolinite (aka 'China Clay') Mineral form of aluminium silicate (Al_2SiO_5); a clay mineral formed from breakdown of feldspar

Karst Permeable limestone landscape lacking runoff; often bare or thin-soiled crags and 'pavements' riddled with solution features such as enlarged joint fractures (clints, grikes), solution pipes and caves

Keel (boat) Shallow-draught, lateen-sailed and oared lighters used to transport coal from upstream staithes along the narrow-fronted upper Tyne estuary to collier barques moored in the outer estuary

Keelmen Skilled and strong operators of Tyneside keel boats

Kerogen Solid insoluble organic material in sediment and sedimentary rock. On heating (≤ 200 degrees centigrade) it partially converts to hydrocarbon. Although amorphous and of no fixed composition, it is classified by origin as lacustrine (algal), marine (planktonic), terrestrial humic (coal) and lipid (pollen, spores, waxy remains).

Kinetic Theory of Gases Gas behaviour and properties defined by randomly interacting and colliding solid particles; the first statistical scientific law, by James Maxwell

L

Lacustrine Pertaining to lakes

Lagerstätte plural **Lagerstätten**: Fossilized *in situ* preservation of an organic community

Lamination Sedimentary texture comprising millimetric lenticular to planar laminae of contrasting grain size

Laurentia Pertaining to the Precambrian oldlands and their Palaeozoic cover at the heart of the St Lawrence Seaway of the eastern North American continent

Laurussia Laurentia plus Scotto-Scandinavia/West Russia

Lava Flow of magma from a volcanic vent and the cooled and solidified product

Laws of Thermodynamics Laws governing the dynamics of heat energy exchange

Leat Channel cut around contours to take water from underground mine workings or natural springs into ore-crushing and sorting plants

Lepidodendron Tree-like Lepidodendrales giant of Carboniferous and Permian forest swamps

Levee Natural, raised, well-drained banks adjacent to river and deltaic channels formed gradually by non-uniform deposition during overbank flow; sediment decreases in grain size and quantity into adjacent floodplains

Level Tunnel at a certain depth driven outwards from a central shaft to intersect and work ore-bearing veins by stoping

Light reflectance The degree of light reflectance measured on a microscope stage of polished sections of certain coal macerals like vitrain and clarain

Lignin (from the Latin root, *lignare*; wood): Plant tissue-stiffener – linked cellulose strands by a polymer made up of phenol groups, as in carbolic acid, with hydroxyl groups attached to a ring-like structure, as in benzene. Made up the thick external bark (like an exoskeleton) of the Lepidodendrales and also the secondary woody cells of the Cordaitales

Lignite (aka Brown Coal) Low-rank, dirty-burning opencast coal

Limestone A calcareous sedimentary rock made up of calcite particles and fossil fragments cemented by calcite or dolomite

Limonite Mixed iron-mineral comprising brown-to-, amorphous, hydrated ferric oxides and hydroxides

Lipid Cellular biomolecules insoluble in water, e.g. fatty acids, waxes, sterols, certain vitamins (A, D, E, and K), mono- and di-glycerides

Liptinite Fine-grained coal maceral remains, originally spores, pollen, cysts, leaf cuticle, resins and waxes as part of kerogen

Lithification Sediment hardening by the precipitation of mostly crystalline mineral matter as natural cementing media holding rock together

Lithosphere Rigid outer 100 km or so of Earth's crust and upper mantle

Little Ice Age Informal and imprecise name for the period after the Medieval Warm Period, roughly from the mid-sixteenth to late-nineteenth centuries; characterized by very cold winters in Northern hemisphere; resulted from a modest global cooling

Lowstand Advent of lowest stable glacial-period sea level

Lepidodendrales/Lycophyte Informal names for extant and extinct spore-bearing members of the class Lycopodiopsida. With dichotomously branching stems bearing simple leaves called microphylls; reproduction by means of spores borne in sporangia. Chief among extinct forms is *Lepidodendron*, the tree-like giant of Carboniferous and Permian forest swamps

M

Maceral Components of bituminous coals analogous to rock-forming minerals; the name cleverly derived by Marie Stopes

Magma Molten silicate liquid below surface

Magnetite Mineral magnetic form of iron (ferric) oxide, Fe_3O_4, the original lodestone; frequently of magmatic origins, low in phosphorus, a much sought-after iron ore

Malachite Green-coloured mineral form of copper carbonate hydroxide ($Cu_2CO_3.2OH$); beautifully banded and often variegated

Mantle Underlies the Earth's crust down to the outer core at *c.*2900 km depth. Comprises ultrabasic (silica-poor) igneous rock and comprises *c.*85% of Earth's volume, the majority in a stiffly plastic state undergoing slow convective motion; a partially molten layer at *c.*100 km depth enables plates to slowly slide over it

Marine band Bed of usually black, often pyritous, organic-rich marine-origin mudrock; emits low levels of gamma radiation; occurs within mostly non-marine coal-bearing strata; contains fossil faunas useful in correlation, including the pelagic goniatites

Medieval (aka Middle Ages) European cultural era from around 500 (i.e. post-Roman) to 1400AD

Mesozoic Era of geological time from the Triassic to Cretaceous periods: 253–66 Ma

Metamorphic rock Recrystallized (but not melted) product of the effects of elevated heat and/or pressure on pre-existing igneous or sedimentary rock

Methane Simple gaseous hydrocarbon (CH_4) making up the majority of natural gas; product of organic breakdown by methanogenesis

Mica Sheet-like silicate minerals, biotite (iron-rich) and muscovite (potassium-rich), both softish with perfect basal cleavage

Milankovitch theory Application of Solar System orbital theory to calculate the amount of incoming solar radiation; applied originally in a comparative manner by James Croll, then most comprehensively and quantitatively by Mikhail Milankovitch; should perhaps always be referred to as Croll-Milankovitch theory

Mineral Crystalline substance of specific chemical composition

Mineralization (in veins) Enrichment in metallic and/or non-metallic content of rock by crystal precipitation in fractures (often faults) from warm, briny circulating waters

Mining (longwall) Method of coal mining appropriate to thick horizontal or gently tilted seams and originating in the Coalbrookdale area of Shropshire. The seam is worked continuously sideways at the same time as an access roadway is kept supported adjacent to it, the working face and roadway course slowly moving laterally in tandem as roof supports are shifted correspondingly and then abandoned

Mining (pillar-and-stall) (in Scots, stoop and room): Method of coal mining whereby a seam was worked by teams or families in rectangular stalls along an access roadway separated by pillars of unmined coal that supported the roof. In Tyneside mines the pillars were subsequently systematically removed from mined-out areas to allow general subsidence

Modernism Post-WW I rejection of the past by creating anew: an explosion of -isms; cubism, surrealism, expressionism, etc.

Moho Sharp compositional boundary between Earth's crust and mantle; after seismologist Andrija Mohorovičić who discovered it

Monazite Uranium/thorium-bearing, rare-earth/phosphate mineral; common as an accessory mineral as small isolated crystals in granitic rocks and sedimentary grains in sandstones; useful in radiometric dating

Monocline A step-like fold in strata, sometimes bounded below its line of maximum inflexion by a reverse or thrust fault

Monsoon Equator-crossing and Coriolis-deflected moist trade winds that follow pressure changes caused by seasonal migration of the intertropical convergence

Mountain-building Events caused by plate subduction, plate collision and/or continent:continent collision, all of which cause crustal thickening by shortening along major thrust faults

Moss Northern English and Scottish Borders vernacular for wetland bog terrain; as in 'moss-troopers' – Border brigand clans (e.g. Kerrs, Armstrongs, Johnstons)

Mud (aka clay) The finest grade of clastic sediment, grains 0.0039–0.0625 mm diameter

Mudrock A silty mudstone

Mudstone A mud-grade sedimentary rock

N

Namurian European stage name (from Wallonia, Belgium) for (roughly) the middle part of the Carboniferous (330 to c.318 Ma)

Nappe Overturned anticline bounded by a sole thrust under its overturned limb

Neo-Romantics Late-1930s–1940s British artists with WW2-induced sombre but emotional Romanticism and a deep attachment to British landscape history

Neptunists Believers (followers of Werner) in the origin of igneous rock as aqueous precipitate from seawater (rather than as intruded or extruded magma, as by the Plutonists); now extinct

Newer Granite Granitic and intermediate plutons and lavas emplaced in northern Britain during Acadian mountain-building

Nodule Lump-like mass of hardish substance; usually microbial-induced growth of chemical precipitate (in the present context particularly iron-bearing siderite, but also flint) occurring in sedimentary rocks

O

Ocean crust The largely basaltic topmost layer of a tectonic plate; sharply overlies the mantle at the oceanic Moho

Ochre Brown, yellowish and red mixtures of iron oxide mineral pigments, goethite, limonite and hematite respectively

Oil Shale Mudstones rich in organic carbon retortable to yield hydrocarbons

Olivine Mineral form of iron and magnesium silicates as the continuously varying series $FeSiO_3$ to $MgSiO_3$; makes up a large proportion of the upper mantle

Ombotrophic bog Mossy bog dependent entirely on rainfall for aqueous sustenance

Oolith Sand-grade, spherical, calcium carbonate grains, like fish-roe; with concentric internal growth rings around a fragmental nucleus, like tiny 'gobstoppers'

Orbital climate theory Explains how the principal periodic perturbations in Earth's orbit – eccentricity (elongation); precession ('wobble' of Earth's spin axis); tilt/obliquity (angle between spin axis and orbital plane – create variations in the magnitude of incoming solar radiation

Orogeny Mountain-building event caused by plate subduction, plate collision or continental collision

Outcrop Bedrock exposed by erosion, protruding into the atmosphere from solid rock underground

Outlier A mass of younger strata surrounded by older, often left behind by scarp retreat or as the core of a synclinal fold

Outwash Sediment deposited outwith a glacial terminus by meltwater channels

Overturned fold A fold pushed over onto one side so that strata on the overturned limb are upside-down; also nappe, recumbent fold

P

Palaeotethys Upper Palaeozoic to Mesozoic equatorial-centred ocean

Palaeozoic Era of time from the Cambrian to Permian periods: 541-253 Ma

Pangea Unitary late-Palaeozoic to mid-Mesozoic supercontinent; formed as Hercynian (Variscan) and Appalachian/Mauritanian mountain-building united Laurentia/Laurussia and Gondwana in the late-Carboniferous/early-Permian

Panthalassa The modern Pacific is a descendant of this Palaeozoic to Mesozoic 'all-enveloping ocean'

Pelagic Free-swimming or drifting marine or freshwater organism

Permeable The condition in a rock when internal pore spaces and microcracks are joined up sufficiently to allow penetration and (usually) slow flow of water under the influence of pressure gradients

Permian Period of geological time 299-253 Ma

Plate Rigid outer 100 km or so of oceanic crust and underlying mantle (together comprising the oceanic lithosphere) that moves upon partially molten asthenosphere

Phloem Living conducting tissue around the inner bark in a tree that transports the water-based sugary products (think maple syrup) of photosynthesis as sap into various non-photosynthesizing parts like roots, bulbs, buds and fruits

Pleistocene Epoch of time 2.6 Ma-11.7 ka

Pliocene Epoch of time 2.6-5.3 Ma

Plume Thermally rising ridges or cylinders of partial melt travelling through the denser and more rigid Earth's mantle, eventually melting as magma sources by decompression near the base of the ocean crust

Pluton (aka batholith) A large three-dimensional mass of intruded plutonic igneous rock

Plutonist Believer in the magmatic origins of igneous rock (rather than as precipitates from seawater, according to the Neptunists)

Porcelain Semi-transluscent and hard fine white pottery; a Chinese invention of a mix of china clay and feldspar fired to the point of vitrification in specially designed very high-temperature (1200–1400 degrees Centigrade) kilns heated by anthracite

Porphyry copper ore Low-grade but voluminous ore of copper sulphides disseminated in intermediate igneous plutons and deposited there by the action of warm to hot fluids associated with subduction-related magma generation and the alteration of primary feldspar to clay minerals of various kinds

Pre-Raphaelite (1848-1880s) British movement against the Industrial Revolution, the sublime and materialism; inspired by 14th-century artists, imitating closely observed nature and rejecting later-Renaissance classically inspired art

Proterozoic The eon of geological time, 2500–541 Ma

Puddling (canals) Production of a muddy mulch to line and make impermeable a canal perimeter cut through permeable sediment or rock

Puddling (iron smelting) The manual collection of decarburized molten iron from a reverberatory fining furnace using tongs to bring together the carbon-free iron into a ball-like mass of wrought iron prior to rolling or drawing

Pyrite Mineral form of iron disulphide, FeS_2; also 'fool's gold'. Not to be confused with copper-bearing chalcopyrite

Q

Quartz Mineral form of silicon dioxide (SiO_2)

Quartzite Metamorphosed sandstone

Quaternary Period of time 2.6 Ma–present

R

Radiometric dating Determination of age from the amount of measured decay products and a knowledge of the decay rate of radioactive substances

Rake Line of excavation along the course of a mineral vein across the countryside; Derbyshire origins

Rayon Artificial yarn made from cellulose

Reed (Phragmites) Hollow-stemmed freshwater wetland colonizer, often in vast beds and forming peats; makes the best thatch

Refugia Diminished remnants of flora left as viable communities under locally favourable conditions during periods of regional climatic deterioration

Renaissance Late-medieval (beginning late-12th century) rediscovery of classical humanistic philosophy; the arts and science fuelled by liberal attitudes towards records of past scholarship in libraries from Egypt to Islamic Spain, but particularly in Florence and Northern Italy

Republic of Letters The idea (still in progress) of a universal community of free-thinking intellectuals untrammelled by state or religious interference; product of the Enlightenment and M. Voltaire

Restoration In England, Scotland and Ireland, the 1660 return of monarchy after failure of the republican-military government of the Commonwealth

Rheic ocean Former Upper Palaeozoic ocean along the line of the Variscan (Hercynian) mountain belt

Rhizome Modified subterranean portion of a plant stem from which roots and shoots emerge at nodes; an enlarged rhizome (aka tuber) stores nutrients for future use

Rift Linear area of subsidence bounded on one or both sides by active normal faults and formed due to crustal stretching (extension)

Ripple Wave-like form of sand moulded by currents and/or waves

Rock Solid natural aggregate of minerals

Rock deformation Action of tectonic compression, extension and shearing that causes rock masses to strain, bend and fracture in response

Romanticism Mid-18th to 20th-century reaction against the rational Enlightenment; emphasizes individual imagination, emotion, sublime landscapes

Roof rock In coal mining, the strata directly overlying a worked coal seam; best as a thick homogenous sandstone

S

Sabkha Arabic word for the coastal salt flats found in extremely hot, arid hell-holes like Kuwait and the Trucial Coast

Sal-ammoniac Ammonium chloride substance used in dyeing and metallurgy

Sandstone A sand-grade sedimentary rock, grains 0.065–2 mm diameter

Schizohaline Aqueous environments subject to marked swings in salinity mode, e.g. brackish to hypersaline; after Bob Folk (i.m.)

Sediment Aggregates of detrital, skeletal or precipitated grains of any size, composition or source

Sedimentary rock Indurated deposits of sedimented detrital, skeletal or precipitated grains

Sedimentary structures Features of sedimentary rocks formed by currents of water or wind, raindrop imprints, desiccation cracks, etc.

Sedimentation The process of sediment deposition

Seed plant Spermatophyte plants that produce seeds either as visible ('naked') forms in the gymnosperms (chiefly the cycads, conifers and the unique gingko) or enclosed within fruits, as in the flowering angiosperms

Serf Person or family indentured for a specific time to another to undertake paid work for them exclusively and as directed

Sgraffito Pottery technique produced by applying a slip coat of liquid clay over the whole of the body of a pot. When it is leather-hard a design is scratched through the slip to reveal the contrasting body of the pot beneath.

Shelly Describes a rock, usually a limestone, comprising mostly calcitic fossil shells

Siderite Mineral form of iron (ferrous) carbonate ($FeCO_3$); as in clay ironstone

Siemens-Martin Open Hearth Self-contained, heat-recycling, smelting furnace for iron ore and/or to change scrap or pig-iron to steel by oxidation of excess carbon

Silica Mineral form of silicon dioxide (SiO_2); as crystalline quartz, flint (amorphous chalcedony) or chert

Silicate Any mineral with bases of silica tetrahedra arranged as fibres, sheets or chains linked by various cations (like sodium, calcium, potassium, magnesium, iron, etc.); includes micas, feldspars, pyroxenes, amphiboles, olivines

Sill Sheet-like slab of igneous rock intruded as magma parallel to stratification of sedimentary 'country' rock

Siltstone A silt-grade sedimentary rock, grains 0.0039–0.0625 mm diameter

Silures Iron Age, Cymru-speaking South Wales tribe and descendants who furiously and persistently resisted Roman, Saxon, Norman and English incursion into their ancestral homelands

Silurian Period of geological time 443–419 Ma; named after the Silures tribe

Slag Waste product of smelting, often as a silicate substance

Slate Strongly compressed mudrock with a pronounced rock (slaty) cleavage

Smithsonite Mineral form of zinc carbonate, $ZnCO_3$

Song Dynasty (China) Dynamic dynastic period 960–1279 AD, corresponding to late-Saxon to Angevin England, Dunkeld Scotland and the free Welsh kingdoms

Sop Three-dimensional, pancake-shaped orebody of hematite in West Cumbria

Sough Adit to drain a working-level in a mine

Sphagnum moss Non-vascular plant notably supporting temperate moist upland bogs and peatlands; fed by rainfall

Sphalerite Mineral form of zinc sulphide (ZnS); previously 'zinc blende'

Spiegeleisen Ferro-manganese 'mirror-like' alloy traditionally used in Bessemer steelmaking to bind and remove impurities and strengthen the steel

Spinning Jenny Multi-spindle spinning frame invented in 1764 by James Hargreaves of Stanhill, Oswaldtwistle, Lancashire

Sporangia Spore-forming, bearing and releasing receptacles in plants

Spore Unicellular dispersal reproductive items from plants; in contrast to a seed they contain no food and germinate into a sporeling – a haploid gametophyte

Steel Hard iron containing a definite and reduced amount of carbon along with other elemental additives needed to give strength, edge and resistance to corrosion

Stigmaria Root systems of *Lepidodendron*, the tree-like Lepidodendrales that dominated Carboniferous forest swamps

Stomata Cellular apertures on leaves and stems whose guard cells appropriately open and close enabling plants to breathe (take in air and water vapour) and respire (pass out water vapour (minus CO_2) and photosynthetic oxygen)

Stope Part of a steeply dipping ore-bearing vein extracted by digging or blasting sideways, upwards or downwards

Strata Group of rock layers – plural of *stratum*

Stratification The presence of bedding in sedimentary rocks

Stratigraphy Study of the succession of strata through geological time

Stratum A bed of sediment or sedimentary rock

Strike The sense of orientation of a tilted bed as a plane in space measured normal to any tilt (dip); a horizontal bed has no strike

Structure The course of strata from structure contours; their dip, strike, folding, faulting and unconformities

Structure contour Contours drawn on the top or bottom of a particular stratum that define structure

Subduction Movement of plate pushed or pulled down into the asthenosphere and lower mantle

Sublime Mid-18th-century relation to a scene that creates a sense of awe, terror or amazement and which 'stuns the soul' with a powerful sense of the dangerous unknowable

Submarine slide Subaqueous sediment flow as a coherent mass downslope

Subsidence Downward movement of the Earth's surface relative to some external reference frame; usually due to tectonics but also sediment compaction and, since the last Ice Age, the loading and unloading by growing or shrinking ice caps and of the effects of global sea-level change

Sulphate Oxidized sulphur; the anion SO_4^{2-}

Sulphur-oxidizing bacteria Bacteria reducing the sulphate ions abundantly present in seawater to hydrogen sulphide; the strain responsible is *Desulfovibrio vulgaris*; provides sulphur that enables precipitation of iron sulphides

Suture In tectonics the suspected, proven or imagined ('hallucinosuture') trace of the join between formerly separated continents or terranes caused by collision or transpression

Syncline Stratal down-bend; a fold with a trough and two inward-tilted limbs passing from adjacent upfolds (anticlines)

T

Tectonic The action of plate stresses to fracture and/or bend rock masses, so giving rise to earthquakes

Terrane (displaced) An area of crust distinct in its nature and age from a now-contiguous terrane, implying distinct provenance and tectonic fusion at some time past

Tethys ocean Former Mesozoic ocean embaying central equatorial Pangea; now marked by Alpine–Himalayan mountain belt

Thermal Sag Broad basin-shaped subsidence of the Earth's surface consequent upon cooling of the asthenosphere after lithospheric stretching

Thrust fault A reverse fault with a low (<10 degrees) tilt

Thrust-and-fold belt A product of 'thin-skinned tectonics'; severe crustal shortening during plate compression whereby the strata of the upper crust are deformed by shallow-dipping thrust faults, each passing upwards into anticlinal folds

Thwitel Small hose-hidden dagger manufactured in Sheffield and first mentioned by Chaucer the 14th century

Till (aka boulder clay) A glacier-ice deposit; usually an ill-sorted, unstratified and variable mixture of clay, stone erratics and sand

Tillite Rock originally deposited as till

Tilt block Segment of crust defining a tilted, fault-bounded crustal segment; a half rift

Trade wind Seasonally strong, meridional, subtropical winds deflected rightwards in each hemisphere by Coriolis force

Transpression Shearing deformation along a bend in a transverse fault producing oblique compression

Transtension Shearing deformation along a bend in a transverse fault producing oblique extension

Transverse fault (aka strike-slip or tear fault) Fracture where rock is displaced along the course (strike) of its fault plane rather than as up or down motion along it in reverse and normal faults

Trap country Volcanic landscapes with step-and-bench topography sculpted from individual lava flows; Scottish usage

Triassic Period of geological time, 253–201 Ma

Trip hammer Massive cog-hammer driven originally (since medieval times) by water-wheel gearing; used in forge and battery work

Trouble Early Modern Lothian mining term; first published use by George Sinclair (1672) for the disruptive effects of dykes and faults displacing coal seams

U

Ultrabasic (ultramafic) igneous rock Dark, dense rock (e.g. peridotite, serpentinite) making up the Earth's mantle; rich in olivine with <45% silica content

Unconformity Relationship between older strata that were tilted or folded and then eroded before deposition of younger strata on top of them; younger strata frequently include detritus of the older; the surface of unconformity represents 'lost' time unrecorded in the local strata

Uniformitarianism Clumsy, ugly and inappropriate term (with 'sectish' implications) coined by William Whewell for James Hutton's logic of 'present causes' as borrowed by Charles Lyell; the root 'uniform' in this context means 'unchanging with time', as used in fluid mechanics; to explain features in the rock record from *vera causa* ('true causes').

Upside-down strata Strata inverted by folding

V

Variscan (aka Hercynian) Late-Carboniferous mountain-building episode associated with closure of the Rheic Ocean and the welding together of the supercontinent Pangea along the line of the mountainous Variscides (Hercynides) from Cantabria to Donetsk

Vascular Tubular connecting and transporting tissues in plants involving root, stem, branch and leaf

Veins Steeply dipping ore-seams more or less continuous along their strike

Victorian/Edwardian Gothic Architectural type; derivative, often trite and heavy; also amusingly jackdaw-gathered; also arrogant colonial-clutter *c*.1860s–1910s

Vitrain (aka Vitrinite) Bright, conchoidal-fracturing coal maceral signifying high rank in the bituminous range

Volcanic Pertaining to volcanoes and the origin of lavas

Volcanic plug or dome Igneous rock intruding the neck of an extinct or resting volcano, sometimes extruded or violently erupted with *nuée ardente* as at Mont Pelée above St Pierre, Martinique in 1902

W

Wagonway Fixed route for self-propelled or animal-pulled wagons along wooden, iron or steel rails

Weathering Physical and chemical alteration of rock and mineral during atmospheric exposure

Wetland More or less permanently water-saturated ground; peatlands, swamps, fens, marshes, bogs, etc.

Whinstone Scots vernacular for dolerite; refers to the gorse (whin) bushes that grow prolifically on its rocky outcrops; most obvious cutting lowland pastoral landscapes like coastal Galloway or Lothian

Witherite Rare mineral form of barium carbonate ($BaCO_3$); dense, white-coloured

X

Xylem Plant connecting tissue for the transport of water and nutrients

Z

Zircon Mineral form of zirconium silicate ($ZrSiO_4$); hard, resistant and eminently useful uranium-bearing mineral common as an accessory in granitic rocks and as a detrital sedimentary grain; widely used nowadays in high-precision radiometric dating

Index